Everyday Mathematics®

The University of Chicago School Mathematics Project

STUDENT REFERENCE BOOK

Mc Graw Hill Education

The University of Chicago School Mathematics Project

Max Bell, Director, *Everyday Mathematics* First Edition
James McBride, Director, *Everyday Mathematics* Second Edition
Andy Isaacs, Director, *Everyday Mathematics* Third, CCSS, and Fourth Editions
Amy Dillard, Associate Director, *Everyday Mathematics* Third Edition
Rachel Malpass McCall, Associate Director, *Everyday Mathematics* CCSS and Fourth Editions
Mary Ellen Dairyko, Associate Director, *Everyday Mathematics* Fourth Edition

Authors
Max Bell
Jean Bell
John Bretzlauf
Amy Dillard
James Flanders
Robert Hartfield
Andy Isaacs
Catherine Randall Kelso
James McBride
Kathleen Pitvorec
Peter Saecker

Writers
Lisa J. Bernstein
Andy Carter
Rosalie A. DeFino
Jeanne Di Domenico
Lila K.S. Goldstein
Jesch Reyes
Elizabet Spaepen
Judith S. Zawojewski

Digital Development Team
Carla Agard-Strickland, Leader
John Benson
Gregory Berns-Leone
Scott Steketee

Technical Art
Diana Barrie, Senior Artist
Cherry Inthalangsy

UCSMP Editorial
Elizabeth Olin
Kristen Pasmore
Molly Potnick

Contributors
Sarah R. Burns
Lance Campbell
Kathryn Flores

www.everydaymath.com

Send all inquiries to:
McGraw-Hill Education
8787 Orion Place
Columbus, OH 43240

ISBN: 978-0-02-138356-6
MHID: 0-02-138356-1

Printed in the United States of America.

3 4 5 6 7 8 9 QVS 20 19 18 17 16

Standards for Mathematical Practice 1

Operations and Algebraic Thinking 35

Number and Operations in Base Ten 63

Games 289

About the *Student Reference Book*

The *Student Reference Book* is a helpful guide to review math concepts and skills, a resource to find the meaning of math terms, and interesting reading when you want to learn about a new topic in mathematics.

A reference book is organized to help readers find information quickly and easily. Dictionaries, encyclopedias, atlases, and cookbooks are examples of reference books. Reference books are not like novels and biographies, which you often read in order from beginning to end. When you read reference books, you look for specific information at the time you need it. Then you read just the pages you need at that time.

You can use this *Student Reference Book* to look up and review information on topics in mathematics. It includes the following information:

- A **table of contents** that lists the topics and gives an overview of how the book is organized
- Essays describing how to use **mathematical practices** to solve problems and show mathematical thinking
- Essays on **mathematical content,** such as algebraic thinking, numbers, operations, fractions, measurement, data, and geometry
- A collection of **photo essays** that show in words and pictures some of the ways that scientists, artists, engineers, and others have used mathematics throughout history or how they use it today
- Directions on how to play **mathematical games** to practice your math skills
- A set of **tables** and **charts** that summarize information, such as a place-value chart, rules for the order of operations, and tables of equivalent values
- An **appendix** that includes directions on how to use a **calculator** to perform various mathematical operations
- A **glossary** of mathematical terms consisting of definitions and some illustrations
- An **answer key** for the **Check Your Understanding** problems
- An **index** to help you locate topics quickly
- **Videos** and **interactive problems** available through the electronic version of this book in the Student Learning Center

About the *Student Reference Book*

As you work in class or at home, you can use the *Student Reference Book* to help you solve problems. For example, when you don't remember the meaning of a word or aren't sure what method to use, you can use the *Student Reference Book* as a tool.

You can look in the **table of contents** or the **index** to find pages that give a brief explanation of the topic. The explanation will often include definitions of important math words and show examples of problems with step-by-step sample solutions.

While reading the text, you can take notes that include words, pictures, and diagrams to help you understand what you are reading. Work through the examples and try to follow each step.

At the end of some of the essays, you will find problems in **Check Your Understanding** boxes. Solve these problems and then check the **answer key** at the back of the book. These exercises will help you make sure that you understand the information you have been reading. Make sense of the problems by comparing your answers with those in the answer key. If necessary, work backward from the sample answers to revise your work.

The **Standards for Mathematical Practice** section includes interesting problems that you can solve. The discussions illustrate how fifth-grade students use the practices to solve these problems.

The world of mathematics is a very interesting and exciting place. Read the **photo essays** or review topics learned in class. The *Student Reference Book* is a great place to continue your investigation of math topics and ideas.

Once you are familiar with the overall structure of the *Student Reference Book,* you can use it to read about different mathematical concepts. As you follow your interests, you will find that your skills as an independent reader and problem-solver will improve.

Standards for Mathematical Practice

Mathematical Practices

Mathematicians, scientists, engineers, and others who use mathematics develop ways of working that help them solve problems. These ways of doing mathematics are called mathematical practices. When you solve problems, you are using mathematical practices.

Look at each picture below.

In what ways might these students be thinking and working as mathematicians?

The next page lists the Standards for Mathematical Practice you will use in *Everyday Mathematics*. Below each standard is a list of goals that can help you understand what it means to use the practices. On the following pages of the *Student Reference Book*, you will see how some fifth-grade students use mathematical practices as they solve problems and reason about mathematics. As you read, think about the ways you can develop these practices to become a more powerful problem solver.

Mathematical Practice 1: Make sense of problems and persevere in solving them.

GMP1.1 Make sense of your problem.

GMP1.2 Reflect on your thinking as you solve your problem.

GMP1.3 Keep trying when your problem is hard.

GMP1.4 Check whether your answer makes sense.

GMP1.5 Solve problems in more than one way.

GMP1.6 Compare the strategies you and others use.

Mathematical Practice 2: Reason abstractly and quantitatively.

GMP2.1 Create mathematical representations using numbers, words, pictures, symbols, gestures, tables, graphs, and concrete objects.

GMP2.2 Make sense of the representations you and others use.

GMP2.3 Make connections between representations.

Mathematical Practice 3: Construct viable arguments and critique the reasoning of others.

GMP3.1 Make mathematical conjectures and arguments.

GMP3.2 Make sense of others' mathematical thinking.

Mathematical Practice 4: Model with mathematics.

GMP4.1 Model real-world situations using graphs, drawings, tables, symbols, numbers, diagrams, and other representations.

GMP4.2 Use mathematical models to solve problems and answer questions.

Mathematical Practice 5: Use appropriate tools strategically.

GMP5.1 Choose appropriate tools.

GMP5.2 Use tools effectively and make sense of your results.

Mathematical Practice 6: Attend to precision.

GMP6.1 Explain your mathematical thinking clearly and precisely.

GMP6.2 Use an appropriate level of precision for your problem.

GMP6.3 Use clear labels, units, and mathematical language.

GMP6.4 Think about accuracy and efficiency when you count, measure, and calculate.

Mathematical Practice 7: Look for and make use of structure.

GMP7.1 Look for mathematical structures such as categories, patterns, and properties.

GMP7.2 Use structures to solve problems and answer questions.

Mathematical Practice 8: Look for and express regularity in repeated reasoning.

GMP8.1 Create and justify rules, shortcuts, and generalizations.

Problem Solving: Make Sense and Keep Trying

When you solve a problem, you work to make sense of the problem, keep trying when the problem is hard, and reflect on your thinking as you work. When you compare your strategy to others, you learn to solve problems in more than one way.

Think about this problem:

Use the numbers in the box below to find all the pairs of factors that make the number sentence true.

3.2	3.4	3.8	6	8

_______ * _______ = A product between 20 and 30

How would you start to solve this problem?

I need to find all possible pairs of numbers in the box with a product between 20 and 30. My answer will be a list of multiplication number sentences.

I have a lot of multiplying to do. I'll make a list of pairs of numbers I need to multiply.

Kayla

3.2 * 3.2	3.2 * 3.4	3.2 * 3.8	3.2 * 6	3.2 * 8
	3.4 * 3.4	3.4 * 3.8	3.4 * 6	3.4 * 8
		3.8 * 3.8	3.8 * 6	3.8 * 8
			6 * 6	6 * 8
				8 * 8

GMP1.1 Make sense of your problem.

Kayla made sense of the problem by listing all the possible pairs of factors.

Kayla uses paper and pencil to multiply. After writing two number sentences with answers, she notices something important.

3.2 * 3.2 = 10.24 3.2 * 3.4 = 10.88

Wait a minute! When both factors are less than 4, the product will be less than 16 because 4 * 4 = 16. So I don't have to try the pairs in the first three columns of my list because they will all be less than 20.

And 6 * 6 is 36, so that's larger than 30. 6 * 8 and 8 * 8 are even larger than 6 * 6, so they don't work either.

Kayla

Kayla deletes the pairs of factors she knows are smaller than 20 or larger than 30. Here is her new list.

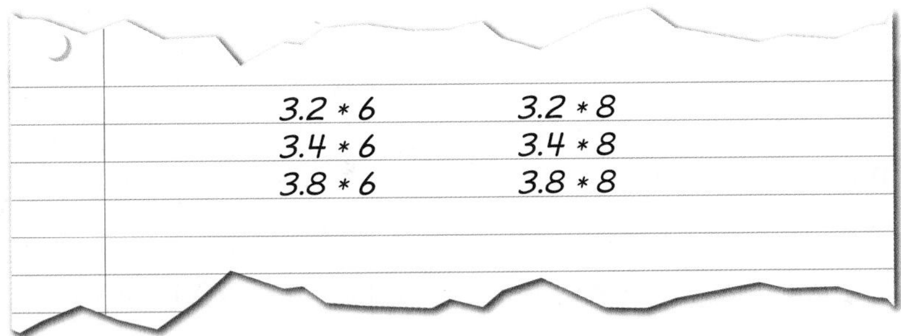

GMP1.2 Reflect on your thinking as you solve your problem.
At first Kayla thought she needed to find every product. But by reflecting on the first two pairs of factors she multiplied, she realized that she didn't have to multiply every pair in the list.

Kayla tries to think of another way to make the work easier. She decides to round and estimate to eliminate some of the pairs. She explains her thinking to her partner, Ava.

"I can round 3.2 * 6 to 3 * 6, so that's about 18.

I can round 3.4 * 6 to 3 * 6, so that's about 18.

I can round 3.8 * 6 to 4 * 6, so that's about 24.

I think that only 3.8 * 6 is between 20 and 30."

Kayla, I've been using a calculator to find the products. When I entered 3.4 * 6, I got a product of 20.4. So even though your estimate was 18, 3.4 * 6 is between 20 and 30.

Ava

Kayla

GMP1.6 Compare the strategies you and others use.
Kayla made estimates, and Ava used a calculator to find exact answers. By comparing their strategies and results, Kayla realized that some of her estimates were not precise enough for this problem.

Kayla thinks about the problem again and explains, "I know from Ava that $3.4 * 6 = 20.4$. So $3.8 * 6$ has to be more than 20.4, too, but it's also less than $4 * 6 = 24$. That means $3.8 * 6$ must work too."

Kayla writes down the products that she knows are between 20 and 30 and a list of the pairs of factors she still needs to check.

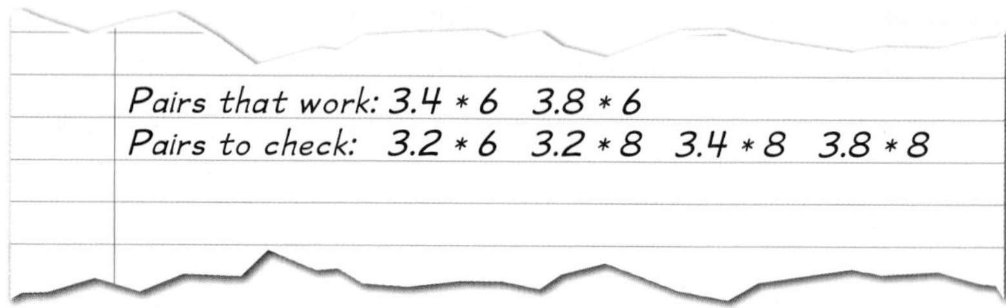

GMP1.3 Keep trying when your problem is hard.
Instead of giving up when her estimates didn't all work, Kayla used what she learned from Ava and tried again. Kayla thought about the size of the products she still needed to check based on the size of the factors and the products she already knew.

Kayla remembers that Ava used the calculator to find $3.4 * 6 = 20.4$. She is pretty sure that $3.2 * 6$ is less than 20, but decides to multiply using paper and pencil as a second strategy to be sure.

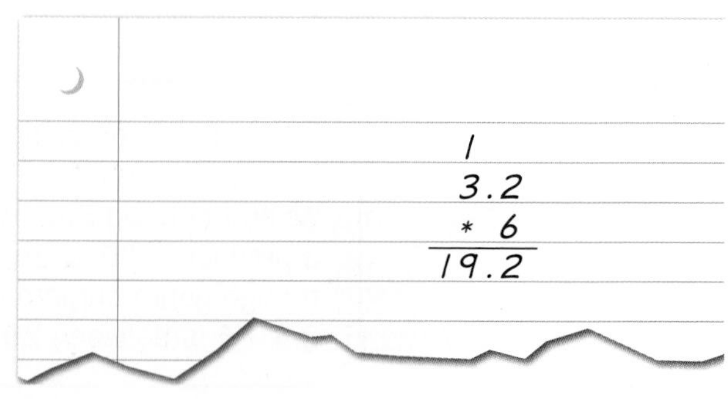

Kayla says, "Doing the multiplication tells me that 3.2 ∗ 6 is less than 20, so I will take it off my list."

Kayla decides to use a calculator for the last 3 pairs. She uses estimation and mental math to check that her answers make sense.

What would you do to finish solving the problem? What strategies and tools can you use?

3.2 ∗ 8 = ?

256 is definitely greater than 30, but 256 doesn't make sense. 3.2 ∗ 8 should only be a little more than 3 ∗ 8 = 24. I need to put the decimal point after the 5. Now 3.2 ∗ 8 = 25.6 makes sense.

3.4 ∗ 8 = ?

3.4 ∗ 8 = 27.2 makes sense. 3 ∗ 8 = 24 and 0.4 ∗ 8 is less than half of 8, or less than 4. So 3.4 ∗ 8 = 24 + a number less than 4. The product will be less than 28.

Kayla uses mental math to check the last pair, 3.8 ∗ 8. She thinks, "3 ∗ 8 = 24 and 0.8 ∗ 8 = 6.4, so 24 + 6 + 0.4 is more than 30. That means this pair doesn't work."

GMP1.4 Check whether your answer makes sense.

Kayla knew that she could make mistakes, even when using a calculator. She used mental math and estimation to be sure that her answers made sense.

Here is Kayla's solution to the problem.

Pairs of factors that have products between 20 and 30:
3.4 ∗ 6   3.8 ∗ 6   3.2 ∗ 8   3.4 ∗ 8

GMP1.5 Solve problems in more than one way.

Kayla used mental math, estimation, paper and pencil, and a calculator to solve the problem.

Mathematical Practice 1: Make sense of problems and persevere in solving them.

Create and Make Sense of Representations

You can use **mathematical representations** to solve problems. Mathematical representations can be numbers, words, pictures, symbols, gestures, tables, graphs, or real objects.

Solve this problem. What representation will you use?

Think about this problem:

$$6\frac{3}{4} + 4\frac{1}{2} = ?$$

Justin uses fraction circle pieces to represent the fractions in the problem.

$$6\frac{3}{4} + 4\frac{1}{2} = 10 + \frac{3}{4} + \frac{1}{2}$$

$$6\frac{3}{4} + 4\frac{1}{2} = 10 + 1\frac{1}{4}$$

$$= 11\frac{1}{4}$$

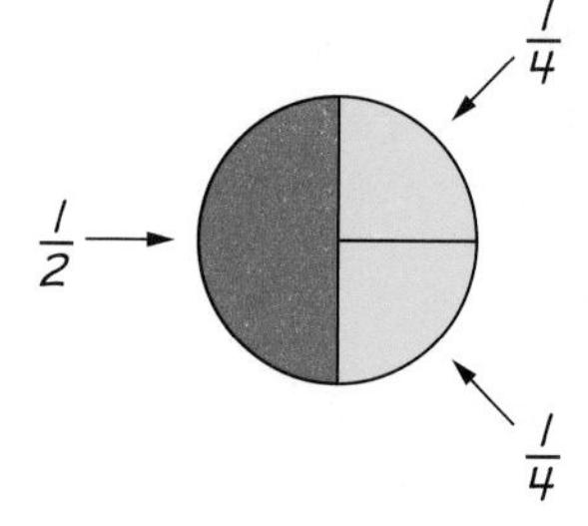

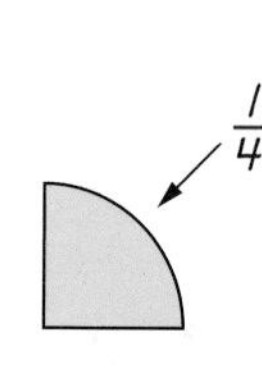

$$1 + \frac{1}{4} = 1\frac{1}{4}$$

Ryan uses a number line to represent the problem.

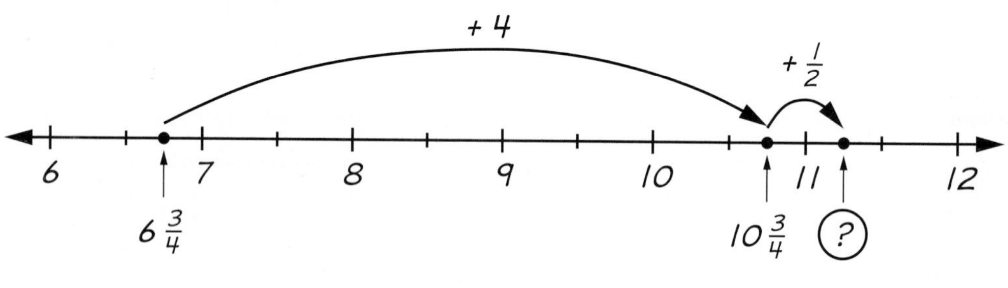

I start at $6\frac{3}{4}$ on my number line and add 4. I end at $10\frac{3}{4}$. I see that Justin's first step shows $10 + \frac{3}{4} + \frac{1}{2}$ for the sum. So I still have $\frac{1}{2}$ to add. I know how to figure out the exact number.

Ryan

+ 4

$+\frac{1}{2} = \frac{1}{4} + \frac{1}{4}$

6 7 8 9 10 11 12

$6\frac{3}{4}$ $10\frac{3}{4}$ $11\frac{1}{4}$

Ryan

GMP2.1 Create mathematical representations using numbers, words, pictures, symbols, gestures, tables, graphs, and concrete objects.

GMP2.2 Make sense of the representations you and others use.
Justin created a representation using fraction circle pieces. Ryan created a number line, but to finish his work he made sense of Justin's representation.

Kayla uses a number sentence.

$$6\frac{3}{4} + 4\frac{1}{2} = 6\frac{3}{4} + 4\frac{2}{4}$$
$$= 10\frac{5}{4}$$
$$= 11\frac{1}{4}$$

Kayla

GMP2.3 Make connections between representations.
Kayla used number sentences and then made connections to Justin's and Ryan's representations. She noticed that their representations showed how $\frac{1}{2}$ equals $\frac{2}{4}$ to make sense of her own solution.

When you create and make sense of different representations, you are reasoning abstractly. When you think about numbers and amounts, you are reasoning quantitatively.

Mathematical Practice 2: Reason abstractly and quantitatively.

Make Conjectures and Arguments

A **conjecture** is a statement that might be true. In mathematics, conjectures are not simply guesses. They are claims based on information or mathematical thinking. **Arguments** in mathematics use mathematical reasoning to show whether a conjecture is true or false. Mathematical arguments can use words, pictures, symbols, or other representations.

Think about this discussion between two students:

Josh makes a conjecture about multiplication.

Josh

He defends his conjecture with this argument:

$9 * 2 = 18$

18 is more than 9 and 2.

$3 * 5 = 15$

15 is more than 3 and 5.

$6 * 9 = 54$

54 is more than 6 and 9.

Josh explains, "Each time I multiply two numbers, the answer I get is larger than both numbers, so my conjecture is true."

Emily

Emily disagrees with Josh's conjecture and explains why.

Here is Emily's argument for why she thinks Josh's conjecture is false:

$\frac{1}{3} * 9 = 3$

Since 3 is less than 9, Josh's conjecture is false.

Josh

Emily

GMP3.1 Make mathematical conjectures and arguments.

Josh made a conjecture and supported it with examples where his conjecture worked. Emily was able to disprove his conjecture with one example where it didn't work. For a conjecture to be true, it must always be true. To show that a conjecture is false, all you need is one example where the conjecture doesn't work.

GMP3.2 Make sense of others' mathematical thinking.

Emily thought about Josh's argument and explained why she disagreed with his thinking. When Josh made sense of Emily's explanation, he decided his conjecture was false.

Mathematical Practice 3: Construct viable arguments and critique the reasoning of others.

Check Your Understanding

Josh decides to revise his conjecture:

When you multiply two numbers, the answer is always larger than at least one of the factors.

Provide an argument for or against Josh's revised conjecture.

Hint: An argument for his conjecture must show it is always true. An argument against his conjecture only needs to show one example where it doesn't work.

Check your answer in the Answer Key.

Create and Use Mathematical Models

You can use mathematics to represent situations or objects in the real world. When you do this, you create a **mathematical model**. Your model might use graphs, drawings, tables, symbols, numbers, diagrams, or words.

When you solve a problem using a model, you analyze the real-world situation, create a model for it, and use the model to answer the question. Your model should help you find an answer that makes sense. If it doesn't, you should revise your model or create a new one that better fits the real-world situation or helps you answer the question.

A fifth-grade class is helping their principal solve this problem:

Principal Pippen plans to paint one wall of a school hallway purple. The school engineer sent Mrs. Pippen this note and information:

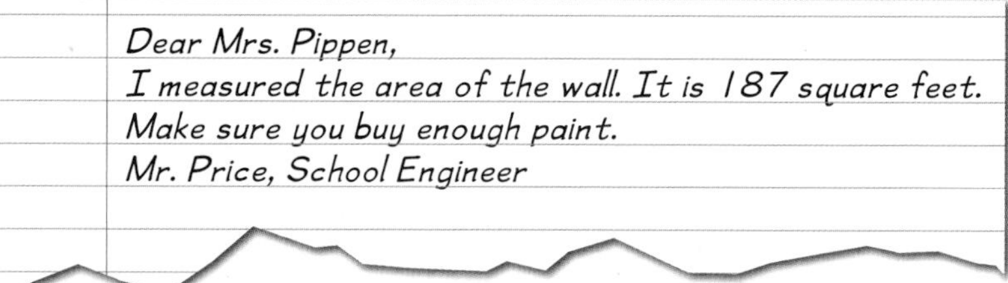
Dear Mrs. Pippen,
I measured the area of the wall. It is 187 square feet.
Make sure you buy enough paint.
Mr. Price, School Engineer

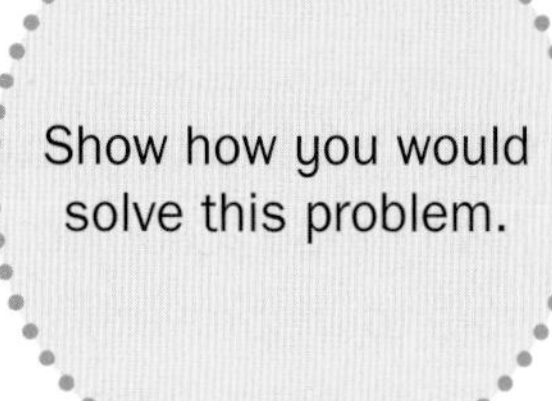

Mrs. Pippen asks the class, "How many pints of paint should I buy?"

The students use models to solve the problem and share their thinking.

Aiden creates a model using numbers:

$187 \div 25 \rightarrow 7 R 12$

Emily makes a drawing of the hallway. She shows her model to the class:

25 ft²	50 ft²	75 ft²	100 ft²	125 ft²	150 ft²	175 ft²	𝍸 𝍸 //

12 more square feet makes 187 ft².

My drawing shows why Aiden used division. I divide the wall into sections of 25 square feet because I know that each pint covers 25 square feet. As I draw each section, I count by 25s. When I get to 175, I use tally marks to show I only need to cover 12 more square feet to get to 187 square feet.

Emily

Justin says he understands Emily's picture, but thinks he can revise it to show how to get an answer for Principal Pippen. He adds to Emily's picture using a green marker. Then Justin explains his additions.

187 square feet

Pints: 1	2	3	4	5	6	7	8
25 ft²	50 ft²	75 ft²	100 ft²	125 ft²	150 ft²	175 ft²	𝍸 𝍸 //

12 more square feet makes 187 ft².

8 pints of paint

Emily's model makes sense to me, but it still doesn't answer the question: How much paint should Principal Pippen buy?

I can label each section of 25 square feet with 1 pint of paint. This way I can see that 7 cans is not enough. Principal Pippen needs to buy 8 cans of paint.

Now I understand why I got a remainder of 12 with my division number model. That's the number of square feet of the wall that wouldn't be covered by the first 7 pints of paint. So Mrs. Pippen would have to buy an extra pint to make sure the whole wall gets painted.

Justin

Aiden

Ava thinks about Emily's and Justin's diagram and draws a table to model the same information. The left column shows the number of pints of paint. The right column shows the square feet covered with each new pint. She gets the same answer: 7 pints would only cover 175 square feet, but 8 pints would leave some paint left over.

Number of Pints	Number of Square Feet
1	25
2	50
3	75
4	100
5	125
6	150
7	175
8	200

GMP4.1 Model real-world situations using graphs, drawings, tables, symbols, numbers, diagrams, and other representations.
These students created different models for the problem. They used number sentences, diagrams, and tables to represent the problem.

GMP4.2 Use mathematical models to solve problems and answer questions.
At first, Aiden wasn't sure how to solve the problem. But after thinking about other students' models, he was able to use his model to answer the question. Justin helped Emily revise her model to clearly show how to reach a solution. These students were modeling with mathematics.

Mathematical Practice 4: Model with mathematics.

Check Your Understanding

Kayla used this model to answer Mrs. Pippen's question.

Make sense of Kayla's model.

1. What does 187 square feet model?
2. What do the 25, 25, 25, and 100 model?
3. What do the 1, 1, 1, and 4 on the right side of the problem model?
4. How can you use this model to answer the question, "How many pints of paint does Principal Pippen need?"
5. How is this model similar to the other models? How is it different?

$$
\begin{array}{rrl}
 & 187\ \text{ft}^2 & \\
- & 25 & 1\ \text{pint} \\
\hline
 & 162 & \\
- & 25 & 1\ \text{pint} \\
\hline
 & 137 & \\
- & 25 & 1\ \text{pint} \\
\hline
 & 112 & \\
- & 100 & 4\ \text{pints} \\
\hline
 & 12\ \text{ft}^2 &
\end{array}
$$

Check your answers in the Answer Key.

Use Tools to Solve Problems

You can use many types of tools to solve problems in mathematics. Examples of tools include pencil and paper, rulers, base-10 blocks, fraction circle pieces, diagrams, tables, charts, graphs, computers, and calculators.

Think about this problem:

The Service Club at Sophia and Aiden's school is helping with a relief effort after a typhoon in Asia. Their goal is to fill a van with boxes of canned food to take to the airport. Sophia and Aiden have to buy the boxes, so they need to figure out how many boxes will fit in the van.

How would you solve this problem? What tools might you use?

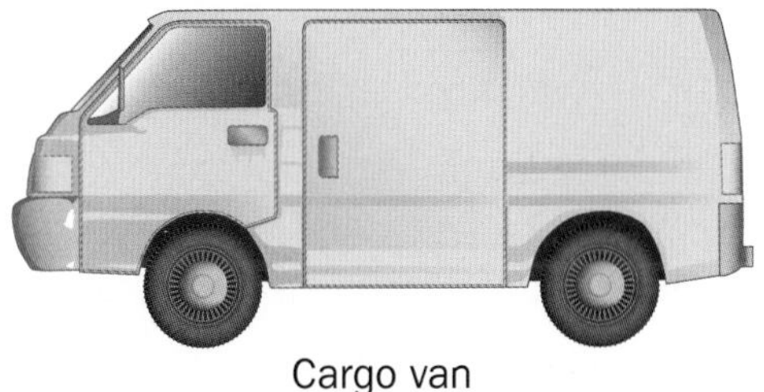

Cargo van

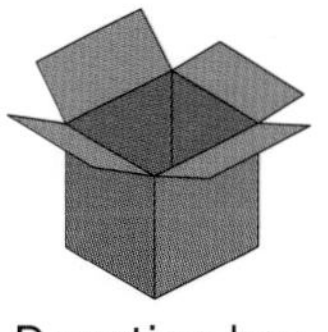

Donation box

Sophia decides to use a meterstick to measure the dimensions of one box and the space in the back of the van. Here are her measurements:

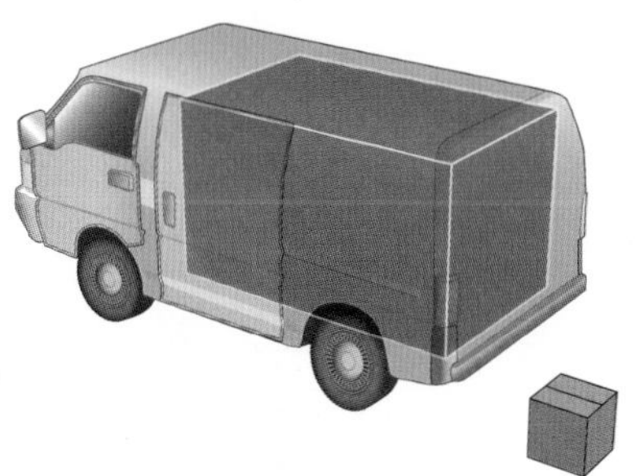

Dimensions of back of cargo van:
Length is 280 cm
Width is 170 cm
Height is 160 cm

Dimensions of donation box:
Each edge is 50 cm.

How can we figure out how many boxes can be packed into the van? What if we try using wooden cubes to represent the boxes?

Aiden

Sophia tries Aiden's idea and uses cubes to model filling the back of the van with boxes.

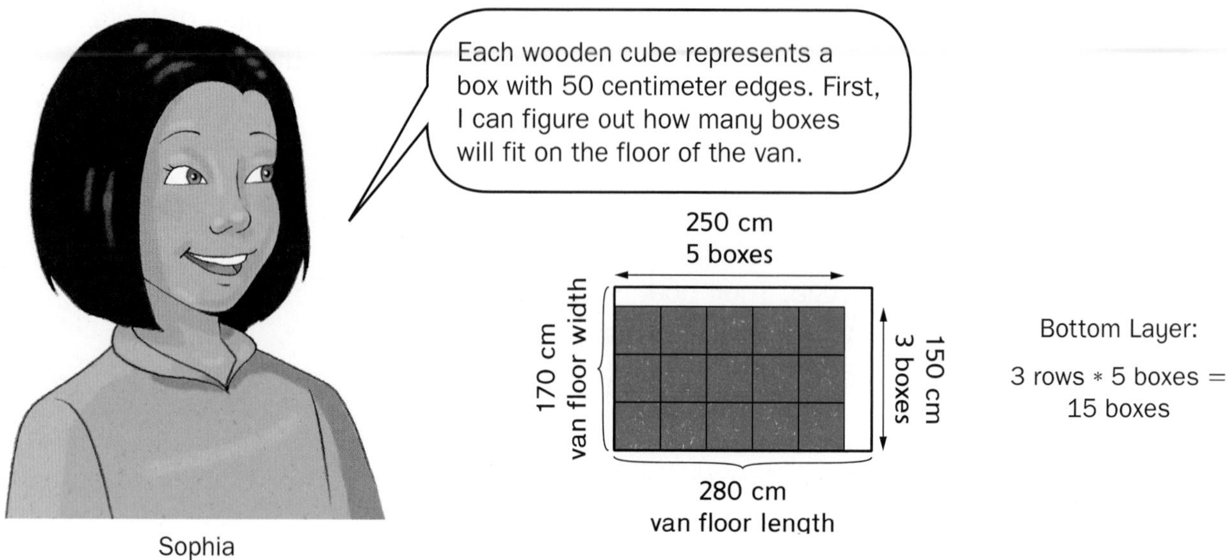

Sophia

Sophia says, "The van is 170 centimeters wide. 3 boxes along the width will be 50 * 3 or 150 centimeters wide. 4 boxes will not fit because 4 boxes are 200 centimeters wide.

"The length of the van is 280 centimeters. So I can fit 5 boxes along the length: 50 * 5 = 250 centimeters long. There is not enough room for another box to fit in the length.

"I fill in the rest of the floor with blocks. There are 3 rows of 5 boxes, so 15 boxes fit in the bottom layer."

Now I have to think about how many layers of boxes will fit.

The height of the van is 160 centimeters, and each box is 50 centimeters tall, so I can make 3 layers: 3 * 50 centimeters = 150 centimeters.

I'll have some space above the boxes, but only 10 centimeters. I can't fit in another layer.

My model shows 3 layers of 15 boxes. That's 45 boxes in all.

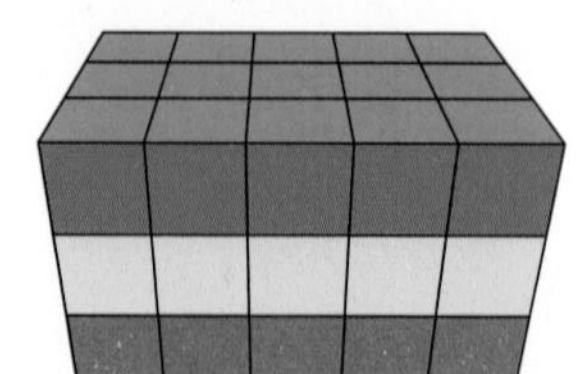

3 layers * 15 boxes = 45 boxes

Sophia

GMP5.1 Choose appropriate tools.

To find the dimensions of the van and the box, Sophia chose to use a meterstick as her tool. She used wooden cubes as tools to figure out how many boxes would fill the van.

Aiden thinks about how the boxes will fit in the back of the van. He makes a drawing and labels it to represent what he pictures in his head. Then Aiden uses number models to find the number of boxes that could fit in the van.

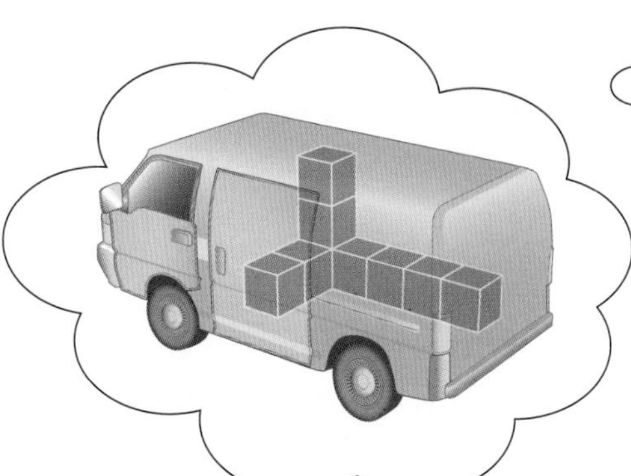

Aiden

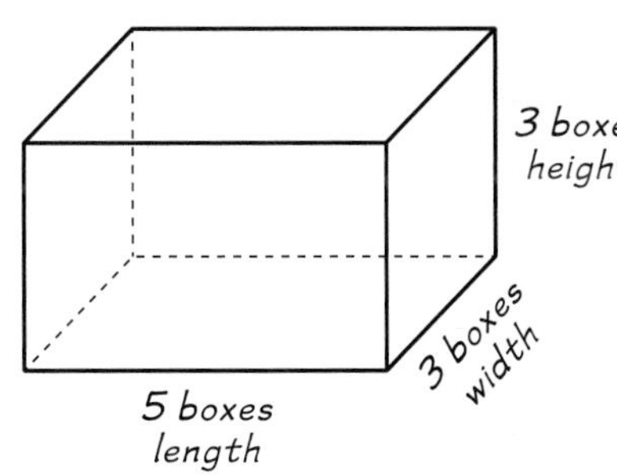

280 cm ÷ 50 cm → 5 R30
5 boxes is the most we can fit along the length.

170 cm ÷ 50 cm → 3 R20
3 boxes is the most we can fit along the width.

160 cm ÷ 50 cm → 3 R10
3 boxes is as high as we can go.

5 * 3 * 3 = 45 boxes in all

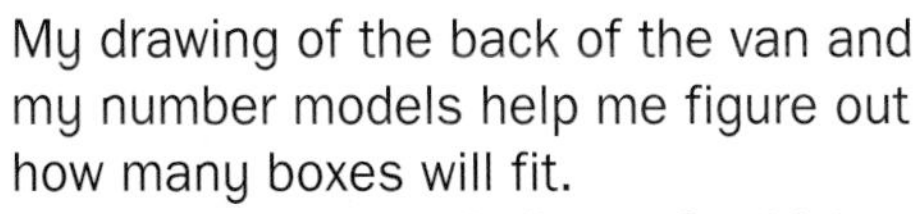

My drawing of the back of the van and my number models help me figure out how many boxes will fit.
My drawing looks similar to Sophia's blocks, and we get the same answer.
We should order 45 boxes.

Aiden

GMP5.2 Use tools effectively and make sense of your results.
Aiden used a drawing of the back of the van and paper and pencil to calculate his results. Aiden and Sophia used their tools effectively to solve the problem. When Aiden compared their strategies and answers, he made sense of the results.

Mathematical Practice 5: Use appropriate tools strategically.

Be Precise and Accurate

Part of doing mathematics is using clear and precise language to explain your solutions and mathematical thinking.

Ryan writes the following rule for deciding which of two fractions is closer to 0:

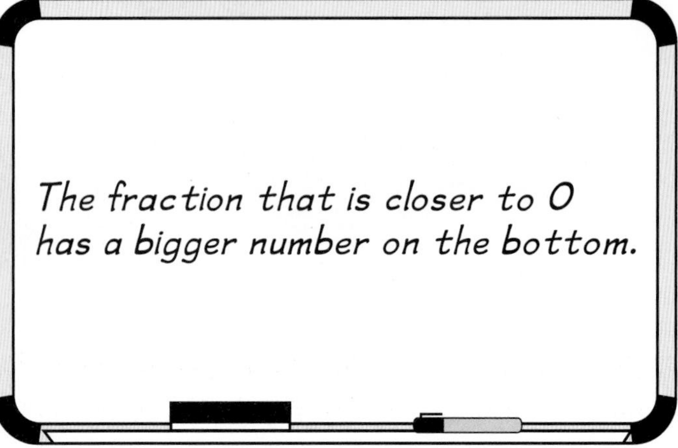

Josh tries to apply Ryan's rule to this problem:

Is $\frac{1}{4}$ or $\frac{3}{8}$ closer to 0?

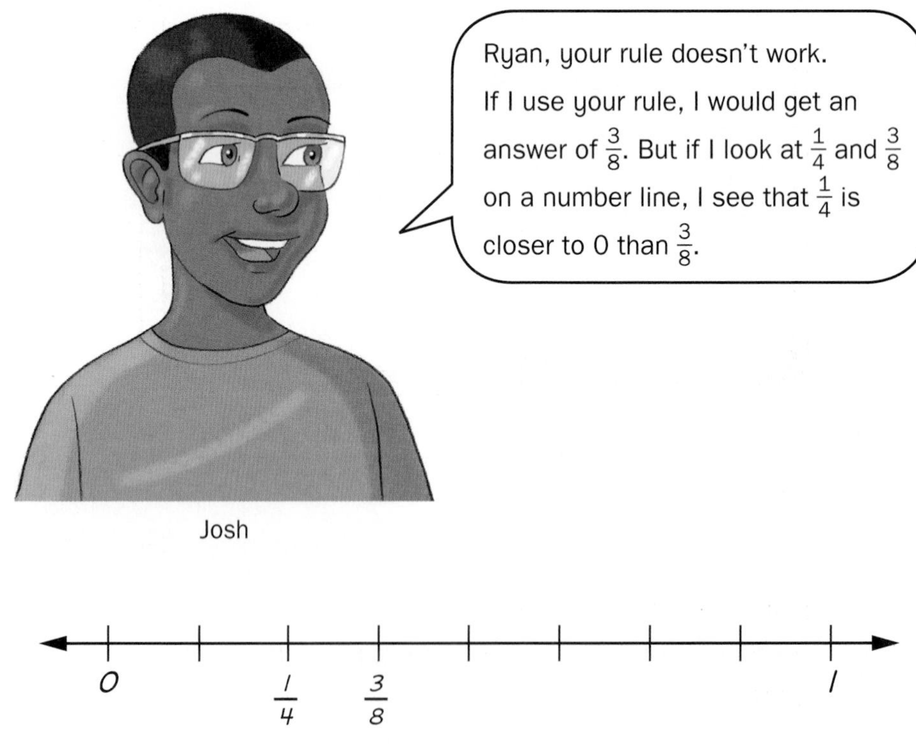

Ryan

Ryan's teacher reminds him that a fraction with a 1 in the numerator is called a **unit fraction.** He asks Ryan to revise his rule using clear mathematical language instead of words like "top" and "bottom."

Ryan revises his rule:

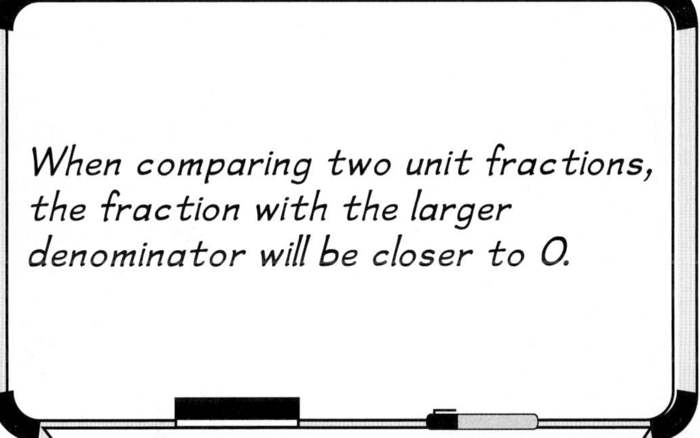

GMP6.1 Explain your mathematical thinking clearly and precisely.
When Ryan revised his rule to include only unit fractions, his language was clearer and more precise.

GMP6.3 Use clear labels, units, and mathematical language.
*When Ryan used the words **numerator**, **denominator**, and **unit fraction** in his revised rule, he was using clear mathematical language.*

It is also important to be precise and accurate when measuring, calculating, or counting.

Think about this problem:

Ava and her little brother Billy want to order a cell phone case to give their mother for her birthday. They need to measure the cell phone so they can get a case that is the right size.

Billy measures the length and width of the cell phone with square-inch tiles:

Ava arranges the tiles so there are no spaces. Billy looks at Ava's measurement and says, "OK. Now I see that we need to order a case that is 2 inches wide and 4 inches long."

Ava thinks about the measurement with square-inch tiles and says, "But look. We can see that a case that is 2 inches wide and 4 inches long will be too small. I can use a ruler to measure the phone with a smaller unit, like a half inch or a quarter inch. That will give us a better measurement for ordering a case."

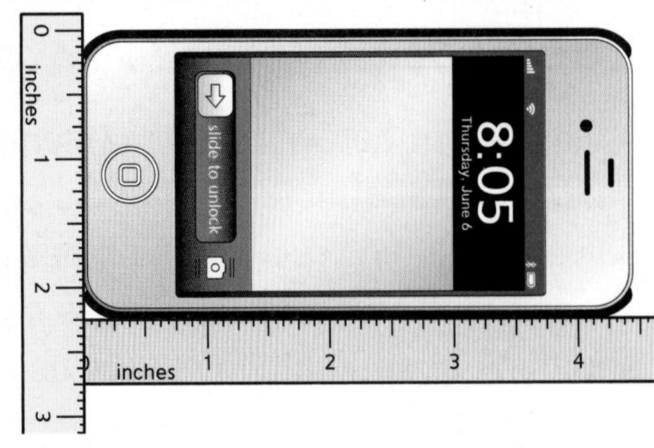

Ava

GMP6.4 Think about accuracy and efficiency when you count, measure, and calculate.
Ava helped Billy measure more accurately when she arranged the tiles so that there were no spaces between them.

GMP6.2 Use an appropriate level of precision for your problem.
Ava chose a smaller unit to measure the phone. Smaller units give more precise measurements. Sometimes more precise measurements are more appropriate when solving a problem.

Mathematical Practice 6: Attend to precision.

Check Your Understanding

1. Choose the units that are appropriate for each situation.

a. Finding the distance from your home to school

city blocks   inches   miles   centimeters

b. Ordering carpet for your bedroom

square inches   square yards   square kilometers

2. How did you decide whether a unit was appropriate for a situation in Problem 1?

Check your answers in the Answer Key.

Look for Structure in Mathematics

An important part of doing mathematics is looking for mathematical structures such as properties, categories, and patterns. Finding relationships between numbers, expressions, shapes, and mathematical ideas can help you solve problems and think about math in new ways.

A **property**, or attribute, of a geometric figure is a mathematical structure. Attributes can help you put shapes into categories.

C B A

Category: Parallelograms

Emily's group is sorting shapes into two categories: "Parallelograms" and "Not Parallelograms."

Emily knows that a parallelogram has four sides and two pairs of parallel sides. She is trying to decide how to classify a square.

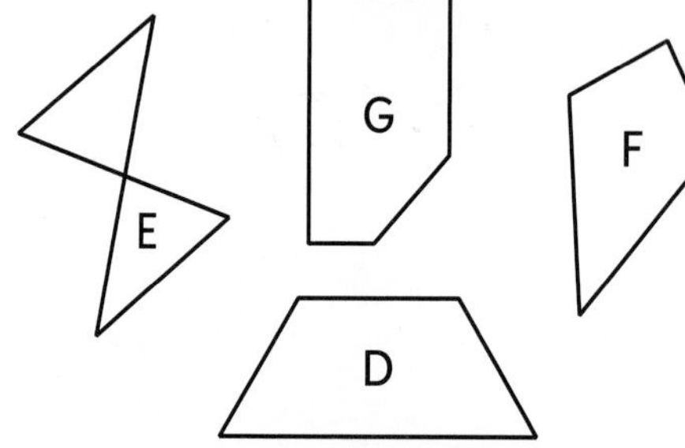

Category: Not Parallelograms

Emily

Ryan

Emily thinks more about the properties of squares and parallelograms and realizes that a square is a *special* parallelogram. Shapes can have more than one name.

GMP7.1 Look for mathematical structures such as categories, patterns, and properties. *At first, Emily didn't notice the properties that a parallelogram and square have in common. She was only thinking about the most common names for these shapes. When Ryan shared his thinking about the properties of squares, he helped Emily think about categories of shapes in a new way.*

Patterns are another kind of mathematical structure. Looking for and using patterns can help you solve problems.

Ava, Kayla, and Justin are helping their teacher solve this problem:

Mr. Bates plans to set up tables for a parent meeting. He has trapezoid-shaped tables that can fit one chair along each short side and two chairs along the long side.

He wants to set up the tables end to end in one long row. How many tables will he need to set up end to end in order to have 38 chairs?

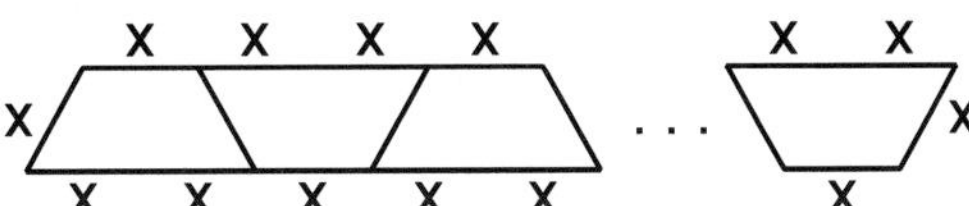

Ava draws 1 table, 2 connected tables, and then 3 connected tables. She labels her drawings with the number of chairs that can fit in each arrangement.

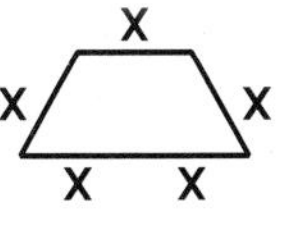

5 chairs

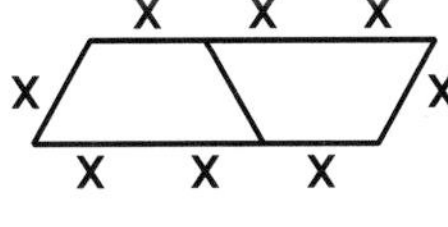

8 chairs

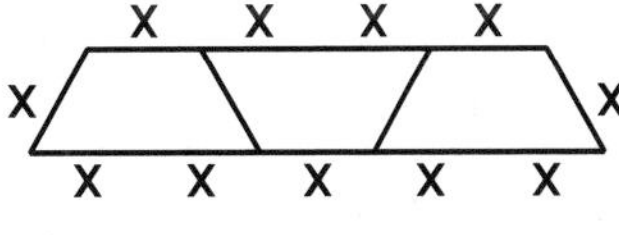

11 chairs

Justin looks carefully at Ava's drawing, and he notices a pattern. Every time they add a table, the total number of chairs increases by 3. He decides to see whether the pattern continues and draws 4 connected tables.

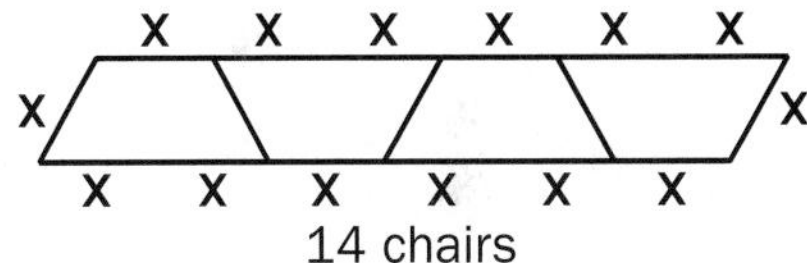

14 chairs

The first table fits 5 chairs. When you put 2 tables together, 10 chairs won't fit. Chairs can't go on the edges that are touching, so 2 tables only fit 8 chairs total. When I add a third table, it only adds 3 more chairs.

Look at my drawing of 4 tables. The pattern keeps going. I think that every time we add another table, it will add 3 more chairs.

Ava

Justin

Kayla realizes that it would take a long time to answer Mr. Bates's question by continuing the pattern until they get to 38 chairs. She makes a table of the numbers they have so far, thinks of the numbers as ordered pairs, and draws a graph.

Tables	Chairs	(x, y)
1	5	(1, 5)
2	8	(2, 8)
3	11	(3, 11)
4	14	(4, 14)

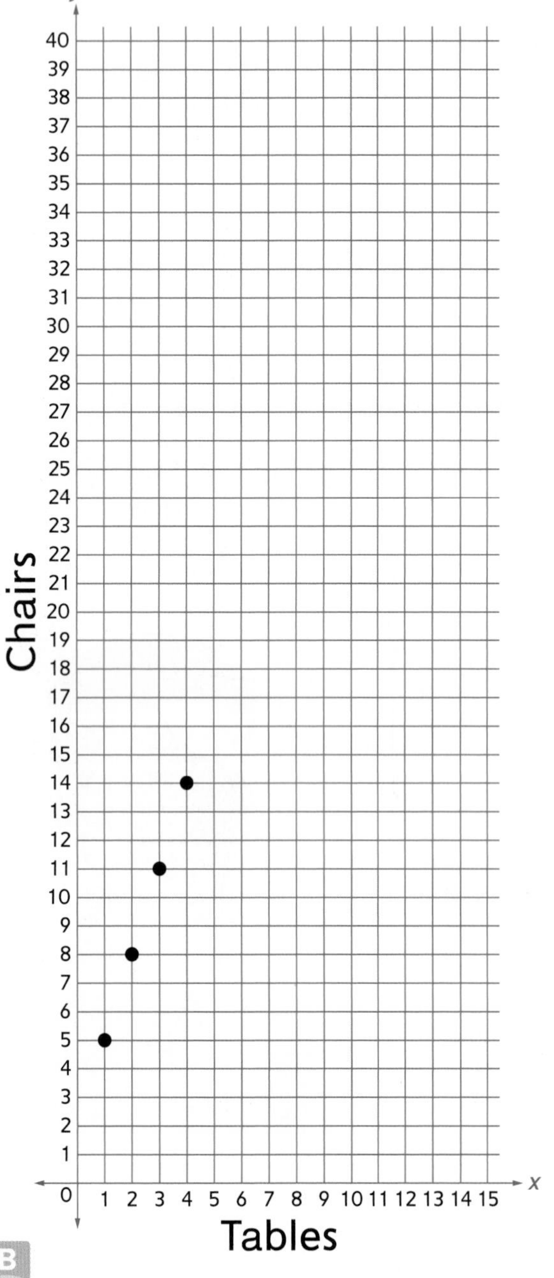

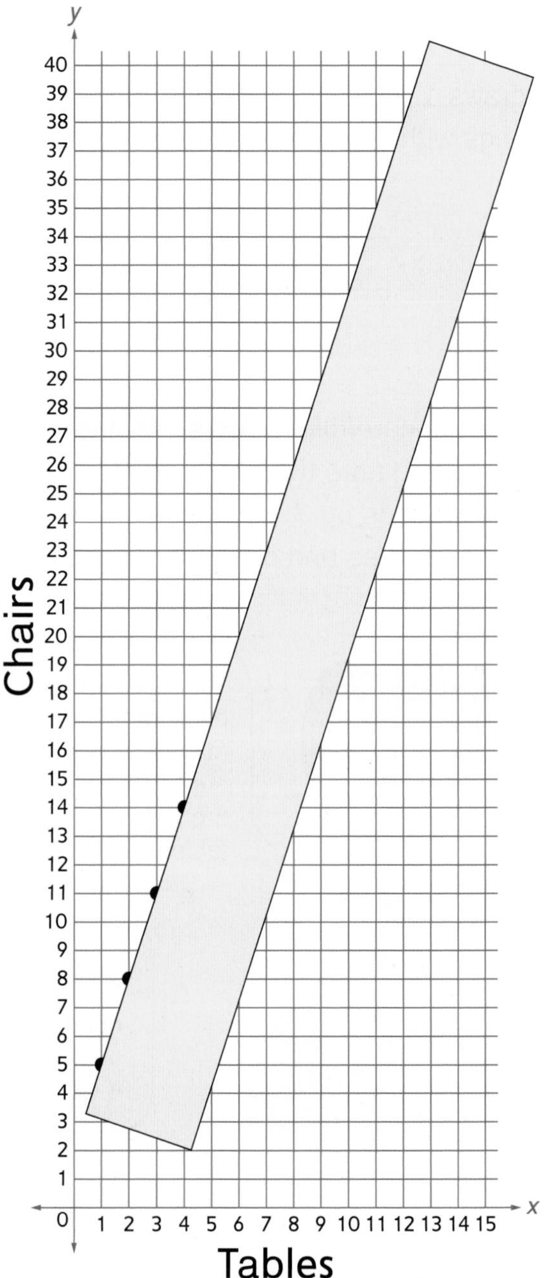

Kayla

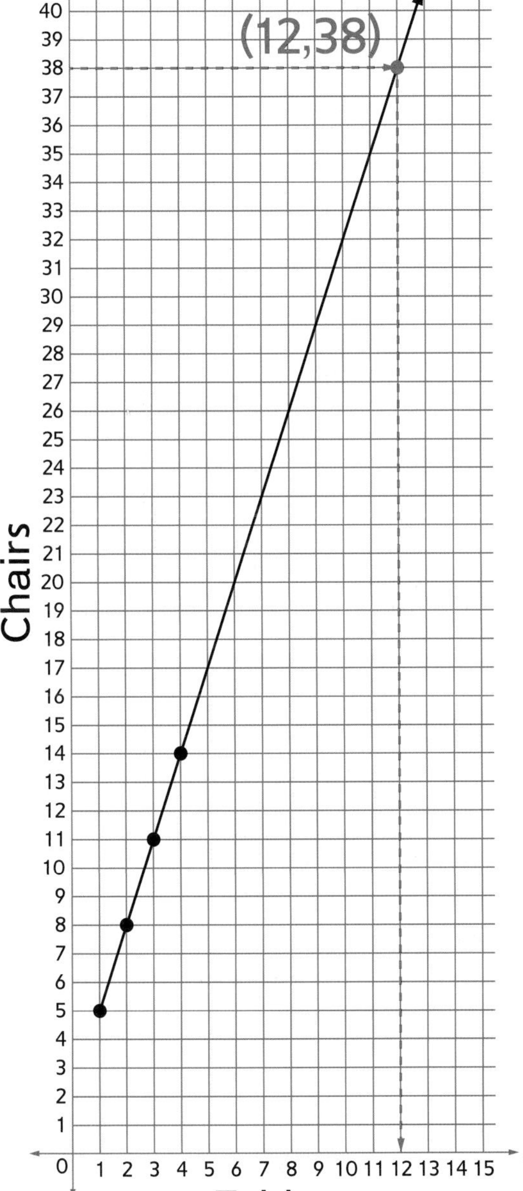

Mr. Bates says, "Kayla, your line is a good idea. It can help you figure out how many chairs will fit with different numbers of tables. Remember that it will only work for whole tables. You can't have half of a table. If you needed 12 chairs, you would have to round up to 4 tables."

Ava uses Kayla's graph to solve the problem.

Ava

Justin

Tables	Chairs	(x, y)
1	5	(1, 5)
2	8	(2, 8)
3	11	(3, 11)
4	14 +3	(4, 14)
5	17 +3	
6	20 +3	
7	23 +3	
8	26 +3	
9	29 +3	
10	32 +3	
11	35 +3	
12	38	

GMP7.2 Use structures to solve problems and answer questions.
Justin, Kayla, and Ava looked for patterns in the numbers in their drawings, table, and graph. They used these patterns to make sense of the problem and solve it.

Mathematical Practice 7: Look for and make use of structure.

Check Your Understanding

1. What properties of shapes D, E, F, and G on page 22 made Emily and Ryan put them in the "Not Parallelograms" category?
2. The school also has square tables that fit one chair on each side. If Mr. Bates uses square tables instead of trapezoid-shaped tables, how many tables would he need to fit 38 chairs?

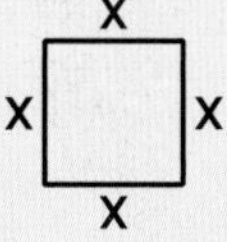

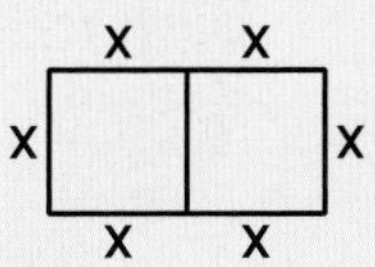

Check your answers in the Answer Key.

Create and Explain Rules and Shortcuts

An important part of doing mathematics is looking for shortcuts, rules, and generalizations to make procedures and operations more efficient, so that when you see a similar problem, you can solve it more easily. When you create a shortcut, rule, or generalization, you need to be able to justify it. For example, when you justify a shortcut, you use mathematical properties to explain how and why it works.

Mr. Bates asks his class to figure out an efficient way to multiply powers of 10. He lists these problems:

10,000 * 10	100 * 1,000	1,000 * 100,000
10,000 * 1,000	100 * 10,000	

Sophia begins to multiply using her calculator.

10,000 * 10

100 * 1,000

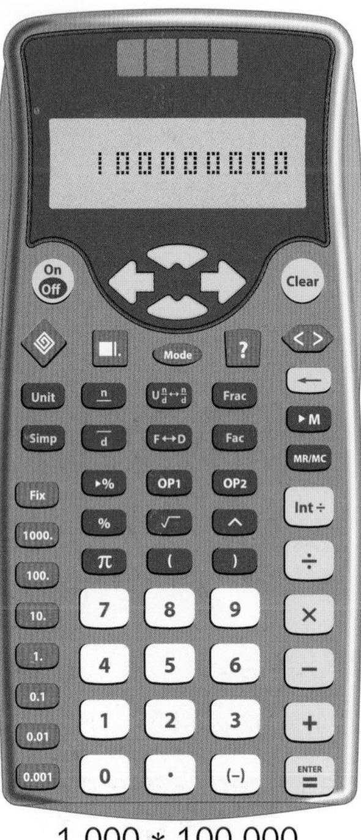

1,000 * 100,000

Sophia

Josh, I just noticed that the total number of zeros in the factors is the same as the number of zeros in the product.

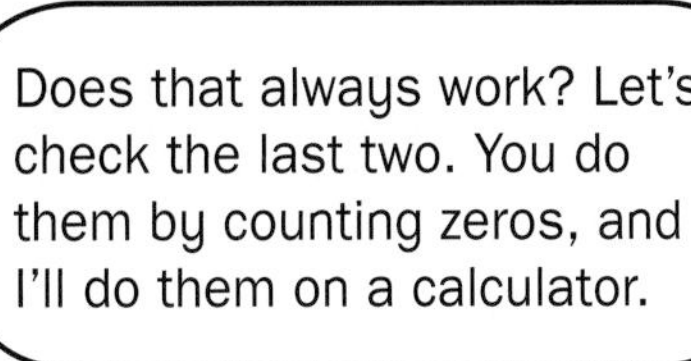

Josh

Josh uses a calculator.

Sophia counts zeros.

10,000 * 1,000 = 10000000

(4 zeros) + (3 zeros) → (7 zeros in all)

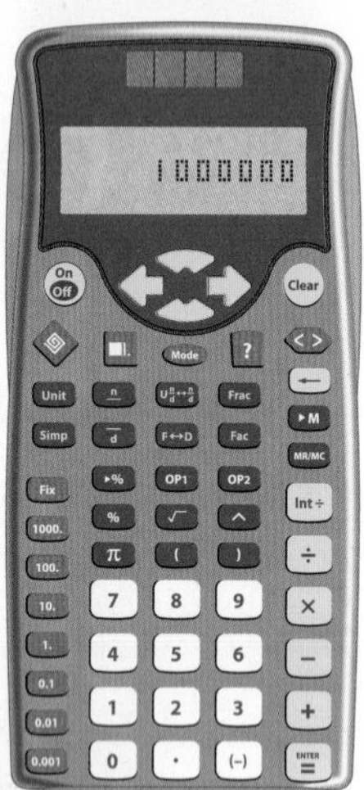

100 * 10,000 = 1000000

(2 zeros) + (4 zeros) → (6 zeros in all)

Sophia

Josh

Mr. Bates asks Josh and Sophia to see if they can explain *why* their shortcut always works.

Josh explains, “Think about place value. A digit in one place represents 10 times what it represents in the place to the right. So each time you multiply a number by 10, the digits move one place to the left, and you put a zero in the ones place.”

$10 * 10 = 100$

$100 * 10 = 1{,}000$

$1{,}000 * 10 = 10{,}000$

$10{,}000 * 10 = 100{,}000$

$100{,}000 * 10 = 1{,}000{,}000$

I get it. We know 1,000 = 10 * 10 * 10.

So you can write the problem 100 * 1,000 as 100 * (10 * 10 * 10). Multiplying 100 by 10 *three* times will move the 1 that started in the hundreds place *3* places to the left. We start with 2 zeros in 100 and need 3 more zeros to show how much larger the product is. That's 5 zeros in all, or 100,000.

Sophia

Mr. Bates asks Sophia and Josh to each write the shortcut on the whiteboard in their own words.

Sophia's Written Shortcut	Josh's Written Shortcut
When you multiply two powers of 10, you: (1) Count the total number of zeros in each of the factors. (2) Write the product as a 1 followed by the total number of zeros you counted.	When you multiply a power of 10 times a power of 10, to figure out the product, you: (1) Write the first factor. (2) Then write more zeros based on the number of zeros in the second factor. (3) Insert the commas.

GMP8.1 Create and justify rules, shortcuts, and generalizations.

Sophia created a shortcut for multiplying powers of 10 by counting zeros. Josh and Sophia justified the rule based on properties of place value.

Mathematical Practice 8: Look for and express regularity in repeated reasoning.

Check Your Understanding

Think about Sophia's and Josh's written rules.

1. How are they alike?

2. How are they different?

3. Do both work? Why or why not?

Check your answers in the Answer Key.

A Problem-Solving Diagram

When you solve problems, you work to make sense of the problem. You reflect on your thinking and keep trying when the problem is hard. As a good problem solver, you always check to see whether your answer makes sense by trying to solve the problem in more than one way and by comparing the strategies you use to those others use.

The diagram below can help you think about problem solving. The boxes in the diagram show the type of things you do when you use mathematical practices to solve problems. The arrows show that you don't always do things in the same order.

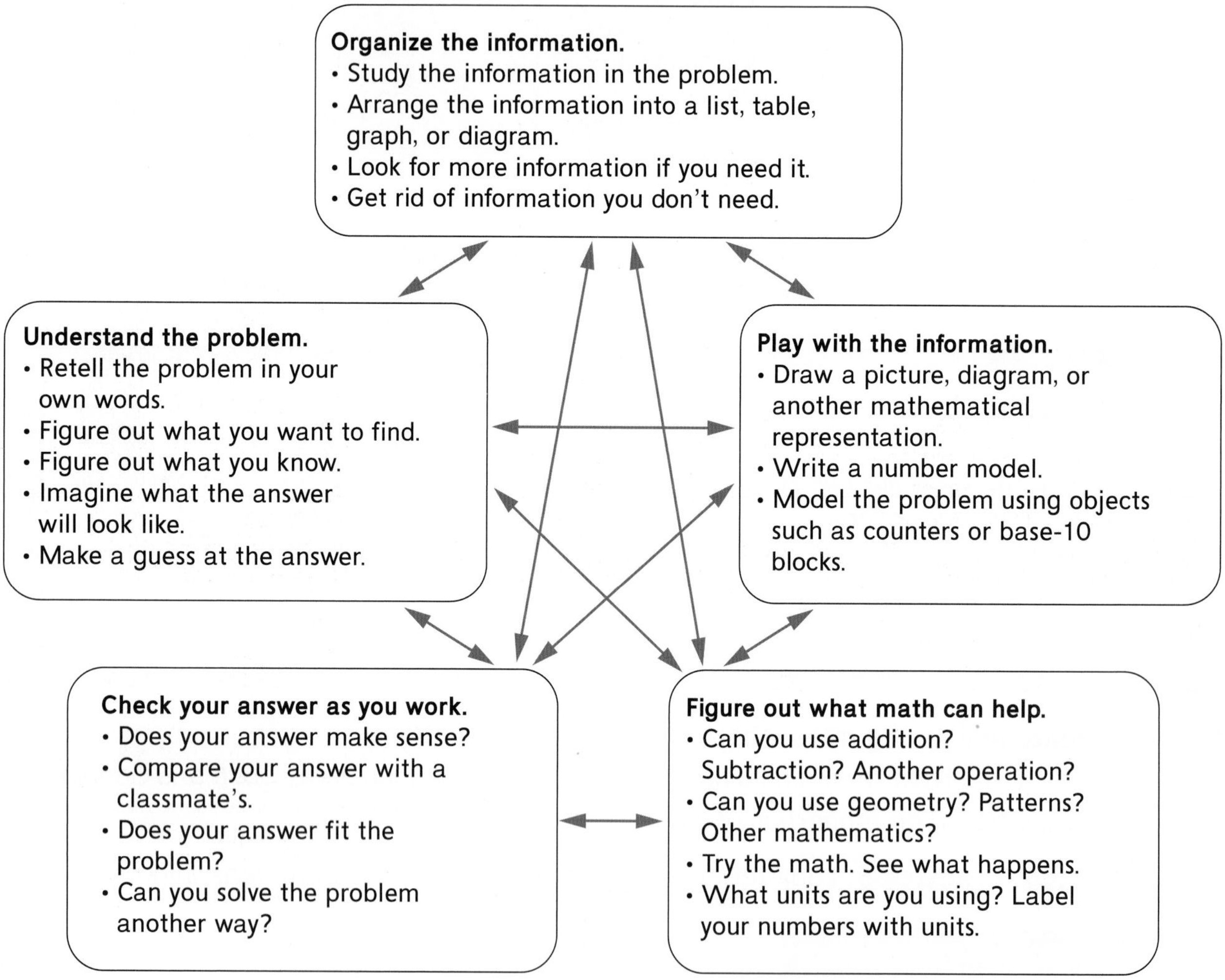

Organize the information.
- Study the information in the problem.
- Arrange the information into a list, table, graph, or diagram.
- Look for more information if you need it.
- Get rid of information you don't need.

Understand the problem.
- Retell the problem in your own words.
- Figure out what you want to find.
- Figure out what you know.
- Imagine what the answer will look like.
- Make a guess at the answer.

Play with the information.
- Draw a picture, diagram, or another mathematical representation.
- Write a number model.
- Model the problem using objects such as counters or base-10 blocks.

Check your answer as you work.
- Does your answer make sense?
- Compare your answer with a classmate's.
- Does your answer fit the problem?
- Can you solve the problem another way?

Figure out what math can help.
- Can you use addition? Subtraction? Another operation?
- Can you use geometry? Patterns? Other mathematics?
- Try the math. See what happens.
- What units are you using? Label your numbers with units.

When you and your classmates solve problems, you can start anywhere in the diagram. You can use different suggestions from different parts of the diagram as you work.

Try this problem and think about how your problem solving fits in the Problem-Solving Diagram:

Here is an up-and-down staircase that is 5 steps tall. How many squares are needed for an up-and-down staircase that is 10 steps tall?

Some fifth graders are using the Problem-Solving Diagram in different ways.

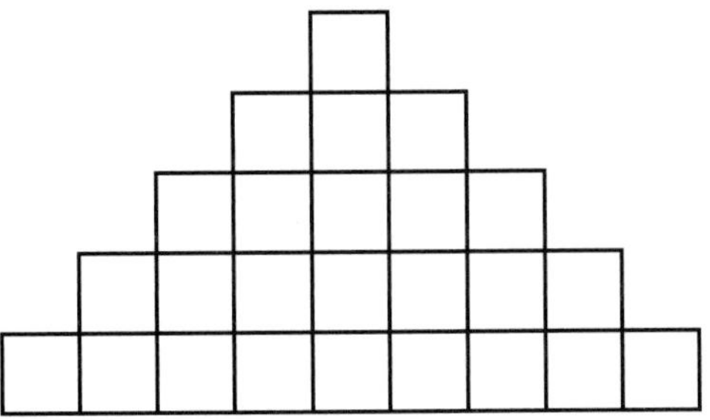

I didn't quite understand the problem, so I began by **playing with the information.** I experimented by drawing a new layer of squares below the bottom layer in the picture.

My picture helped me realize that each new bottom layer has two more squares than the layer above.

Sophia

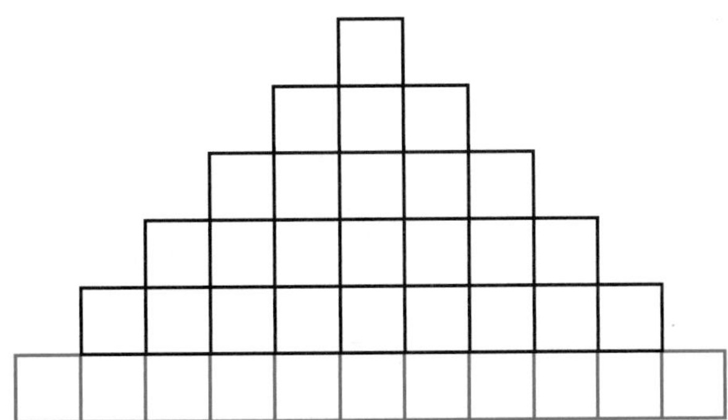

Ava

Emily and I started by **trying to understand the problem**. We told each other the problem in our own words, and then tried to imagine how many squares might be in the 10-step staircase.

We imagined that the 10-step staircase would be *a whole lot more* than the 5-step staircase, because the new staircase would be more than twice as wide. We would need to add lots and lots of new squares.

I looked at the picture and noticed right away that there was a pattern. There are two more squares in each layer as you go from top to bottom. So I **tried to figure out what math can help** by writing a number sentence that shows adding the number of squares in each layer and seeing what happens.

1 + 3 + 5 + 7 + 9 + 11 + 13 + 15 + 17 + 19 + 21 + 23 + ???

Ryan

The trouble was that I didn't know where to stop. I decided to **organize the information**. I went back and started to write the number of steps in the staircase below the addends from that same layer. Then I knew to stop at 19. Here's what it looked like after I got more organized.

1 + 3 + 5 + 7 + 9 + 11 + 13 + 15 + 17 + 19 ~~+ 21 + 23 + ???~~

1 2 3 4 5 6 7 8 9 10

Sophia

Can you explain to me what all your numbers mean?

Ryan

OK, I'll put labels on my work. That will help us **figure out what math can help**. I'll finish by adding the number of squares in each layer up to 10 layers.

Adding the number of squares in each layer:

1 + 3 + 5 + 7 + 9 + 11 + 13 + 15 + 17 + 19 ~~+ 21 + 23 + ???~~

The layer number is the same as the number of steps:

1 2 3 4 5 6 7 8 9 10

The answer is the sum of the number of squares in all 10 layers: 100

I **checked the answer** by using a calculator. The sum is 100 squares.

Ryan

Emily

$$1 + 3 + 5 + 7 + 9 + 11 + 13 + 15 + 17 + 19$$

(1 + 9 = 10, 3 + 7 = 10, 11 + 19 = 30, 13 + 17 = 30)

$$10 + 10 + 5 + 30 + 30 + 15 = 100 \text{ squares}$$

Using the Problem-Solving Diagram

When you use the Problem-Solving Diagram to think about how you solve problems, you'll find that you solve problems in different ways and in different orders.

Check Your Understanding

Use the Problem-Solving Diagram to think about how the fifth-grade students solved the problem on pages 4–7. Describe how their process fits the diagram.

Operations and Algebraic Thinking

Algebra

Algebra is a part of mathematics that uses mathematical statements to describe patterns, express numerical relationships, and model real-world situations. Algebra is like arithmetic, but algebra uses letters, blanks, question marks, and other symbols to stand for quantities that are not known or that vary. By learning how to write and use these mathematical statements, you will be able to solve problems more easily.

You use algebra when you write number models to represent real-world situations to help solve problems. Figuring out missing numbers and working backwards from what you know are types of algebraic thinking.

Example

Solve: $n + \frac{1}{3} = 1$

Think about what you know.

You can think about fraction circles.

1 red circle is the same as 3 orange pieces.

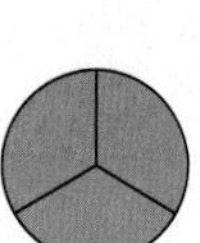

$1 = \frac{3}{3} = \frac{1}{3} + \frac{1}{3} + \frac{1}{3}$

So, n must equal $\frac{1}{3} + \frac{1}{3} = \frac{2}{3}$.

$n = \frac{2}{3}$

Did You Know?

Algebra often involves solving number problems with one or more numbers that are not known. The missing numbers are called the "unknowns." In early times, words were used for the unknowns, as in "Eight times some number equals eight hundred." Then in the late 1500s, François Viète began using letters, as in $8 * x = 800$, to stand for unknown quantities. Viète's invention made solving number problems much easier and led to many discoveries in mathematics and science.

Variables and Unknowns

The letters or other symbols that you sometimes see in number sentences are called **variables.** Variables can be used in multiple ways.

Variables Can Be Used to Stand for Unknown Numbers. In the equation $8 * x = 800$, the variable x stands for an unknown number. The variable could stand for any number, but in this case only one number makes the number sentence true. Finding the number that makes a number sentence true is called "solving the number sentence." For this problem, the number 100 makes the number sentence true because $8 * 100 = 800$. Symbols such as question marks or blanks are also used for unknown numbers.

Variables Can Be Used to Stand for Quantities that Change or Vary. The number model $c = n * 15 + 2$ can be used to find the total cost of ordering a group of tickets if concert tickets cost $15 each and there is a $2 handling fee for ordering a group of tickets online. The variables in the number model are c which stands for the total cost of the tickets and n which stands for the number of tickets in an order. The cost and number of tickets vary from order to order.

Variables Can Be Used to State Properties of the Number System. **Properties** of the number system are statements that are true for all numbers.

Property	Example of the Property
$0 + a = a$	$0 + 5 = 5$
$1 * a = a$	$1 * 3.5 = 3.5$
$a + b = b + a$	$10 + 8 = 8 + 10$
$a * b = b * a$	$\frac{1}{2} * 4 = 4 * \frac{1}{2}$
$a = a$	$47.5 = 47.5$

Variables are often used in statements to describe properties. For example, the following statement is a property: *Any number multiplied by 1 is equal to itself.* In the table to the right, this property is written as $1 * a = a$ where the variable a stands for *any number*.

Variables Can Be Used in Formulas. Formulas are used in everyday life, in science, in business, and in many other situations as an easy way to describe relationships. For example, the formula for the perimeter of a square is $p = 4 * s$, where p is the perimeter and s is the length of a side. The formula for the area of a square is $A = s^2$, where A is the area.

Variables Can Be Used to Express Rules or Functions. Function machines and "What's My Rule?" tables have rules that tell you how to get the "out" numbers from the "in" numbers. These rules can be written using variables. For example, a "What's My Rule?" table might have the rule, "$* 3$," which means you triple the "in" number to get the "out" number. This rule can be written as $y = 3 * x$ using variables.

Rule
$y = 3 * x$

in	out
x	y
0	0
1	3
2	6
3	9
...	...

In some rules and formulas, numbers and variables are written next to each other without an operation symbol between them. This means that the variable and the number are to be multiplied. For example, the rule $y = 3 * x$ can also be written $y = 3x$.

Variables Can Be Used in Computers and Calculators. Variables are used in computer spreadsheets, which makes it possible to apply formulas quickly and efficiently. Computer programs are made up of a series of "commands" that contain variables. These commands look very much like number sentences that contain variables.

Expressions

An **expression** is a group of mathematical symbols that represents a number. An expression may include numbers, variables, operation symbols, and grouping symbols, but does *not* include relation symbols (such as =, >, or <).

An expression like (10 + 7) * 6, which use numbers and not variables is a *numerical expression.* An expression that contains a variable, such as c + 5, is an *algebraic expression.*

Examples

All of the following are expressions:

$\frac{3}{4} * 9$ (15 * *a*) + 4 10^4 [12 ÷ (6 − 3)] + 2 *n* * 4.1 23 − 0.98

Evaluating Expressions

To evaluate something is to find out what it is worth. To **evaluate a numerical expression,** perform the operations in the order described below.

1. Do operations inside parentheses or other grouping symbols first.
2. Calculate all expressions with exponents. (For example: $10^2 = 10 * 10 = 100$)
3. Multiply and divide in order, from left to right.
4. Add and subtract in order, from left to right.

Examples

Evaluate the expression: $8.7 * 10^4$	**$8.7 * 10^4$**
Calculate all expressions with exponents.	8.7 * 10,000
Then multiply.	87,000
Evaluate the expression: 4 + (15 * 3)	**4 + (15 * 3)**
Do operations inside parentheses first.	4 + 45
Then add.	49

To evaluate an algebraic expression, first replace each variable with a given value.

Example

Evaluate the algebraic expression *k* * 4 when *k* = 5.

	***k* * 4**
Replace *k* with 5. Then multiply.	5 * 4 = 20

If *k* = 5, then *k* * 4 is equal to 20.

Number Sentences

Number sentences are made up of mathematical symbols.

Mathematical Symbols				
Digits	Variables	Operation Symbols	Relation Symbols	Grouping Symbols
0, 1, 2, 3, 4, 5, 6, 7, 8, 9	a b n x l w A P ? □	+ − × * / ÷	= ≠ ≈ < > ≤ ≥	( ) [ ] { }

A **number sentence** must contain numbers (or variables) and a **relation symbol.** Number sentences that contain the = symbol are called **equations.** Number sentences that contain symbols such as ≠, <, or >, are called **inequalities.**

A number sentence that has no variables is either **true** or **false.** You may need to perform operations on one or both sides of the relation symbol to see whether it is true or false.

Examples

Determine whether each number sentence is true or false.

$200 + 50 + 8 = 258$

This number sentence is **true** because $200 + 50 + 8$ *is* equal to 258.

$5 < \frac{1}{5}$

This number sentence is **false** because 5 *is not* less than $\frac{1}{5}$. 5 is greater than $\frac{1}{5}$.

$30 * 60 > 20 * 80$

This number sentence is **true** because $30 * 60 = 1{,}800$, $20 * 80 = 1{,}600$, and 1,800 *is* greater than 1,600.

In some number sentences, a **variable,** such as a letter or question mark, takes the place of one or more of the numbers. You cannot tell whether a number sentence such as $9 * ? = 54$ is true or false until you know what number replaces the variable.

Solving Number Sentences

If a number used in place of a variable makes the number sentence true, that number is called a **solution of the number sentence.** For example, the number 6 is a solution of the number sentence $9 * ? = 54$ because the number sentence $9 * 6 = 54$ is true. When you are asked to solve a number sentence, you are being asked to find its solution(s).

Example

Solve the number sentence: $15 - n = 14.25$.

Think: What number could replace n to make the number sentence true?
The difference between 15 and 14.25 is 0.75. Try 0.75.

Is $15 - 0.75 = 14.25$ a true number sentence? Yes, so the solution is 0.75.

Relations

A relation tells how two things compare. The table below shows the most common relations that compare numbers and lists their symbols.

Symbol	Relation
$=$	is equal to
$\neq$	is not equal to
$\approx$	is approximately equal to
$<$	is less than
$>$	is greater than
$\leq$	is less than or equal to
$\geq$	is greater than or equal to

Note *Reminder:* When writing $>$ or $<$, be sure the arrow tip points to the smaller number.

Equations

A number has many different names. For example, the expressions 16, $10 + 6$, $4 * 4$, and $1.6 * 10$ are different names for the same number (sixteen). Expressions that name the same number are called **equivalent names.** Expressions that name the same number are equal.

One way to state that two things are equal is to write a number sentence using the $=$ symbol. Any two of the expressions 16, $10 + 6$, $4 * 4$, and $1.6 * 10$ are equal because they all name the same number. So we can write many true number sentences using the $=$ symbol: $4 * 4 = 1.6 * 10$, $10 + 6 = 16$, and so on.

Number sentences that contain the $=$ symbol are called **equations.** An equation that does not contain a variable is either true or false.

Examples

Here are some true and false equations:

$17.9 = 5.3 + 12.6$	$12 = 12$	$(4 * 100) + (5 * 10) = 405$	$10^3 = 1{,}000$
True	True	False	True

Check Your Understanding

Write true or false for each equation.

1. $7 * 80 = 80 * 7$ **2.** $(330 / 10) + 2 = 32$ **3.** $\frac{1}{4} + \frac{1}{4} = \frac{2}{8}$

4. $9{,}872 = 9{,}000 + 800 + 70 + 2$ **5.** $\frac{7}{8} = \frac{7}{8}$ **6.** $54 = 15 + 51 - 2$

Check your answers in the Answer Key.

Inequalities

Number sentences that do not contain the = symbol are called **inequalities.** Like equations, inequalities may be true or false.

Examples

Here are some inequalities:

$\frac{7}{8} + \frac{5}{6} < 1$	$26.9 > 26.9 - 4.7$	$3 * 10 \neq 10^3$
False	True	True

Note You can often use estimation or mathematical reasoning to decide whether an inequality is true or false. For example, by thinking "$\frac{7}{8}$ is close to one, and $\frac{5}{6}$ is close to one," you can estimate that $\frac{7}{8} + \frac{5}{6}$ is close to two. It would not make sense for the sum to be less than one, so you can reason that the number senence $\frac{7}{8} + \frac{5}{6} < 1$ is false.

The symbols ≤ and ≥ combine two meanings. ≤ means "is less than *or* equal to"; ≥ means "is greater than *or* equal to."

Examples

Here are some true and false inequalities:

$5 \leq 5$ True	$300 \geq 350$ False	$\frac{3}{4} + \frac{2}{3} \geq 3$ False
$35 \geq 28.9 + 15.3$ False	$48 * 643 \leq 35{,}000$ True	$5 * 100 \leq 500$ True

Inequalities on the Number Line

For any pair of numbers on the number line, the number to the left is less than the number to the right.

Example

Use the number line to complete the number sentence. 

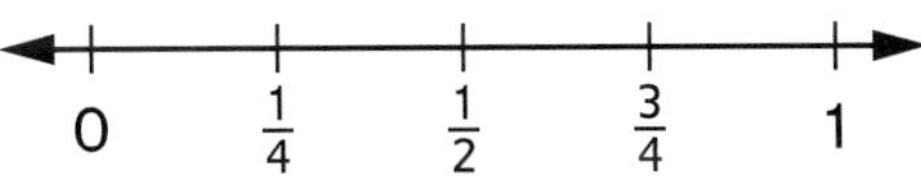

$\frac{1}{2}$ is to the left of $\frac{3}{4}$, so $\frac{1}{2}$ is less than $\frac{3}{4}$. $\frac{1}{2} < \frac{3}{4}$

Example

Use the number line to complete the number sentence. 27.7 ____ 27.5

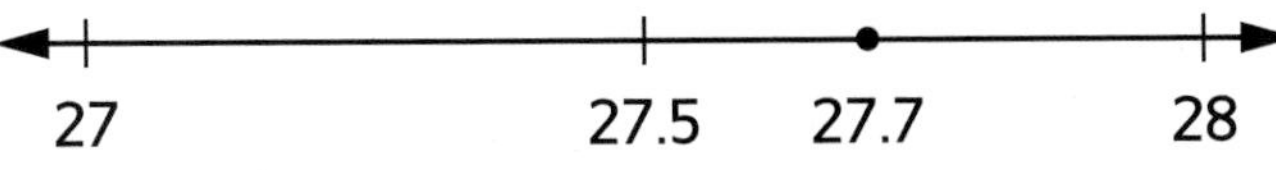

27.7 is to the right of 27.5, so 27.7 is greater than 27.5. 27.7 > 27.5

Grouping Symbols

Grouping symbols, such as parentheses ( ), brackets [ ], and braces { }, make the meaning of a number sentence or expression clear. When there are grouping symbols in a number sentence or expression, the operations inside the grouping symbols are always done first.

Examples

Evaluate the following expression: (17 – 4) * 3

The parentheses tell you to subtract 17 – 4 first.	(17 – 4) * 3
Then multiply by 3.	13 * 3
The answer is 39.	39

Evaluate the following expression: 17 – (4 * 3)

The parentheses tell you to multiply 4 * 3 first.	17 – (4 * 3)
Then subtract 12 from 17.	17 – 12
The answer is 5.	5

Example

Insert parentheses to make the following number sentence true:
5 * 1 + 8 = 45

There are two possible places to insert parentheses.
Try each possibility.

Is (5 * 1) + 8 = 45 true?

With parentheses around 5 * 1, 5 is multiplied by 1 before 8 is added. This results in 13, not 45, so the number sentence is false.	(5 * 1) + 8 5 + 8 13

Is 5 * (1 + 8) = 45 true?

With parentheses around 1 + 8, 1 and 8 are added before their sum is multiplied by 5. This results in 45, which makes the number sentence true.	5 * (1 + 8) 5 * 9 45

Parentheses should be inserted around 1 + 8, making the number sentence 5 * (1 + 8) = 45.

Which three different relation symbols can you use to make this number sentence true?
(17 – 4) * 3 ____ 17 – (4 * 3)

Check Your Understanding

Insert parentheses to make each number sentence true.

1. 25 – 15 + 10 = 0 **2.** 100 = 9 + 1 * 10 **3.** 4 + 1 / 10 – 5 = 1

Check your answers in the Answer Key.

In some situations, more than one set of grouping symbols is needed. When one set of parentheses is inside another set of parentheses, they are called *nested parentheses.* The operations in the innermost parentheses are carried out first.

Example

Solve the following number sentence: 100 – ((3 + 7) ∗ 4) = **?**

There are two sets of parentheses.

Do the operation within the innermost parentheses first. Start by adding 3 and 7.	100 – ((3 + 7) ∗ 4) = ?
Then the outer parentheses tell you to multiply 10 and 4.	100 – (10 ∗ 4) = ?
Finally, subtract 40 from 100.	100 – 40 = ?
The answer is 60.	= 60

100 – ((3 + 7) ∗ 4) = **60**

When parentheses are nested, one set of parentheses is sometimes replaced with brackets or braces.

Example

Evaluate the following expression: 20 ∗ [5 + (18 ÷ 6)]

There are two sets of grouping symbols: brackets and parentheses.

The parentheses are inside of the brackets, so divide 18 by 6 first.	20 ∗ [5 + (18 ÷ 6)]
Then the brackets tell you to add 5 and 3.	20 ∗ [5 + 3]
Finally, multiply 20 by 8.	20 ∗ 8
The answer is 160.	160

20 ∗ [5 + (18 ÷ 6)] = **160**

Check Your Understanding

Evaluate each expression.

1. 15 – (6 ÷ 2) **2.** 22 + ((10 + 5) / 5) **3.** 5 ∗ {(3 / 3) – (2 / 2)}

State whether the number sentences are true or false.

4. 2 ∗ [3 ∗ (4 ∗ 5)] = [(2 ∗ 3) ∗ 4] ∗ 5

5. (4 + 2) / (1 + 2) = 4 + (2 / 1) + 2

Check your answers in the Answer Key.

Writing Expressions and Number Sentences

Number Models

Writing an expression or number sentence is one way to model a situation. A number sentence or an expression that describes a situation is called a **number model.** Often two or more number models can fit a given situation.

Example

Write number models that describe each problem situation.

Problem	Number Models	
	Expression	Number Sentence
Zaria walked $\frac{3}{4}$ mile on Monday and $1\frac{1}{2}$ miles on Tuesday. How many miles did she walk all together?	$\frac{3}{4} + 1\frac{1}{2}$	$\frac{3}{4} + 1\frac{1}{2} = m$
Jonathan had $20 and spent $12.89 on a DVD. How much money does he have left?	$20 − $12.89	$20 − $12.89 = *r* or $12.89 + *r* = $20
A school has 6,825 unit cubes. If the cubes are separated into bags of 25, how many bags will the school need to store all of the unit cubes?	6,825 ÷ 25	6,825 ÷ 25 = *b* or 25 ∗ *b* = 6,825
An elephant at the zoo weighs about 3 tons. There are 2,000 pounds in 1 ton. About how much does the elephant weigh in pounds?	3 ∗ 2,000	3 ∗ 2,000 ≈ *w*

Number models can help you figure out which operations to use to solve problems. They can also help you report your answer after you have solved a problem.

Chris Willig

Recording Calculations

Expressions and number sentences can be used to record calculations with numbers and to summarize solutions to problems. If you used more than one operation to solve a problem, you may need to use grouping symbols to show which operation you performed first.

Example

Write an expression to record the calculations.

Subtract the sum of 10 and 15 from 30.

Think: What operation is performed first?

In order to find the sum of 10 and 15, you must add 10 and 15. $10 + 15$

Then subtract the sum from 30. Use parentheses to show which operation you complete first. $30 - (10 + 15)$

The expression $30 - (10 + 15)$ shows that the sum of $10 + 15$ is subtracted from 30.

Example

Solve the problem and write a number sentence to summarize your solution. Report your solution using units.

Problem: Find the area of a rectangle that is 2 m long and $1\frac{1}{2}$ m wide.

Solution: Use the area formula $A = l * w$ and multiply 2 m by $1\frac{1}{2}$ m to get 3 m^2.

Number sentence: $2 \text{ m} * 1\frac{1}{2} \text{ m} = 3 \text{ m}^2$, or $2 * 1\frac{1}{2} = 3$

The area is 3 m^2.

Note You can write expressions and number sentences with units, as in $2 \text{ m} * 1\frac{1}{2} \text{m} = 3 \text{ m}^2$, or without units, as in $2 * 1\frac{1}{2} = 3$. Either way, make sure to keep track of the units you are working with so you know how to report your answer.

Check Your Understanding

Write a number model for each problem.

1. Mr. Delavan's class is going on a field trip. The cost is $50 for the bus rental plus $4.50 for each of his 26 students for food. What is the total cost?
2. Mariana has $2\frac{1}{3}$ cups of flour. She needs $3\frac{1}{2}$ cups of flour for a recipe. How much more flour does she need?

Check your answers in the Answer Key.

Interpreting Expressions

You can often interpret, or make sense of, expressions without evaluating them. In order to interpret an expression, think about the meaning of the operations in the expression.

Example

Expression	Sample Interpretation
10 + 5	The value of the expression is 5 more than 10, or 10 more than 5.
200 − 19	The value of the expression is 19 less than 200.
4 * 36	The value of the expression is 4 times as much as 36, or 36 times as much as 4.
600 ÷ 2	The value of the expression is half as much as 600, or 600 divided into 2 equal groups.
7 + (4 * 3)	The value of the expression is 7 more than the product of 4 and 3.

You can also think about the meaning of operations to compare two expressions without evaluating them.

Example

Determine whether each expression is greater than or less than 8.

$8 - 5$ Subtracting 5 from 8 will decrease the value of the expression, so $8 - 5 < 8$.

$8 + 1$ Adding 1 to 8 will increase the value of the expression, so $8 + 1 > 8$.

$8 \div 4$ Dividing 8 by 4 is like splitting 8 into 4 equal parts. Part of 8 is less than 8, so $8 \div 4 < 8$.

$2\frac{1}{2} * 8$ Multiplying 8 by $2\frac{1}{2}$ is like making $2\frac{1}{2}$ groups of 8. More than one group of 8 will be greater than 8, so $2\frac{1}{2} * 8 > 8$.

Example

Compare the expressions 2 * (1,568 + 427) and 1,568 + 427.

The expression 2 * (1,568 + 427) contains the expression 1,568 + 427, but multiplies it by 2.

So the value of 2 * (1,568 + 427) is twice as much as the value of 1,568 + 427.

When you interpret mathematical expressions, think carefully when multiplying or dividing with numbers less than 1.

Example

Predict whether the product will be greater or less than 4.

$\frac{1}{2} * 4$ Multiplying by $\frac{1}{2}$ is the same as taking half of 4, or splitting 4 into 2 equal parts and taking one of the parts. Part of 4 is less than 4, so $\frac{1}{2} * 4 < 4$.

$\frac{1}{2}$ of 4 cherries is 2 cherries.

Example

Predict whether the quotient will be greater or less than 4.

$4 \div 0.1$ Dividing by 0.1 is the same as asking: "How many tenths are in 4?" Since there are 10 tenths in 1 whole, there are 4 groups of 10 tenths in 4. Four groups of 10 is more than 4, so $4 \div 0.1 > 4$.

A square is one whole.

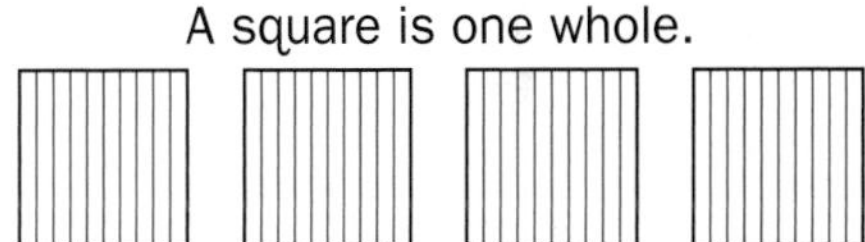

There are 40 groups of 0.1 in 4.

Check Your Understanding

Use <, >, or = to make the following number sentences true.
Try to complete the problems by interpreting the expressions instead of calculating.

1. 9,876 + 1,234 ____ 9,876 − 1,234 **2.** $\frac{7}{8} + \frac{1}{2}$ ____ $\frac{7}{8} + \frac{1}{3}$

3. 18 / 3 ____ $18 * \frac{1}{3}$

Check your answers in the Answer Key.

Using Inverse Operations to Check

Addition and subtraction are called **inverse operations** because one operation "undoes" the other. For example, if you add 5 to a starting amount, you can return to the starting amount by subtracting 5.

Similarly, multiplication and division are inverse operations because division undoes multiplication and multiplication undoes division. If you multiply a starting amount by 5, you can return to the starting amount by dividing by 5.

Note When you use fact families to solve a problem or check your work, you are using inverse operations. For example, the fact family $4 * 5 = 20$, $5 * 4 = 20$, $20 \div 4 = 5$, and $20 \div 5 = 4$ shows the relationship between multiplication and division.

You can use inverse operations to check your work with expressions and number sentences.

Example

Check that $14.56 - 2.31 = 12.25$.

Since 2.31 was subtracted from 14.56, adding 2.31 to the difference of 12.25 should bring the total back to 14.56.

$$\begin{array}{r} 12.25 \\ +\ \ 2.31 \\ \hline 14.56 \end{array}$$

Since $12.25 + 2.31 = 14.56$, it is true that $14.56 - 2.31 = 12.25$.

Example

Check that $149 * 16 = 2{,}384$.

The number sentence $149 * 16 = 2{,}384$ means that it takes 149 groups of 16 to make 2,384. Dividing 2,384 by 16 should give a quotient of 149.

$$\begin{array}{r|r} 16\overline{)2{,}384} & \\ -\ 1{,}600 & 100 \\ \hline 784 & \\ -\ 640 & 40 \\ \hline 144 & \\ -\ 144 & 9 \\ \hline 0 & 149 \end{array}$$

Since $2{,}384 \div 16 = 149$, it is true that $149 * 16 = 2{,}384$.

Example

Check that $4 \div \frac{1}{2} = 8$.

The number sentence $4 \div \frac{1}{2} = 8$ means that there are 8 [$\frac{1}{2}$s] in 4, so use multiplication to check.

$$8 * \frac{1}{2} = \frac{8}{2}$$
$$\frac{8}{2} = 4$$

Since $8 * \frac{1}{2} = 4$, it is true that $4 \div \frac{1}{2} = 8$.

Some Properties of Arithmetic

Certain facts are true of all numbers. Some of them are obvious, such as "every number equals itself," but others are less obvious. Since you have been working with numbers for years, you probably already know most of these facts, or **properties,** but you may not know their mathematical names.

The Identity Properties

The sum of any number and 0 is that number. For example, $15 + 0 = 15$. The **additive identity** is 0. Using variables, you can write this as $a + 0 = a$, where a is any number.

$a + 0 = a$

$0 + a = a$

The product of any number and 1 is that number. For example, $75 * 1 = 75$. The **multiplicative identity** is 1. Using variables, you can write this as $a * 1 = a$, where a is any number.

$a * 1 = a$

$1 * a = a$

The Commutative Properties

When two numbers are added, the order of the numbers does not change the sum. For example, $8 + 5 = 5 + 8$. This is known as the **Commutative Property of Addition.** Using variables, you can write this as $a + b = b + a$, where a and b are any numbers.

$a + b = b + a$

When two numbers are multiplied, the order of the numbers does not change the product. For example, $7 * 2 = 2 * 7$. This is known as the **Commutative Property of Multiplication.** Using variables, you can write this as $a * b = b * a$, where a and b are any numbers.

$a * b = b * a$

The Associative Properties

When three numbers are added, it makes no difference which two are added first. For example, $(3 + 4) + 5 = 3 + (4 + 5)$. This is known as the **Associative Property of Addition.** Using variables, you can write this as $(a + b) + c = a + (b + c)$, where a, b, and c are any numbers.

$(a + b) + c = a + (b + c)$

When three numbers are multiplied, it makes no difference which two are multiplied first. For example, $(3 * 4) * 5 = 3 * (4 * 5)$. This is known as the **Associative Property of Multiplication.** Using variables, you can write this as $(a * b) * c = a * (b * c)$, where a, b, and c are any numbers.

$(a * b) * c = a * (b * c)$

The Distributive Property

When you use an area model or partial products to multiply, you use the **Distributive Property.** The Distributive Property relates multiplication and addition.

		5	8
*			6
$6 * 50 \rightarrow$	3	0	0
$6 * 8 \rightarrow$ +		4	8
	3	4	8

For example, when you solve $6 * 58$ with partial products, you think of 58 as $50 + 8$, and multiply each part by 6.

The Distributive Property says: $6 * (50 + 8) = (6 * 50) + (6 * 8)$.

Using variables, you can write this as $a * (b + c) = (a * b) + (a * c)$.

Example

Show how the Distributive Property works by finding the area of the rectangle at the right in two different ways.

One way: Find the total length of the rectangle $(3 + 4)$ and multiply that by the width (5).

$5 * (3 + 4) = 5 * 7$
$= 35$

Another way: Find the area of each smaller rectangle, and then add the areas.

$(5 * 3) + (5 * 4) = 15 + 20$
$= 35$

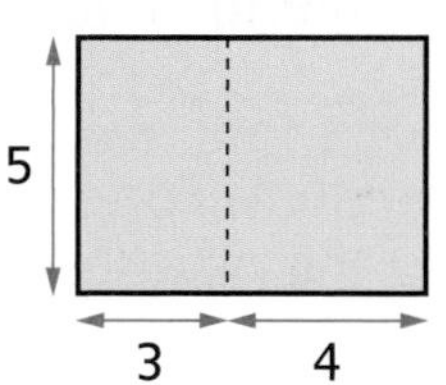

Both methods show that the area of the largest rectangle is 35 square units.

$5 * (3 + 4) = (5 * 3) + (5 * 4)$

Note The Distributive Property works with subtraction too. For example, you can solve $6 * 18$ by thinking about $6 * (20 - 2)$:

$6 * (20 - 2)$
$= (6 * 20) - (6 * 2)$
$= 120 - 12$
$= 108$
$6 * 18 = 108$

The Distributive Property also works with mixed numbers. Think of each mixed number as a whole number plus a fraction.

Example

Use the area model to multiply $5\frac{1}{3}$ by 4.

Think of $5\frac{1}{3}$ as $5 + \frac{1}{3}$. Multiply each part of $5\frac{1}{3}$ by 4.

Find the area of each smaller rectangle.

$5 * 4 = 20$

$\frac{1}{3} * 4 = \frac{4}{3}$, or $1\frac{1}{3}$

Add the smaller areas: $20 + 1\frac{1}{3} = 21\frac{1}{3}$.

The area model shows that $5\frac{1}{3} * 4 = 21\frac{1}{3}$.

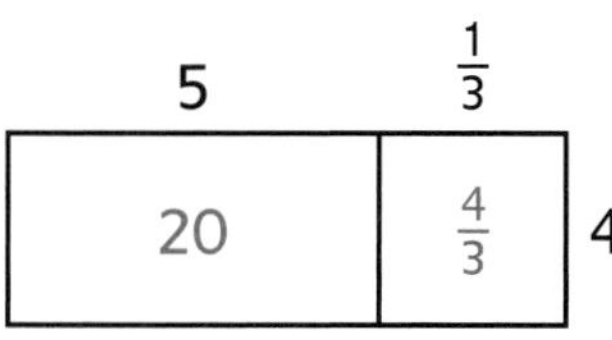

Number Patterns

When numbers are listed in a particular order according to a pattern, they form a **sequence.** There is often a rule that you can use to figure out the numbers in the list. For example, if you start from 0 and use the rule "Add 2," you get the sequence 0, 2, 4, 6, 8, and so on. The numbers in a sequence are sometimes called **terms.**

NOTE: In previous grades, you may have used **Frames-and-Arrows diagrams** to show number patterns. In a Frames-and-Arrows diagram, the frames hold the numbers, and the arrows show the path from one frame to the next. Each diagram has a rule box. The rule in the box tells how to get from one frame to the next. The numbers in the frames are the terms.

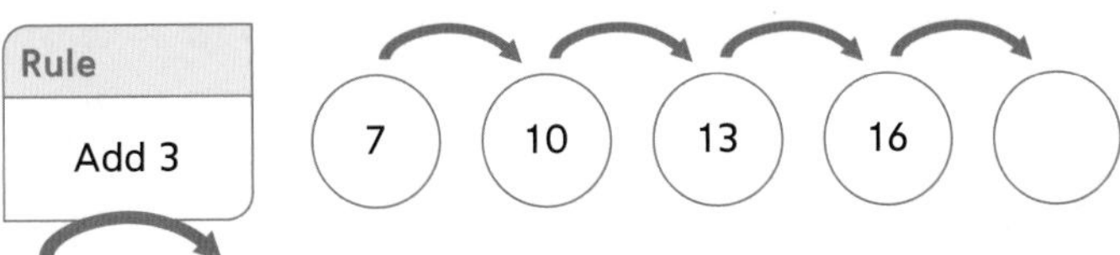

You can use a given rule to continue a sequence.

Example

Use the rule to find the next three terms in the sequence below.

Rule: Multiply by 2 **Sequence:** 2, 4, 8, . . .

Apply the rule to the last term in the sequence:

8 ∗ 2 = 16, so the fourth number in the sequence is 16.

16 ∗ 2 = 32, so the fifth number in the sequence is 32.

32 ∗ 2 = 64, so the sixth number in the sequence is 64.

With the next three terms, the sequence is: 2, 4, 8, 16, 32, 64.

You can also use the numbers in a sequence to find a rule. Make sure that the rule works for all of the terms that you are given.

Example

Find a rule for the following sequence: 99, 88, 77, 66, . . .

Compare the terms in order:

88 is 11 less than 99.

77 is 11 less than 88.

66 is 11 less than 77.

Each term is 11 less than the term before it, so a rule is "Subtract 11."

Identifying Relationships Between Patterns

When two number patterns are compared, you can often identify relationships between them by comparing the **corresponding terms** of each sequence. Corresponding terms are pairs of terms that are in the same position in each sequence.

You can easily compare corresponding terms by making a table showing both patterns.

Example

Compare the two sequences below.

Rule: + 2 **Sequence:** 0, 2, 4, 6, . . .

Rule: + 4 **Sequence:** 0, 4, 8, 12, . . .

Make a table with a column for each pattern. Each row gives a pair of corresponding terms.

The first pair of corresponding terms is 0 and 0. There are many operations that could relate these two terms.

The second pair of corresponding terms is 2 and 4. 4 is 2 more than 2 (2 + 2 = 4), and 4 is 2 times as much as 2 (2 * 2 = 4).

The third pair of corresponding terms is 4 and 8. 8 is 4 more than 4 (4 + 4 = 8), and 8 is 2 times as much as 4 (4 * 2 = 8).

The fourth pair of corresponding terms is 6 and 12. 12 is 6 more than 6, and 12 is 2 times as much as 6 (6 * 2 = 12).

Rule: + 2	Rule: + 4
0	0
2	4
4	8
6	12

All of the terms in the second sequence are 2 times as much as the corresponding terms in the first sequence. This makes sense because the rule + 4 adds 2 times as much as the rule + 2.

One way to describe the relationship between two patterns is with a rule. Think of this rule as a function machine that uses terms of the first sequence for "in" numbers and gives the terms of the second sequence as "out" numbers. You can think of this rule as an in/out relationship rule.

Note Remember that the in/out relationship rule is different from the rules for each column. To find a rule for each column, look down each column for a pattern or relationship. To find the in/out relationship, look across the rows.

Example

Write a rule that relates the two patterns in the table.

Notice that the rule for the first pattern, + 2, gives the list of "in" numbers moving down the column on the left.

The rule for the second pattern, + 4, gives the list of "out" numbers moving down the column on the right.

Looking across each row, all of the "out" numbers are 2 times as much as the "in" numbers.

So, an in/out relationship rule is "* 2."

in Rule: + 2	out Rule: + 4
0	0
2	4
4	8
6	12

Function Machines and "What's My Rule?" Problems

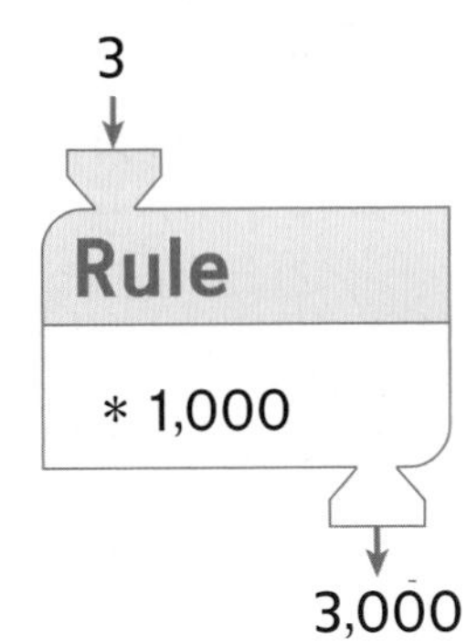

A **function machine** in *Everyday Mathematics* is an imaginary machine that takes in numbers, uses a rule to change those numbers, and then gives out new numbers.

Here is a function machine with the rule "* 1,000." The machine will multiply any number put into it by 1,000, then give out the result.

If you put 3 into this "* 1,000" machine, it will multiply 3 * 1,000. The number 3,000 will come out. If you put 4.7 into this machine, it will multiply 4.7 * 1,000. The number 4,700 will come out.

If you put n, an unknown number, into the machine, it will multiply $n * 1{,}000$. The number $n * 1{,}000$ will come out.

To keep track of what goes in and what comes out, you can organize the "in" and "out" numbers in a table.

in	out
3	3,000
4.7	4,700
0.5	500
. . .	. . .
n	$n * 1{,}000$

In previous grades, you solved many problems with function machines. You had to find the "out" numbers, the "in" numbers, or a rule that fit the given "in" and "out" numbers. In *Everyday Mathematics*, these are called "What's My Rule?" problems.

Example

Find the "out" numbers.

The rule is * 2 ÷ 3, so the "in" number is first multiplied by 2, then the result is divided by 3.

9 goes in: 9 * **2** = 18 and 18 ÷ **3** = 6, so 6 comes out.

12 goes in: 12 * **2** = 24 and 24 ÷ **3** = 8, so 8 comes out.

15 goes in: 15 * **2** = 30 and 30 ÷ **3** = 10, so 10 comes out.

18 goes in: 18 * **2** = 36 and 36 ÷ **3** = 12, so 12 comes out.

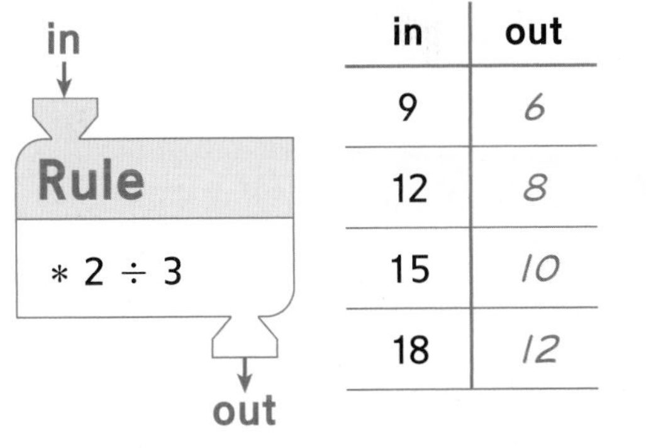

in	out
9	*6*
12	*8*
15	*10*
18	*12*

Example

Find the "in" numbers.

The rule is to multiply the "in" number by 12, so the "out" numbers must be 12 times more than the "in" numbers.
12 came out, so 12 ÷ 12, or 1, went in.
24 came out, so 24 ÷ 12, or 2, went in.
36 came out, so 36 ÷ 12, or 3, went in.

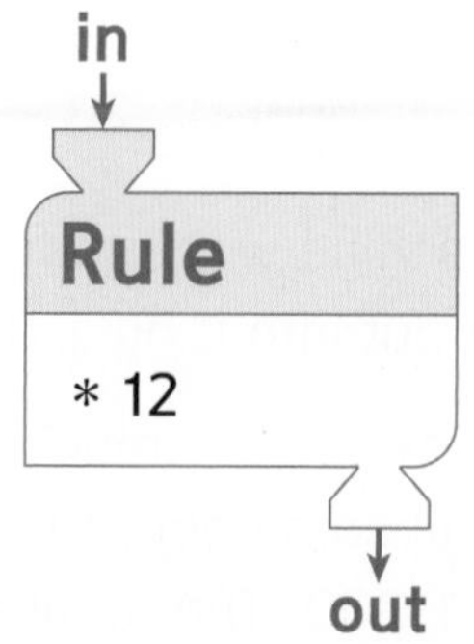

in	out
1	12
2	24
3	36

Example

Use the table to find a rule.

When you have a table of "in" and "out" numbers, there may be several different rules that will give the same pairs of numbers. Make sure that your rule works for all of the "in" and "out" numbers in your table.

A rule that works for the table shown here is ÷ 2.

Check: 10 ÷ 2 = 5
20 ÷ 2 = 10
30 ÷ 2 = 15

Another rule that works for the same table is $* \frac{1}{2}$.

Check: $10 * \frac{1}{2} = 5$
$20 * \frac{1}{2} = 10$
$30 * \frac{1}{2} = 15$

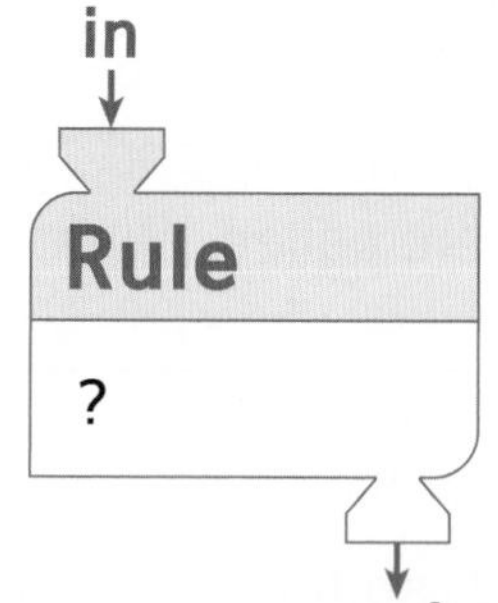

in	out
10	5
20	10
30	15

Check Your Understanding

Find a rule.

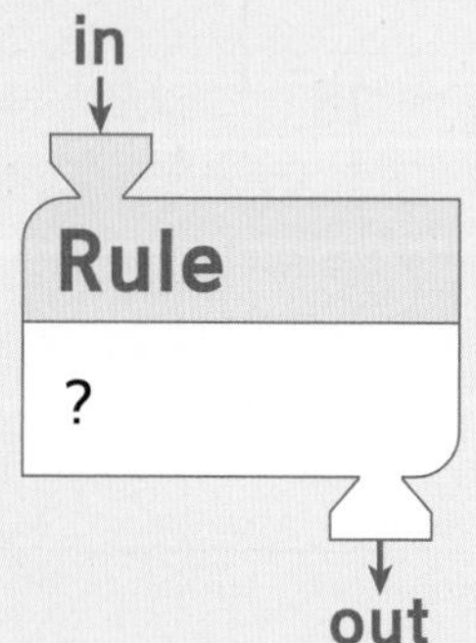

in	out
100	25
28	7
16	4
5	$1\frac{1}{4}$

Check your answer in the Answer Key.

Rules, Tables, and Graphs

Relationships between variables can be shown by rules, by tables, or by graphs.

Example

Lauren's brother earns $12 per hour. Use a rule, a table, and a graph to show the relationship between how many hours Lauren's brother works and how much he earns.

Rule: $12 ∗ Number of hours worked = Earnings

Table: Make an in/out table for a function machine with the rule "∗ 12."

Hours Worked x	Earnings ($) y
0	0
1	12
2	24
3	36
...	...

Graph: To draw a graph, plot the number pairs from the table: (0, 0), (1, 12), (2, 24), (3, 36).

Plot each number pair as a point on a coordinate grid.

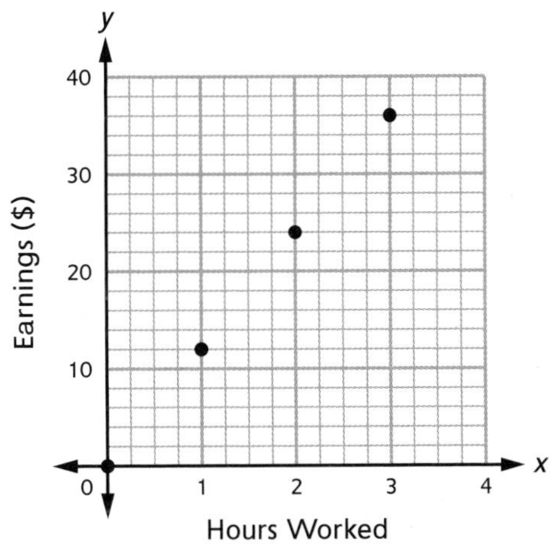

Connecting points on a grid with a straight line can help you see how patterns and relationships continue.

Draw this line to complete the graph.

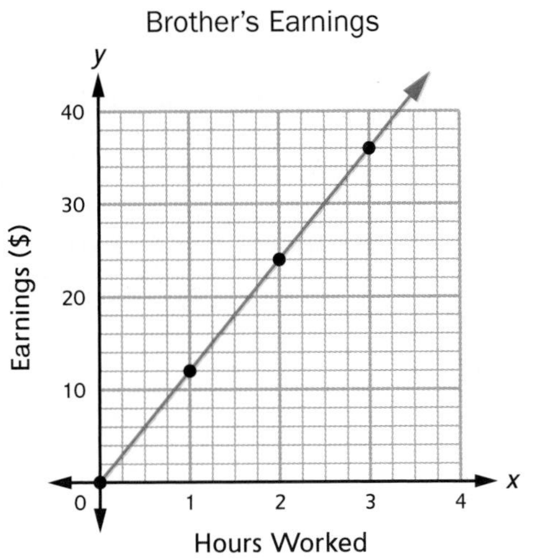

Did You Know?

The size of your head (the distance around your head) increases as you get older (up to age 18). The relationship between head size and age is hard to describe with a rule, so it can be shown by a table or graph.

Age (years)	Average Head Size (cm)
0	34
0.5	43
1	46
2	48.5
4	50
6	51
8	52
10	53
15	55
18	55.5

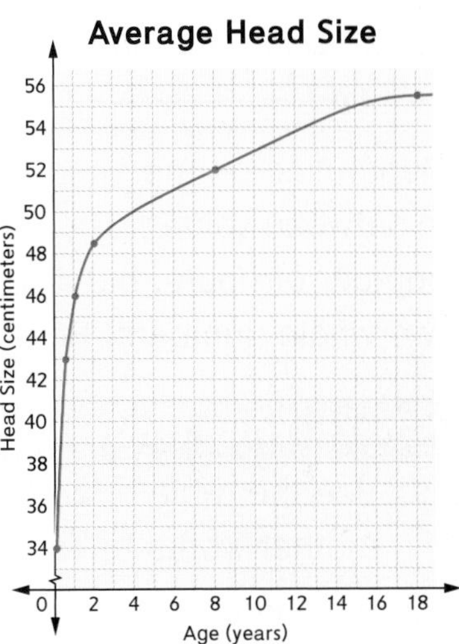

Example

Use the table, the graph, and the rule to find how much Lauren's brother earns if he works for 2 hours and 30 minutes.

One way to use the table to find the earnings is to think of 2 hours and 30 minutes as 2 hours + $\frac{1}{2}$ hour. For 2 hours, Lauren's brother earns \$24. For $\frac{1}{2}$ hour, he earns half of \$12, or \$6. Lauren's brother earns \$24 + \$6 = \$30.

Another way to use the table is to note that $2\frac{1}{2}$ hours is halfway between 2 hours and 3 hours, so Lauren's brother's earnings are the amount halfway between \$24 and \$36, which is \$30.

Hours Worked x	Earnings (\$) y
0	0
1	12
2	24
3	36
...	...

To use the graph, first find $2\frac{1}{2}$ hours on the horizontal axis. Then go straight up to the graph line. Turn left and go across to the vertical axis. You end at \$30.

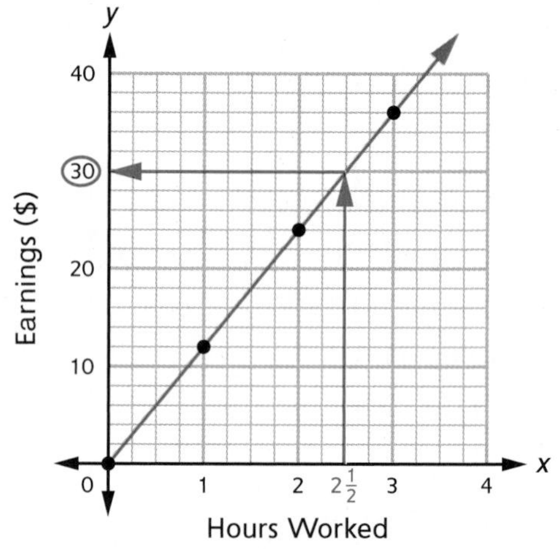

You can also use the rule to find Lauren's brother's earnings:

\$12 $*$ Number of hours worked = Earnings

$\$12 * 2\frac{1}{2} = ?$

$12 * 2\frac{1}{2} = 12 * (2 + \frac{1}{2})$

$12 * 2 = 24 \qquad 12 * \frac{1}{2} = 6 \qquad 24 + 6 = 30$

Lauren's brother earns \$30 in $2\frac{1}{2}$ hours. The rule, table, and graph all give the same answer.

Depending on the situation, it may or may not make sense to use a graph line to find in-between values. For example, it makes sense to use a graph line to find a fraction of a number of hours, but it would not make sense to consider a number of people in between two whole numbers. Be sure to think carefully about the situation when using a graph line.

Check Your Understanding

1. Use the table or graph above to find the number of hours that Lauren's brother works to earn \$18.
2. How much will Lauren's brother earn in 10 hours?

Check your answers in the Answer Key.

Computer-Generated Art

Computer-generated art is created when a person develops a computer program, runs the program, and then selects the best works from all those generated by the computer. This is quite different from **computer-aided art,** which involves using computer software to create original art or to manipulate images that have been loaded onto the computer. There is incredible variety in computer-generated art, both in the way in which it is developed and in the types of images created.

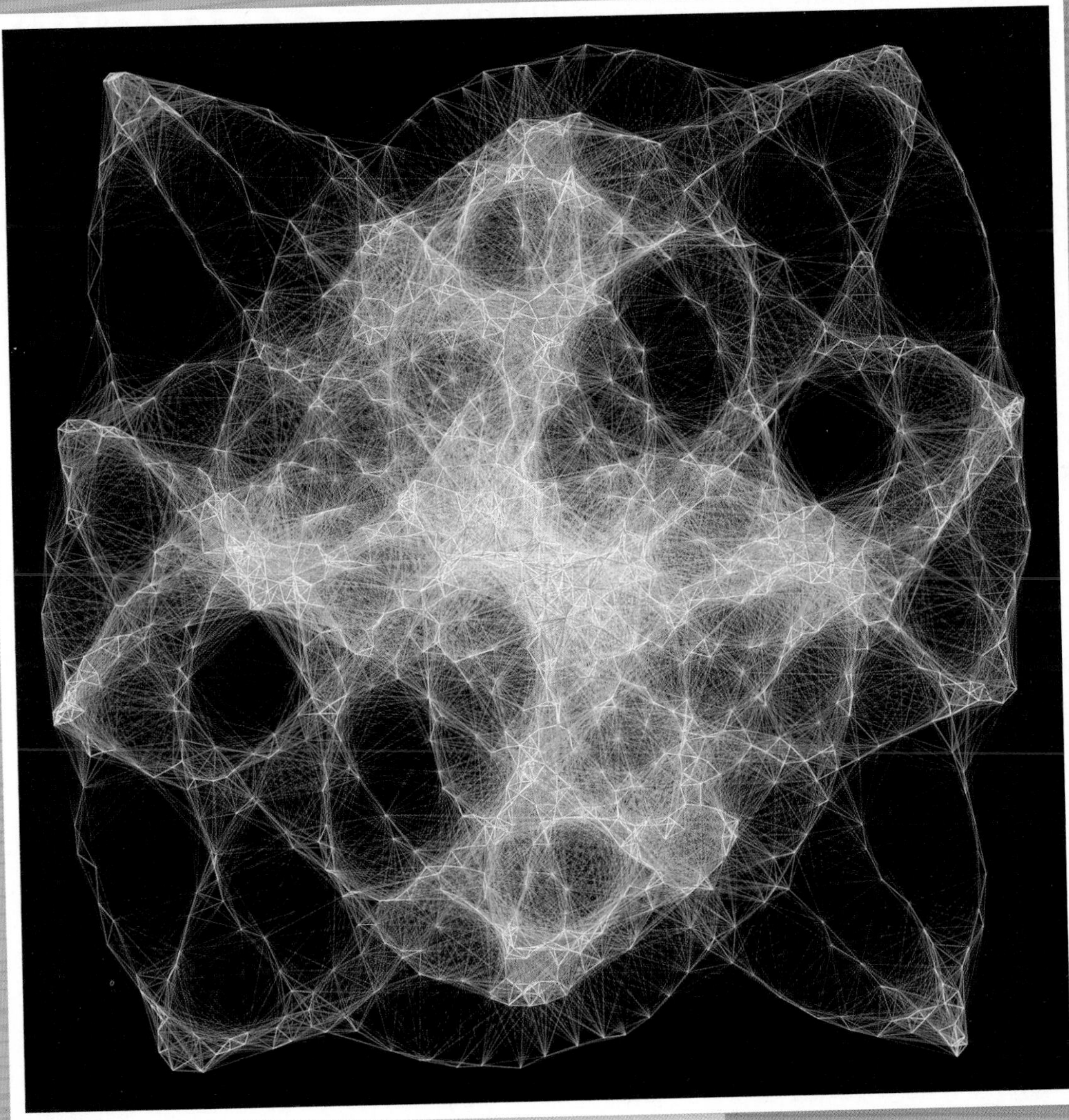

PASIEKA/Science Photo Library/Getty Images

This computer-generated creation is called a Lissajous curve. The computer is used to graph the possible solutions of complicated equations.

Optical Art

Computer-generated optical art uses the repetition of simple forms and colors to create the illusion of movement. Optical art, also called *Op art*, tricks the eye by creating a variety of visual effects such as foreground-background confusion and an exaggerated sense of depth.

This image was created by starting with the largest circle. Each successive circle is two units smaller in diameter than the previous one, with the center moved 45 degrees up and to the left.

This image was created by starting with the largest square. Successive squares were created with each new square 90% of the size of the one before it. Each new square was then rotated counterclockwise and inscribed inside the preceding square. Finally, every second and third square was deleted so the squares would not touch each other.

This image was created by inscribing triangles inside triangles. Starting with the largest triangle, successive triangles were rotated 5 degrees clockwise and then reduced in size.

Transformations

Any image that is drawn on the coordinate grid, or plane, is made up of points that can be named as ordered pairs. Some computer programs generate art by moving each ordered pair (x, y) on a coordinate grid to a new location (x', y') according to a rule. Such a rule is called a "transformation of the plane."

Computers can make simple transformations of objects, as shown below.

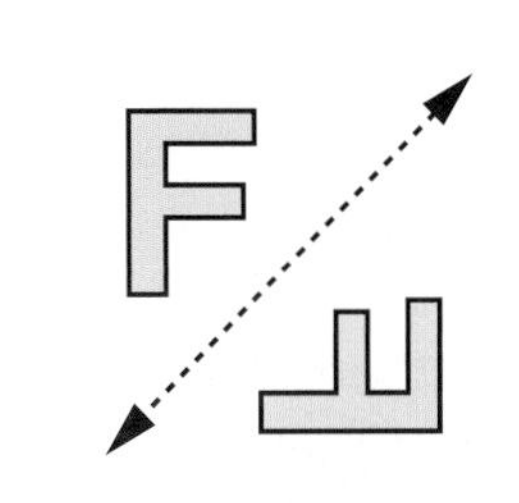

A **reflection** moves a figure by "flipping" it over a line.

A **rotation** moves a figure by "turning" it around a point.

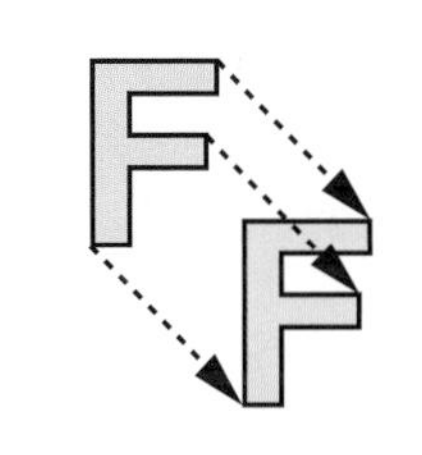

A **translation** moves a figure by "sliding" it to a new location.

Computers can use transformations to make much more complicated works of art.

This image was created using a special type of rule called a Möbius transformation, which moves circles into other circles.

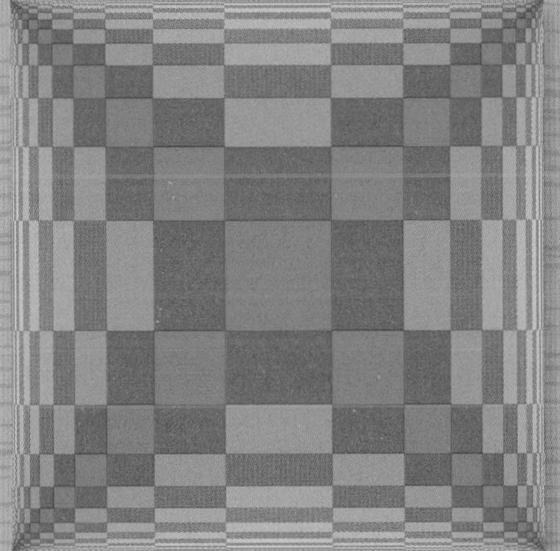

The transformation that created this image moves grid lines on the plane to new locations. The original grid squares become rectangles of many different shapes and sizes.

This image was created using rotation transformations.

Fractals

Fractals are figures that result from the repetition, or iteration, of a pattern. *Iteration* is the application of the same operation over and over again. One of the most exciting developments in computer-generated art is the use of fractals to create extremely detailed images. Like the figures below, fractals are created from many iterations of a self-similar pattern.

This fractal was created by putting points through a special function machine that applies its rule over and over again. The color of a point in the image depends on what happens when the function machine applies its rule to that point over and over. Look for the self-similar patterns that are repeated in the figure.

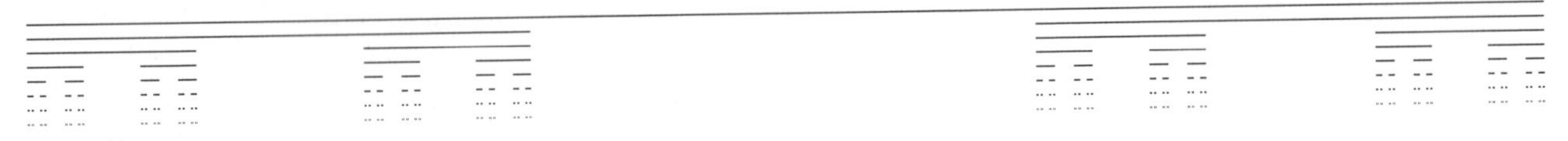

The term *fractal* was introduced in 1975 by Benoît Mandelbrot, but the idea of fractals has been around for hundreds of years. In 1883, Gregor Cantor introduced the *Cantor set*. As shown above, the Cantor set begins with a line segment divided into thirds with the middle third removed. Removing this section creates two new line segments where the operation can be repeated over and over. The iterations of the self-similar pattern in the Cantor set are now recognized as a fractal.

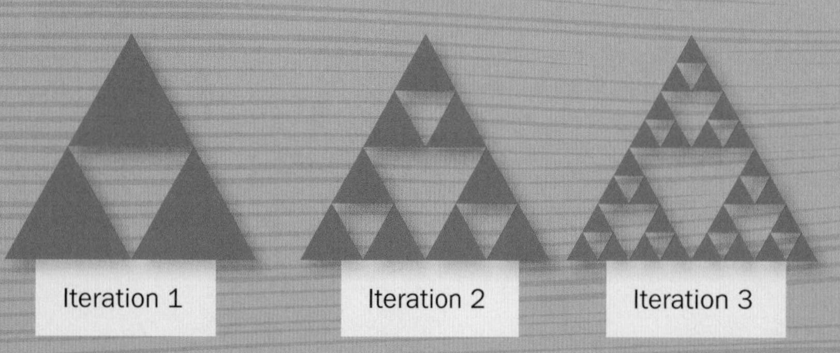

The *Sierpiński triangle* is a simple fractal that leads to an amazing result. This fractal begins by dividing an equilateral triangle into four congruent triangles and removing the center triangle. This action is then iterated by doing the same thing to the three remaining triangles. Here are the first three iterations of the Sierpiński triangle.

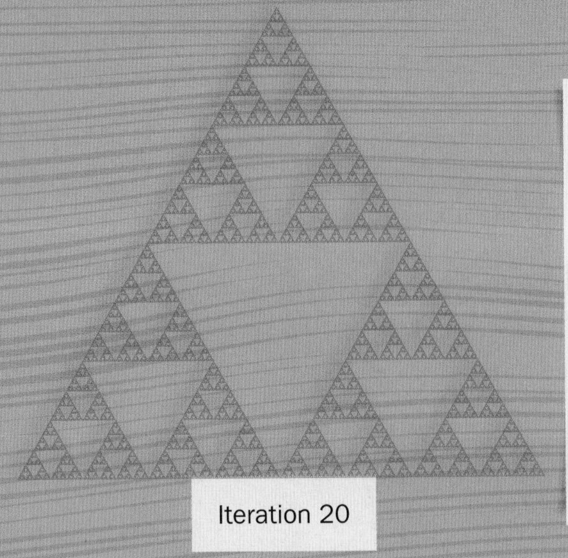

Iteration 20

This is the 20th iteration of the Sierpiński triangle. Look at what is happening to the fractal's *area*, which appears in blue. When the center triangles are removed in each iteration, the area of the entire figure (shown in blue) decreases. The *perimeter* of the fractal is the sum of the perimeters of all the triangles in the figure. So as new triangles appear in each iteration, the perimeter of the fractal increases. If the computer could iterate this pattern an infinite number of times, you would see an amazing result: a figure that has no blue area, but an infinite perimeter.

The *Heighway dragon* is a simple fractal that you can draw. Begin with a line segment. For each iteration, rotate a copy (shown in red) of the previous figure by 90° clockwise and attach it to the end of the previous iteration (shown in black).

How many "dragons" can you find in this fractal?

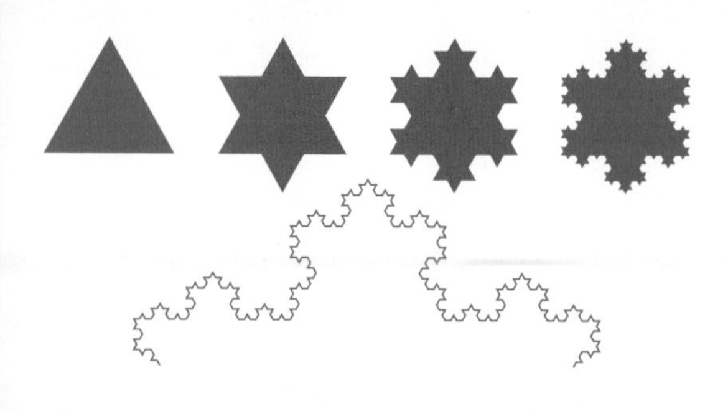

The *Koch snowflake* is another simple fractal that can be easily made and iterated using a computer. In this series, every iteration adds a smaller equilateral triangle to each side of the previous triangle. Each smaller triangle has sides that are $\frac{1}{3}$ the length of the sides of the larger triangle. After several iterations, the image looks like a snowflake.

Fractal landscapes like this one can be created using simple mathematical formulas that take up very little computer memory. From "dragon" patterns to landscape scenery, fractals are playing an ever-increasing role in the development of computer-generated art.

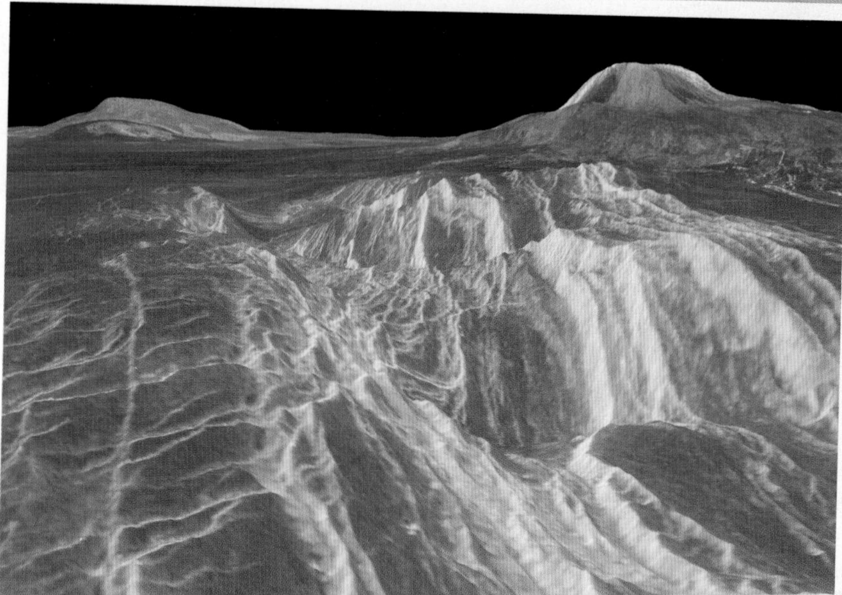

More complex fractals can be used to create landscapes that have the appearance of natural terrain. Fractal landscapes like the one above are now being used in movie production. The first use of such a fractal-generated landscape in a film was in 1982 in the movie *Star Trek II: The Wrath of Khan*.

Number and Operations in Base Ten

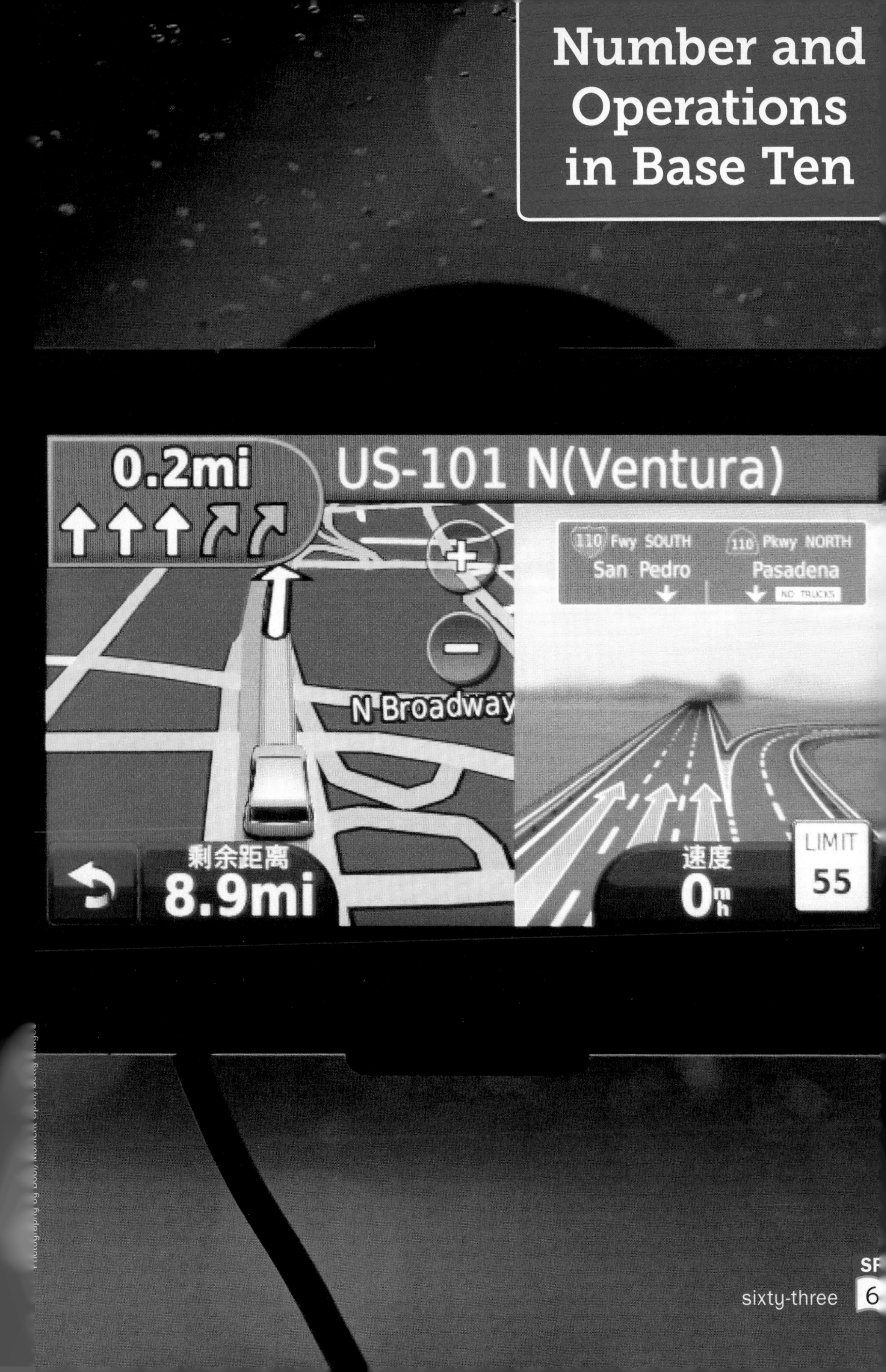

Uses of Numbers

It is hard to live even one day without using or thinking about numbers. Numbers are used on clocks, calendars, license plates, rulers, scales, and so on.

- Numbers are used for counting.

Examples

Students sold 129 tickets to the school play.
The first U.S. Census, taken in 1790, counted 3,929,214 people.
The population of Malibu, California is 15,272.

- Numbers are used for measuring.

Examples

Ivan swam the length of the pool in 34.5 seconds.
The package is 31 inches long and weighs $4\frac{3}{8}$ pounds.

- Numbers are used to show where something is in a reference system.

Examples

Situation	Type of Reference System
Normal room temperature is 21°C.	Celsius temperature scale
David was born on June 22, 2006.	Calendar
The time is 2:39 P.M.	Clock time
Detroit is located at 42°N and 83°W.	Earth's latitude and longitude system

- Numbers are used to compare measures or counts.

Examples

The cat weighs $\frac{1}{2}$ as much as the dog.
There were 4 times as many boys as girls at the game.

- Numbers are used for identification and as codes.

Examples

phone number: (709) 555-212 ZIP code: 60637
driver's license number: M286-423-2061
bar code (used to identify product and manufacturer): 9 780021 308088
car license plate: HTX 585

(t)Fotosearch/Getty Images; (c)Jill Braaten/McGraw-Hill Education; (bc)McGraw-Hill Education; (b)Iconotec/Glow Images

Kinds of Numbers

The **counting numbers** are the numbers used to count things. The set of counting numbers is 1, 2, 3, 4, and so on.

The **whole numbers** are any of the numbers 0, 1, 2, 3, 4, and so on. The whole numbers include all of the counting numbers and the number zero (0).

Counting numbers are useful for counting, but they often do not work for measures. Some measures fall between whole numbers. **Fractions** and **decimals** were invented to keep track of such in-between measures.

Fractions are often used in recipes for cooking and for measures in carpentry and other building trades. Decimals are used for almost all measures in science and industry. Money amounts are usually written as decimals.

Did You Know?

In 1498, Johann Widmann wrote the first book that used + and – signs. Traders and shopkeepers had used both of the signs long before this. They used + to show they had too much of something. They used – to show they had too little of something.

Example

The banana weighs 5.8 ounces.

The recipe calls for $1\frac{1}{2}$ cups of flour.

BANANA BREAD

- $\frac{1}{2}$ cup butter
- 1 cup sugar
- 2 eggs, beaten
- 4 bananas, crushed
- $1\frac{1}{2}$ cups flour
- 1 tsp baking soda
- $\frac{1}{2}$ tsp salt
- $\frac{1}{2}$ tsp vanilla

Negative numbers are used to describe numbers that are less than zero.

Example

On the Celsius scale, a temperature of 10 degrees below zero is written as −10°C.

A depth of 147 feet below sea level is written as −147 feet.

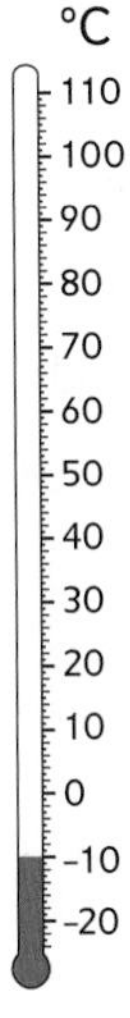

Negative numbers are also used to indicate changes in quantities.

Example

A weight loss of $7\frac{1}{2}$ pounds can be recorded as $-7\frac{1}{2}$ pounds.

Place Value for Whole Numbers

Any number, no matter how large or small, can be written using one or more of the **digits** 0, 1, 2, 3, 4, 5, 6, 7, 8, and 9. A **place-value chart** shows how much each digit in a number is worth. The **place** for a digit is its position in the number. The **value** of a digit is how much it is worth according to its place in the number.

Study the place-value chart below. Look at the numbers that name the places. As you move from right to left across the chart, each number is **10 times as large as the number to its right.**

$*$ 10 $*$ 10 $*$ 10 $*$ 10

10,000s ten thousands	1,000s thousands	100s hundreds	10s tens	1s ones

1 [10] = 10 [1s] 10 = 10 * 1

1 [100] = 10 [10s] 100 = 10 * 10

1 [1,000] = 10 [100s] 1,000 = 10 * 100

1 [10,000] = 10 [1,000s] 10,000 = 10 * 1,000

As you move from left to right across the chart, each number is **$\frac{1}{10}$ the size of the number to its left.** Finding $\frac{1}{10}$ of a number is the same as dividing that number by 10.

$\frac{1}{10}$ of $\frac{1}{10}$ of $\frac{1}{10}$ of $\frac{1}{10}$ of

10,000s ten thousands	1,000s thousands	100s hundreds	10s tens	1s ones

1 [1] = $\frac{1}{10}$ of 10 1 = 10 / 10

1 [10] = $\frac{1}{10}$ of 100 10 = 100 / 10

1 [100] = $\frac{1}{10}$ of 1,000 100 = 1,000 / 10

1 [1,000] = $\frac{1}{10}$ of 10,000 1,000 = 10,000 / 10

Example

10,000s ten thousands	1,000s thousands	100s hundreds	10s tens	1s ones
3	8	9	0	5

The number 38,905 is shown in the place-value chart above.

The value of the 3 is 30,000 or 3 * 10,000.

The value of the 8 is 8,000 or 8 * 1,000.

The value of the 9 is 900 or 9 * 100.

The value of the 0 is 0 or 0 * 10.

The value of the 5 is 5 or 5 * 1.

38,905 is read as "thirty-eight thousand, nine hundred five."

In larger numbers, groups of 3 digits are separated by commas. Commas help identify the thousands, millions, billions, trillions, and so on.

Example

The number 135,246,015,808,297 is shown in the place-value chart.

trillions				billions				millions				thousands				ones		
100	10	1	,	100	10	1	,	100	10	1	,	100	10	1	,	100	10	1
1	3	5	,	2	4	6	,	0	1	5	,	8	0	8	,	2	9	7

Read from left to right. Read "trillion" at the first comma. Read "billion" at the second comma. Read "million" at the third comma. And read "thousand" at the last comma. This number is read as "135 **trillion,** 246 **billion,** 15 **million,** 808 **thousand,** 297."

Check Your Understanding

Read each number to yourself. What is the value of the 5 in each number?

1. 25,308 **2.** 74,546,002 **3.** 643,057 **4.** 2,450,609

Solve the number riddles.

5. I am the smallest 7-digit number you can write with the digits 8, 4, 8, 7, 3, 2, and 1. What number am I?

6. I have 4 digits: two 9s and two 3s. One 9 is worth 90. One of my 3s is worth 300. The other 3 is worth 10 times as much as 300. What number am I?

Check your answers in the Answer Key.

Powers of 10

Numbers like 10, 100, and 1,000 are called **powers of 10.** They are numbers that can be written as products of 10s.

100 can be written as $10 * 10$ or 10^2.
1,000 can be written as $10 * 10 * 10$ or 10^3.

The smaller, raised digit is an **exponent.** The number before the exponent is the **base.** The exponent tells how many times the base is used as a factor. For powers of 10, 10 is the base, so the exponent tells how many 10s are multiplied.

Note 10^2 is read "10 to the second power" or "10 squared." 10^3 is read "10 to the third power" or "10 cubed." 10^4 is read "10 to the fourth power."

A number written using place value, like 1,000, is in **standard notation.** A number written with an exponent, like 10^3, is in **exponential notation.**

The chart below shows powers of 10 from ten through one billion.

Powers of Ten		
Standard Notation	Product of 10s	Exponential Notation
10	10	10^1
100	10∗10	10^2
1,000 (1 thousand)	10∗10∗10	10^3
10,000	10∗10∗10∗10	10^4
100,000	10∗10∗10∗10∗10	10^5
1,000,000 (1 million)	10∗10∗10∗10∗10∗10	10^6
10,000,000	10∗10∗10∗10∗10∗10∗10	10^7
100,000,000	10∗10∗10∗10∗10∗10∗10∗10	10^8
1,000,000,000 (1 billion)	10∗10∗10∗10∗10∗10∗10∗10∗10	10^9

Did You Know?

Raised numbers have been used to indicate powers since at least 1484.

Note Some calculators have special keys such as [x^y] or [^] for entering numbers using exponential notation. See the appendix for more information.

Example

1,000 ∗ 1,000 = ?

Use the table above to write 1,000 as 10 ∗ 10 ∗ 10.

$$1{,}000 * 1{,}000 = (10 * 10 * 10) * (10 * 10 * 10)$$
$$= 10^6$$
$$= 1 \text{ million}$$

So, 1,000 ∗ 1,000 = 1 million.

Example

1,000 millions = ?

Write 1,000 ∗ 1,000,000 as (10 ∗ 10 ∗ 10) ∗ (10 ∗ 10 ∗ 10 ∗ 10 ∗ 10 ∗ 10). This is a product of nine 10s or 10^9.

1,000 millions = 1 billion

Renaming Numbers Using Exponential Notation

The population of the world is about 7 billion people. The number 7 billion can be written as 7,000,000,000 or as $7 * 10^9$.

The number 7,000,000,000 is written in **standard notation.**

The number $7 * 10^9$ is written using **exponential notation.** $7 * 10^9$ is read "seven times ten to the ninth power."

10^9 is the product of 10 used as a factor 9 times.

$10^9 = 10 * 10 * 10 * 10 * 10 * 10 * 10 * 10 * 10$

$= 1,000,000,000$

$= 1$ billion

So, $7 * 10^9 = 7 * 1,000,000,000$.

$= 7,000,000,000$

$= 7$ billion

The world's population is about $7 * 10^9$.

Did You Know?

A number written in *scientific notation* is the product of a number that is at least 1 and less than 10 and a power of 10. For example, 8,200,000 is $8.2 * 10^6$ in scientific notation. Changing numbers from standard notation to scientific notation often makes them easier to write and work with. Calculator displays only show a limited number of digits. When an answer doesn't fit on a calculator display, different calculators use different symbols to display the answer using scientific notation.

Examples

Write each number using exponential notation.

5,000,000 = ?

$5,000,000 = 5 * 1,000,000$

$1,000,000 = 10 * 10 * 10 * 10 * 10 * 10 = 10^6$

So, $5,000,000 = 5 * 10^6$.

240,000 = ?

$240,000 = 24 * 10,000$

$10,000 = 10 * 10 * 10 * 10 = 10^4$

So, $240,000 = 24 * 10^4$.

Examples

Write in standard notation.

$4 * 10^3 = ?$

$10^3 = 10 * 10 * 10 = 1,000$

So, $4 * 10^3 = 4 * 1,000 = 4,000$.

$5.6 * 10^7 = ?$

$10^7 = 10 * 10 * 10 * 10 * 10 * 10 * 10 = 10,000,000$

So, $5.6 * 10^7 = 5.6 * 10,000,000 = 56,000,000$.

Check Your Understanding

Write each number in standard notation.

1. $8 * 10^6$ **2.** $76 * 10^4$ **3.** $1.2 * 10^7$ **4.** $4.9 * 10^2$

Write each number using exponential notation.

5. 500 **6.** 44,000 **7.** 900,000,000

Check your answers in the Answer Key.

Expanded Form

The number 481,926 is written in **standard notation,** the most common way of writing a number. Standard notation is also called **standard form.** When a number is written as the sum of the values of each digit, it is written in **expanded form.** There are several ways to represent a number in expanded form.

Note Based on patterns with powers of 10, mathematicians agree that $10^0 = 1$. Each power of 10 is $\frac{1}{10}$ of the next power.

$\frac{1}{10}$ of ($10^1 = 10$

$\frac{1}{10}$ of ($10^2 = 100$

$10^3 = 1{,}000$

10^0 must be $\frac{1}{10}$ of 10^1.
$\frac{1}{10}$ of 10 is 1, so $10^0 = 1$.

Example

The number 481,926 is shown in the place-value chart below.

100,000s hundred thousands	10,000s ten thousands	1,000s thousands	100s hundreds	10s tens	1s ones
4	8	1	9	2	6

Write the number using words to show the place names:
4 hundred thousands + 8 ten thousands + 1 thousand + 9 hundreds + 2 tens + 6 ones

Write the number using the value of each digit in standard notation:
400,000 + 80,000 + 1,000 + 900 + 20 + 6

Write the number using equal-groups notation:
4 [100,000s] + 8 [10,000s] + 1 [1,000] + 9 [100s] + 2 [10s] + 6 [1s]

Write the number as a sum of products of powers of 10:
(4 * 100,000) + (8 * 10,000) + (1 * 1,000) + (9 * 100) + (2 * 10) + (6 * 1)

Write the number as a sum of products using exponential notation:
$(4 * 10^5) + (8 * 10^4) + (1 * 10^3) + (9 * 10^2) + (2 * 10^1) + (6 * 10^0)$

Check Your Understanding

Write each number in standard notation.

1. (8 * 10,000) + (3 * 1,000) + (4 * 100) + (5 * 1)

2. $(2 * 10^6) + (6 * 10^5) + (7 * 10^4) + (3 * 10^3) + (9 * 10^2) + (5 * 10^1) + (2 * 10^0)$

Write each number in expanded form.

3. 8,744 **4.** 1,456,900

Check your answers in the Answer Key.

Comparing Numbers and Amounts

When two numbers or amounts are compared, there are two possible results: They are equal, or they are not equal because one is larger than the other. Different symbols are used to show that numbers and amounts are equal or not equal.

- Use an *equal sign* (=) to show that the numbers or amounts *are equal*.
- Use a *not-equal sign* ($\neq$) to show that they are *not equal*.
- Use a *greater-than symbol* ($>$) or a *less-than symbol* ($<$) to show that they are *not equal* and to show which is larger.

Examples

Symbol	=	$\neq$	$>$	$<$
Meaning	"equals" or "is the same as"	"is not equal to"	"is greater than"	"is less than"
Examples	$\frac{1}{2} = 0.5$ 4 cm = 40 mm $2 * 5 = 9 + 1$	$2 \neq 3$ $4 * 3 \neq 9$ 1 m $\neq$ 100 mm	$9 > 5$ 16 ft 9 in. $>$ 15 ft 11 in. $10^3 > 100$	$3 < 5$ 98 minutes $<$ 3 hours $3 * (3 + 4) < 5 * 6$

When you compare amounts that include units, use the same unit for both amounts to find which is larger.

Example

Compare 30 yards and 60 feet.

The units are different—yards and feet. You can change yards to feet, then compare.
1 yd = 3 ft, so 30 yd = $30 * 3$ ft, or 90 ft.

Now compare feet. 90 ft $>$ 60 ft

So, 30 yd $>$ 60 ft.

0 1 2 3 4 5 6 7 8 9 10 11 12 13 14 15 16 17 18 19 20 21 22 23 24 25 26 27 28 29 30 31 32 33 34 35 36

0 1 2 3 4 5 6 7 8 9 10 11 12

Or change feet to yards, and then compare.
1 ft = $\frac{1}{3}$ yd, so 60 ft = $60 * \frac{1}{3}$ yd, or 20 yd.

Now compare yards. 30 yd $>$ 20 yd

So, 30 yd $>$ 60 ft.

Multiples

When you skip count by a number, your counts are the **multiples** of that number. Since you can always count further, lists of multiples can go on forever.

Example

Find multiples of 3, 5, and 10.

Multiples of 3: 3, 6, 9, 12, 15, 18, 21, 24, 27, 30, 33, 36, ...

Multiples of 5: 5, 10, 15, 20, 25, 30, 35, 40, 45, 50, 55, ...

Multiples of 10: 10, 20, 30, 40, 50, 60, 70, 80, 90, 100, 110, 120, ...

Note The three dots (...) mean that a list can go on in the same way forever.

Common Multiples

A number is a **common multiple** of two or more numbers if it is a multiple of each of those numbers.

Example

Find common multiples of 2 and 3.

Multiples of 2: 2, 4, **6,** 8, 10, **12,** 14, 16, **18,** 20, 22, **24,** ...
Multiples of 3: 3, **6,** 9, **12,** 15, **18,** 21, **24,** 27, **30,** 33, **36,** ...

Common multiples of 2 and 3: 6, 12, 18, 24, ...

Note Common multiples are useful when finding common denominators of fractions.

Least Common Multiples

The **least common multiple** of two numbers is the smallest number that is a multiple of both numbers.

Example

Find the least common multiple of 6 and 8.

Multiples of 6: 6, 12, 18, **24**, 30, 36, 42, **48**, 54, ...
Multiples of 8: 8, 16, **24**, 32, 40, **48**, 56, ...
24 and 48 are common multiples. 24 is the smallest common multiple.
24 is the smallest number that can be divided evenly by both 6 and 8.

24 is the least common multiple of 6 and 8.

Check Your Understanding

Find at least two common multiples of each pair of numbers.

1. 4 and 12 **2.** 6 and 10 **3.** 9 and 15 **4.** 6 and 14

5. Are the multiples of 10 the same as the powers of 10? Why or why not?

Check your answers in the Answer Key.

Factors of a Counting Number

One way to identify **factors** of a counting number is to make a **rectangular array,** an arrangement of objects in rows and columns that form a rectangle. Each row has the same number of objects. Each column has the same number of objects. A multiplication **number model** can represent a rectangular array.

Note When you find factors of a counting number, the factors must also be counting numbers. The counting numbers are 1, 2, 3, and so on.

Example

Find factors of 15.

Begin by making a rectangular array with 15 red dots.
This array has 3 rows with 5 dots in each row.
$3 * 5 = 15$ is a number model for this array.
3 and 5 are **factors** of 15.
15 is the **product** of 3 and 5.

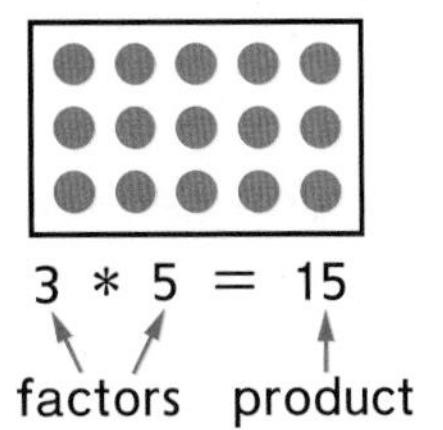

3 and 5 are a **factor pair** for 15.

Counting numbers can have more than one factor pair. 1 and 15 are another factor pair for 15 because $1 * 15 = 15$.

To test whether a counting number *m* is a factor of another counting number *n*, divide *n* by *m*. If the result is a counting number and the remainder is 0, then *m* is a factor of *n*.

Examples

4 is a factor of 12 because 12 / 4 gives 3 with a remainder of 0.

6 is *not* a factor of 14 because 14 / 6 gives 2 with a remainder of 2.

One way to find all the factors of a counting number is to find all the factor pairs for that number.

Example

Find all the factors of the number 24.

Write all multiplication number sentences that equal 24. You can draw a factor rainbow to check that you have found all of the factors.

Number Sentences	Factor Pairs
$24 = 1 * 24$	1, 24
$24 = 2 * 12$	2, 12
$24 = 3 * 8$	3, 8
$24 = 4 * 6$	4, 6

Factor Rainbow

1 2 3 4 6 8 12 24

The factors of 24 are 1, 2, 3, 4, 6, 8, 12, and 24.

Check Your Understanding

List all the factors of each number.

1. 15 **2.** 8 **3.** 28 **4.** 36 **5.** 11 **6.** 100

Check your answers in the Answer Key.

Divisibility

When one counting number is divided by another counting number and the quotient is a counting number with a remainder of 0, then the first number is **divisible by** the second number. If the quotient has a non-zero remainder, then the first number is *not divisible by* the second number. A counting number is divisible by all of its **factors.**

For some counting numbers, even large ones, it is possible to test for divisibility without dividing.

Here are divisibility tests that you can use instead of dividing:

- All counting numbers are **divisible by 1.**
- Counting numbers with a 0, 2, 4, 6, or 8 in the ones place are **divisible by 2.** They are the **even numbers.**
- Counting numbers with 0 in the ones place are **divisible by 10.**
- Counting numbers with 0 or 5 in the ones place are **divisible by 5.**
- If the sum of the digits in a counting number is divisible by 3, then the number is **divisible by 3.**
- If the sum of the digits in a counting number is divisible by 9, then the number is **divisible by 9.**
- A counting number divisible by both 2 and 3 is **divisible by 6.**

Note Division by 0 is *never* permitted.

Note We say that the number 0 is divisible by any counting number *m* because 0 / *m* → 0 R0. For example, 0 / 2 → 0 R0

Examples

Find some numbers 216 is divisible by.

216 is divisible by:

2 because the 6 in the ones place is an even number.
3 because the sum of its digits is 2 + 1 + 6 = 9, and 9 is divisible by 3.
9 because the sum of its digits is divisible by 9.
6 because it is divisible by both 2 and 3.
216 is not divisible by 10 or 5 because it does not have a 0 or a 5 in the ones place.

Check Your Understanding

Which numbers below are divisible by 2? By 3? By 5? By 6? By 9? By 10?

705 4,470 616 621 14,580

Check your answers in the Answer Key.

Prime and Composite Numbers

A **prime number** is a counting number greater than 1 that has exactly two *different* factors: 1 and the number itself. A prime number is divisible only by 1 and itself.

Note The only factor of 1 is 1 itself, so it has exactly one factor. The number 1 is neither prime nor composite.

A **composite number** is a counting number that has more than two different factors.

Examples

11 is a prime number because its only factors are 1 and 11.

4 is a composite number because it has more than two different factors. Its factors are 1, 2, and 4.

Every composite number can be renamed as a product of prime numbers. This is called the **prime factorization** of that number. One way to find the prime factorization of a composite number is to make a **factor tree.** First write the number. Then below the number, write any two factors whose product is that number. Repeat the process for these two factors. Continue until all the factors are prime numbers.

Note The prime factorization of a prime number is that number. For example, the prime factorization of 11 is 11.

Example

Find the prime factorization of 24.

No matter which two factors are used to start the tree, the tree will always end with the same prime factors.

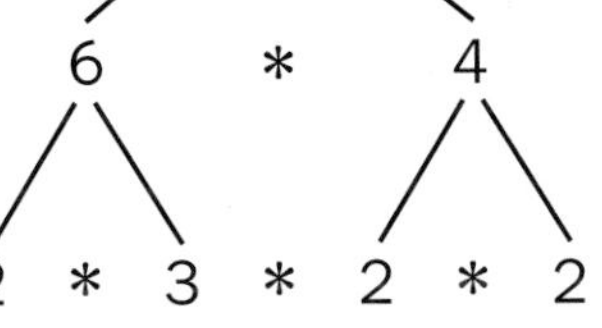

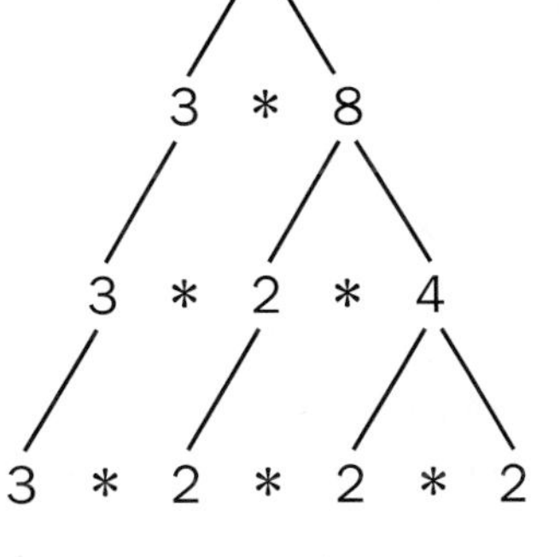

24 = 2 * 2 * 2 * 3

The prime factorization of 24 is 2 * 2 * 2 * 3.

Check Your Understanding

Find the prime factorization of each number. You may use a factor tree to help you.

1. 15 **2.** 20 **3.** 40 **4.** 36 **5.** 17 **6.** 100

Check your answers in the Answer Key.

Estimation

An **estimate** is a number that is close to an exact answer. You estimate when you use a nearby number that can help you make sense of a situation or help you make an approximate calculation. Useful estimates are based on reasoning.

Example

The school play will be performed in 1 month, 3 weeks, and 4 days.

When a girl says, "The play will be performed in about 2 months," she's using an estimate based on knowing that there are about 4 weeks in a month.

Estimation is useful when finding the exact answer is not necessary and it is more efficient to estimate. The estimate needs to be close enough for the purpose.

Example

A boy has a $20 bill to pay for these DVDs. He wonders if he has enough money.

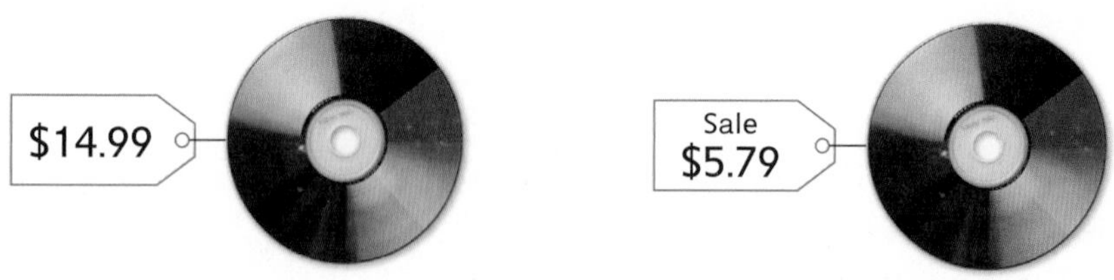

The boy thinks, "$14.99 is only a penny away from $15 and with the $5 from the $5.79, I'm already at $20. The rest makes the total go over $20, so I cannot buy both DVDs."

Using estimation, the boy figures out that he does not have enough money to buy both DVDs.

Did You Know?

Sometimes estimates are used to make sense of a situation when the exact number cannot be known. For example, scientists can estimate the number of fish in a lake by counting a sample and using that information to estimate the total number of fish in the lake. It is impossible to know the exact number because the scientists cannot see all the fish and the fish move about, entering and leaving the lake at different times.

Estimation is also useful when you need to find an exact answer to a calculation. You can make an estimate when you start working on a problem to help you understand the problem better. You can also use your estimate after you calculate to check whether your answer makes sense. If your answer is not close to your estimate, then you need to use another method of estimation or check your work, correct it, and check whether the new answer makes sense.

Example

249 + 345 + 185 = **?**

Estimate: 249 is close to 250, 345 is close to 350, and 185 is close to 200.

250 + 350 + 200 = 800

Calculate:

$$\begin{array}{r} {}^{1} \\ 249 \\ 345 \\ +\ 185 \\ \hline \not{679} \end{array}$$

Check: The numbers used to make the estimate were very close to the original numbers. So, 679 seems too far away from the estimate of 800. Check the work again.

Correct the work:

$$\begin{array}{r} {}^{1}{}^{1} \\ 249 \\ 345 \\ +\ 185 \\ \hline 779 \end{array}$$

Check again: There was an error before. 779 seems more reasonable because it is closer to the estimate of 800.

Front-End Estimation

A common way to produce an **estimate** is to keep the digit in the highest place value and replace the rest of the digits with zeros. This is called **front-end estimation.**

Note Sometimes *front-end estimation* is called *leading-digit estimation.*

Examples

A girl saved $219.

The digit in the highest place value in $**2**19 is **2** in the hundreds place.

So the front-end estimate is $**2**00.

The girl says, "I saved more than 2 hundred dollars."

At its farthest, Earth is 405,500 kilometers away from the moon.

The digit in the highest place value in **4**05,500 km is **4** in the hundred-thousands place.

So the front-end estimate is **4**00,000 km.

A boy says, "Earth is about 400,000 kilometers from the moon."

You can also estimate answers to calculations by using front-end estimation.

Examples

How much will 6 pens cost if the price is 74¢ per pen?

The digit in the highest place value in 74¢ is the 7 in the tens place.
Use 70¢.
Calculate: 6 * 70¢ = 420¢, or $4.20
Estimate: The 6 pens will cost a little more than $4.20.

A boy adds 4.2 + 21.35. His answer is 21.77. Is he correct?

Find front-end estimates for each number in the problem:
Use 4 as a front-end estimate for 4.2.
Use 20 as a front-end estimate for 21.35.
Calculate: 4 + 20 = 24
Estimate: The exact answer is more than 24. The boy's answer is incorrect. He should look at the addition again to correct his work.

Note Whenever you use front-end estimation to add, your resulting estimate is always less than the exact answer. This is because all of the digits, other than the first digit, are replaced with zeros. So the numbers you use to calculate the estimate are always less than those in the original problem.

Check Your Understanding

Use front-end estimation to decide whether the answers are correct.

1. Emily added 837 + 273 + 704 and got 1,054.

2. Luis said that 973 / 36 is 102.

Check your answers in the Answer Key.

Rounding

Rounding is another way to make sense of numbers in situations and to estimate an answer for a calculation. Whole numbers are often rounded to the nearest multiple of a **power of 10.** (Numbers such as 1,000, 100, and 10 are powers of 10.) You can round by using or thinking about number lines. You can also round by reasoning about the numbers or by using a shortcut.

Rounding Using Number Lines

When using a number line to round a given number:

- Decide what place you are rounding to. Think of that place as a power of 10. Find the two multiples of that power of 10 that the number you are rounding is between.
- Sketch the number line showing the two multiples and the halfway point.
- Place the number to be rounded on the number line.
- You will be able to see which of the two multiples the number is closer to. Round the number to the closer multiple.

Round 4,611 to the nearest 1,000.

Find the two multiples of 1,000 that 4,611 is between.

4,000 and 5,000 are multiples of 1,000.

4,611 is more than 4,000 and less than 5,000.

Sketch a number line showing 4,000 and 5,000 and the point halfway between.

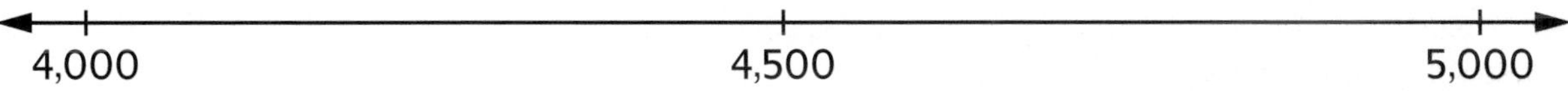

Then estimate where 4,611 is on the number line. Label it.

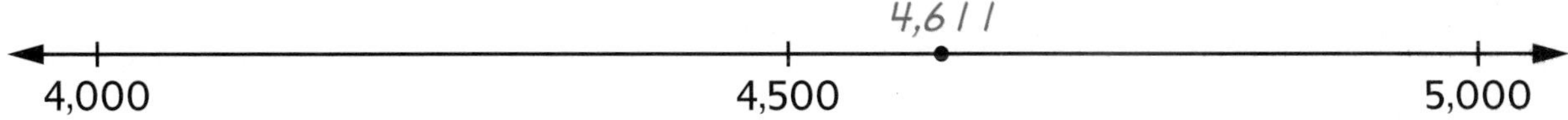

Think: Is 4,611 closer to 4,000 or 5,000?

4,611 is closer to 5,000. 4,611 rounded to the nearest thousand is 5,000.

Example

Round 4,611 to the nearest 100.

Find the two multiples of 100 that 4,611 is between.
4,600 and 4,700 are multiples of 100. 4,611 is more than 4,600 and less than 4,700.
Sketch a number line showing 4,600 and 4,700. Mark the halfway point and 4,611.

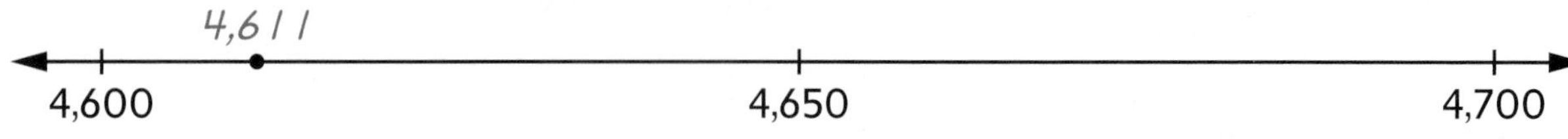

Think: Is 4,611 closer to 4,600 or 4,700?

4,611 is closer to 4,600 than 4,700. 4,611 rounded to the nearest hundred is 4,600.

The next examples show steps you can use to round a number. Your thinking as you follow these steps is similar to your thinking as you use a number line to round.

Examples

A. Round 4,538 to the nearest hundred.

B. Round 26,781 to the nearest thousand.

C. Round 32,951 to the nearest hundred.

Step 1: Write the number you are rounding and underline the digit in the place you are rounding to.	**Step 2:** Replace all numbers to the right of the underlined digit with 0. This is the **lower number.**	**Step 3:** Add 1 to the digit in the place you are rounding to. This is the **higher number. Note:** If the sum is 10, write 0 in the place you're rounding to and add 1 to the digit to its left.	**Step 4:** Ask: *Is the number I am rounding closer to the lower number?* If yes, round down to the lower number. *Is the number I am rounding closer to the higher number?* If yes, round up to the higher number.
A. 4,538	4,500	4,600	4,538 is closer to 4,500 than 4,600. 4,538 *rounds down* to 4,500.
B. 26,781	26,000	27,000	26,781 is closer to 27,000 than 26,000. 26,781 *rounds up* to 27,000.
C. 32,951	32,900	33,000	32,951 is closer to 33,000 than 32,900. 32,951 *rounds up* to 33,000.

Note **Steps 1, 2, and 3** help you figure out what two multiples the number is between.

Note **Step 4** is based on the halfway point. If the number being rounded is less than halfway to the higher number, round to the lower number. If the number being rounded is more than halfway to the higher number, round to the higher number.

A Rounding Shortcut

Once you practice rounding numbers by using or thinking about the number line, you may notice patterns to help you decide when to round down and when to round up. Here is one shortcut:

- Find the digit in the place you are rounding to.
- Identify the possible numbers to round to.
- Look at the digit to the right of the place you are rounding to.
- If the digit to the right is less than 5, round down. If it is 5 or greater, round up.

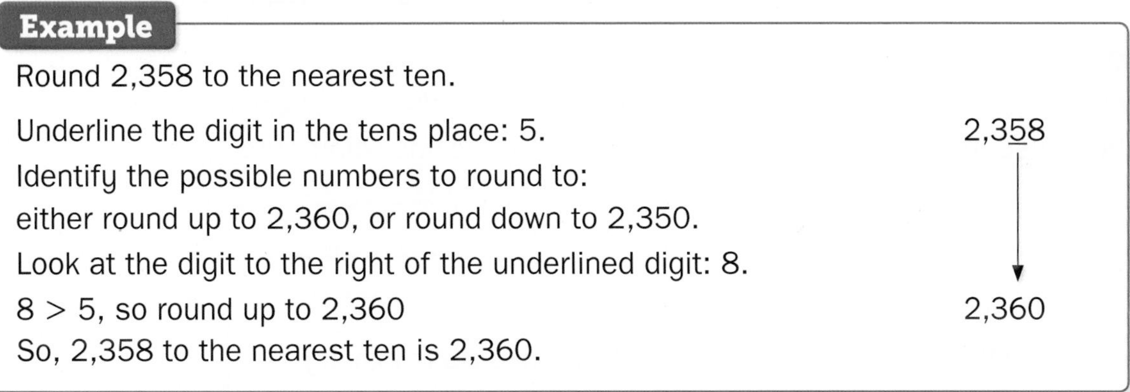

Example

Round 2,358 to the nearest ten.

Underline the digit in the tens place: 5. — $2{,}3\underline{5}8$

Identify the possible numbers to round to:
either round up to 2,360, or round down to 2,350.

Look at the digit to the right of the underlined digit: 8.

$8 > 5$, so round up to 2,360 — 2,360

So, 2,358 to the nearest ten is 2,360.

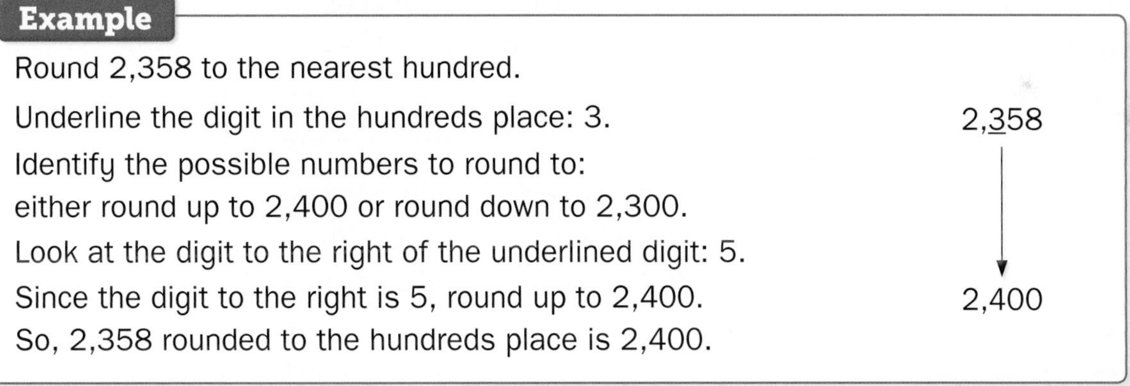

Example

Round 2,358 to the nearest hundred.

Underline the digit in the hundreds place: 3. — $2{,}\underline{3}58$

Identify the possible numbers to round to:
either round up to 2,400 or round down to 2,300.

Look at the digit to the right of the underlined digit: 5.

Since the digit to the right is 5, round up to 2,400. — 2,400

So, 2,358 rounded to the hundreds place is 2,400.

Example

Round 2,358 to the nearest thousand.

Underline the digit in the thousands place: 2.

Identify the possible numbers to round to:
either round up to 3,000 or round down to 2,000.

Look at the digit to the right of the underlined digit: 3.

3 < 5, so round down to 2,000.

So, 2,358 rounded to the nearest thousand is 2,000.

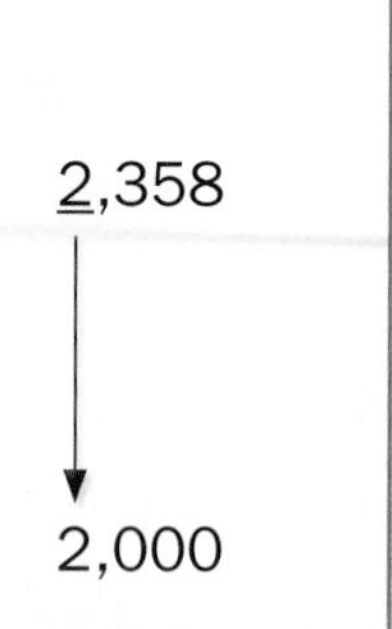

Rounding Halfway Numbers

When the number being rounded is exactly halfway between the higher and lower numbers, use the real-world or problem situation to decide whether to round up or down.

Example

Estimate the product: 66 ∗ 55 = ?

66 is closer to 70 than 60, so round 66 up to 70. Since 55 is exactly halfway between 50 and 60, use the problem situation to decide whether to round up or down. Since the first factor was rounded up, it makes sense to round 55 down to 50 to get a closer estimate.

Estimate: 70 ∗ 50 = 3,500

When there is nothing in the problem situation or any real-world information to help you decide whether to round up or down, the "rounding up" rule usually applies. That means you round up to the higher multiple or number.

Check Your Understanding

Round 25,296 to the nearest

1. hundred **2.** ten thousand **3.** ten **4.** thousand

5. Estimate the sum. $150 + $676 + $212 + $398

6. Estimate the product. 75 * 79

Check your answers in the Answer Key.

Close-but-Easier Numbers

When calculating, some numbers are easier to mentally compute than others. Selecting **close-but-easier numbers** can help you estimate the answers for calculations.

Example

Estimate 42 ∗ 2,488 = ?

One way: Think of an easier mental math problem that can help you estimate: 4 ∗ 25 = 100.

Write the two factors as close-but-easier numbers:

42 is close to **4**0.

2,488 is close to **2,5**00.

Mentally calculate: 40 ∗ 2,500 = ?

40 ∗ **2,5**00 = (**4** ∗ 10) ∗ (**25** ∗ 100)
= (**4** ∗ **25**) ∗ (10 ∗ 100)
= (**100**) ∗ (1,000)
= **100**,000

42 ∗ 2,488 is about 100,000.

Another way: You might choose a different easier mental math problem to make an estimate: 4 ∗ 2 = 8.

Write the two factors as close-but-easier numbers:

42 is close to **4**0.

2,488 is close to **2**,000.

NOTE: Both factors are rounded down, so the product will be more than the estimate.

Mentally calculate: 40 ∗ 2,000 = ?

40 ∗ **2**,000 = (**4** ∗ 10) ∗ (**2** ∗ 1,000)
= (**4** ∗ **2**) ∗ (10 ∗ 1,000)
= (**8**) ∗ (10,000)
= **8**0,000

42 ∗ 2,488 must be more than 80,000.

Note When multiplying two **powers of 10,** the number of zeros in the factors is the same as the number of zeros in the product. Example: 100 ∗ 10 = 1,000. There are 2 zeros in the first factor and one zero in the second factor, and then 3 zeros in the product.

Example

Estimate 7,690 / 35 = ?

One way: Think of an easier mental math problem that can help you estimate: 75 / 3 = 25 because 3 * 25 = 75.

Write the **dividend** and **divisor** as close-but-easier numbers:

The dividend, 7,690, is close to **7,5**00.

The divisor, 35, is close to **3**0.

Mentally calculate: **7,5**00 / **3**0 = ?

$$
\begin{aligned}
\mathbf{7{,}5}00 / \mathbf{3}0 &= (\mathbf{75} * 100) / (\mathbf{3} * 10) \\
&= (\mathbf{75} / \mathbf{3}) * (100 / 10) \\
&= (\mathbf{25}) * (10) \\
&= \mathbf{25}0
\end{aligned}
$$

7,690 / 35 is about 250.

Another way: You might think of a different mental math problem that is easy to divide: 80 / 40 = 2.

Write the dividend and divisor as close-but-easier numbers:

7,690 is close to **8,0**00.

35 is close to **40.**

Mentally calculate: **8,0**00 / **40** = ?

$$
\begin{aligned}
\mathbf{8{,}0}00 / \mathbf{40} &= (\mathbf{80} * 100) / (\mathbf{40} * 1) \\
&= (\mathbf{80} / \mathbf{40}) * (100 / 1) \\
&= (\mathbf{2}) * (100) \\
&= \mathbf{2}00
\end{aligned}
$$

7,690 / 35 is about 200.

Note A shortcut for mental division with some whole numbers: When you have trailing zeros (zeros at the end of the number) in both the divisor and the dividend, you can use extended multiplication facts to think of an equivalent problem that is easier to divide. The answer to the equivalent problem will be the same as the answer to the original problem.

$7{,}500/30 = \frac{7{,}500}{30} = \frac{(750 * 10)}{(3 * 10)}$

Equivalent problem: **750/3**

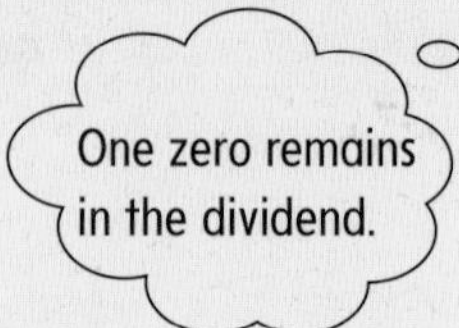

75 / **3** = **25**, so **750** / **3** will be 10 times as much, or **250**.

So, **7,500** / **30** = **250**.

Check Your Understanding

Use close-but-easier numbers to estimate the answer to each problem.

1. 4,941 / 17 = ? **2.** 263 * 47 = ? **3.** 3,044 / 16 = ?

Check your answers in the Answer Key.

Addition Methods

Partial-Sums Addition

Partial-sums addition is used to find sums mentally or with paper and pencil.

To use partial-sums addition:

- Add the value of the digits in each place separately.
- Add the partial sums.

Use an estimate to check whether the answer is reasonable.

Note To make your written work more efficient, you don't need to write the steps shown in green because steps such as estimating or adding the 1s, 10s, and 100s can be done using mental math.

Example

348 + 177 = **?**

Estimate: 348 is close to 350, and you can round 177 to 200.

An estimated sum is 350 + 200 = 550.

			3	4	8
		+	1	7	7
Add the 100s.	300 + 100 →		4	0	0
Add the 10s.	40 + 70 →		1	1	0
Add the 1s.	8 + 7 →			1	5
Add the partial sums.			5	2	5

348 + 177 = **525**

The answer is reasonable because it is close to the estimate of 550.

Did You Know?

In 1642, at age 21, Blaise Pascal invented one of the first mechanical adding machines. The machine had 8 dials and was operated with a pointed, pen-shaped tool. Pascal invented the machine as an aid to his father, who was a tax collector.

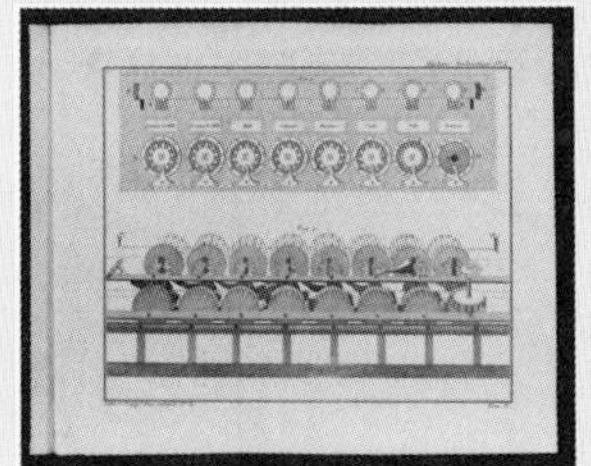

Column Addition

Column addition can be used to find sums with paper and pencil. To add numbers using column addition:

- Add the numbers in each place-value column. Write each sum in its column.
- If the sum of any column is a 2-digit number, make a trade with the column to the left.

Use an estimate to check whether the answer is reasonable.

Example

359 + 289 = **?**

Estimate: 359 is close to 350, and 289 is close to 300.

An estimated sum is 350 + 300 = 650.

		100s	10s	1s
		3	**5**	**9**
	+	**2**	**8**	**9**
Add the numbers in each column.		5	13	18
Trade 10 ones for 1 ten.				
Move 1 ten to the tens column.		5	14	8
Trade 10 tens for 1 hundred.				
Move the 1 hundred to the hundreds column.		6	4	**8**

359 + 289 = **648**

The answer is reasonable because it is close to the estimate of 650.

U.S. Traditional Addition

U.S. traditional addition was taught to most adults in the United States. In this method you add from right to left one column at a time, making trades mentally as you go. Partial sums are not recorded.

Did You Know?

Writing numbers above the addends in U.S. traditional addition is sometimes called "carrying." Numbers that are carried, such as the blue 1 in the tens column in Step 1 below, are sometimes called "carry marks" or "carries."

Example

248 + 187 = **?**

Estimate: 248 is close to 250, and you can round 187 to 200.

An estimated sum is 250 + 200 = 450.

Step 1 Add the ones.

```
    1
  2 4 8
+ 1 8 7
-------
      5
```

8 ones + 7 ones = 15 ones = 1 ten and 5 ones

Step 2 Add the tens.

```
  1 1
  2 4 8
+ 1 8 7
-------
    3 5
```

1 ten + 4 tens + 8 tens = 13 tens = 1 hundred and 3 tens

Step 3 Add the hundreds.

```
  1 1
  2 4 8
+ 1 8 7
-------
  4 3 5
```

1 hundred + 2 hundreds + 1 hundred = 4 hundreds

248 + 187 = **435**

The answer is reasonable because it is close to the estimate of 450.

This method also works with larger numbers.

Note **Addends** are numbers that are added. In 8 + 4 = 12, the numbers 8 and 4 are addends.

Example

7,945 + 8,438 = **?**

Estimate: 7,945 is close to 8,000, and 8,438 is close to 8,400.

An estimated sum is 8,000 + 8,400 = 16,400.

Step 1 Add the ones.

```
      1
  7 9 4 5
+ 8 4 3 8
---------
        3
```

Step 2 Add the tens.

```
      1
  7 9 4 5
+ 8 4 3 8
---------
      8 3
```

Step 3 Continue through the thousands.

```
   1   1
   7 9 4 5
+  8 4 3 8
----------
 1 6 3 8 3
```

7,945 + 8,438 = **16,383**

The answer is reasonable because it is close to the estimate of 16,400.

Opposite-Change Rule

Here is the **opposite-change rule**: If you subtract a number from one addend and add the same number to the other addend, the sum is the same.

Use this rule to make an addition problem easier by changing either of the addends to a number that has 0 in the ones place. Make the *opposite change* to the other addend.

Note **Addends** are numbers that are added. In the example below, 59 and 26 are addends.

Example

59 + 26 = **?**

Estimate: 59 rounds to 60, and 26 rounds to 30.

An estimated sum is 60 + 30 = 90.

One way: Add and subtract 1, because 59 is just one away from 60, making the mental math easier.

$$\begin{array}{r} 59 \\ +\ 26 \\ \hline \end{array} \quad \begin{array}{c} \text{(add 1)} \\ \text{(subtract 1)} \end{array} \quad \begin{array}{r} 60 \\ +\ 25 \\ \hline 85 \end{array}$$

Another way: Subtract and add 4 because 26 is 4 away from 30, making the mental math easier.

$$\begin{array}{r} 59 \\ +\ 26 \\ \hline \end{array} \quad \begin{array}{c} \text{(subtract 4)} \\ \text{(add 4)} \end{array} \quad \begin{array}{r} 55 \\ +\ 30 \\ \hline 85 \end{array}$$

59 + 26 = **85**

The answer is reasonable because it is close to the estimate of 90.

Check Your Understanding

Add. Estimate to check whether your answers are reasonable.

1. $\begin{array}{r} 355 \\ +\ 532 \\ \hline \end{array}$ **2.** $\begin{array}{r} 46 \\ +\ 87 \\ \hline \end{array}$ **3.** $\begin{array}{r} 277 \\ +\ 44 \\ \hline \end{array}$ **4.** $\begin{array}{r} 678 \\ +\ 345 \\ \hline \end{array}$ **5.** 329 + 534 **6.** 751 + 79

Check your answers in the Answer Key.

Subtraction Methods

Trade-First Subtraction

To use **trade-first subtraction,** compare each digit in the top number with each digit below it and make any needed trades before subtracting.

- If each digit in the top number is greater than or equal to the digit below it, subtract separately in each column.
- If any digit in the top number is less than the digit below it, make a trade with the digit to the left before doing any subtraction. Mark the problem to show each trade.

Use an estimate to check whether the answer is reasonable.

Note If you can keep track of the places in your head, you don't need to draw lines between columns.

Example

Subtract 275 from 463 using the trade-first method.

Estimate: You can round 463 to 500 and 275 to 300.

An estimated difference is $500 - 300 = 200$.

	100s	10s	1s
	4	6	3
−	2	7	5

Look at the 1s place.

$3 < 5$, so you need to make a trade.

	100s	10s	1s
		5	13
	4	~~6~~	~~3~~
−	2	7	5

So trade 1 ten for 10 ones.

Mark the problem to show the trade. Now look at the 10s place.

$5 < 7$, so you need to make a trade.

	100s	10s	1s
		15	
	3	~~5~~	13
	~~4~~	~~6~~	~~3~~
−	2	7	5
	1	8	8

So trade 1 hundred for 10 tens.

Now subtract in each column.

$463 - 275 =$ **188**

The answer is reasonable because it is close to the estimate of 200.

Same-Change Rules

Here are some **same-change rules** for subtraction problems:

- If you add the same number to both numbers in the problem, the answer is the same.
- If you subtract the same number from both numbers in the problem, the answer is the same.

One way to use this rule is to change the second number in the problem to a number that has 0 in the ones place to make the mental math easier. Make the *same change* to the first number. Then subtract.

Example

83 − 27 = **?**

Estimate: 83 rounds to 80, and 27 rounds to 30.

An estimated difference is 80 − 30 = 50.

One way: Add 3 to change the second number to 30.

$$\begin{array}{r} 83 \\ -\ 27 \\ \hline \end{array} \quad \begin{array}{l} \text{(add 3)} \\ \text{(add 3)} \end{array} \quad \begin{array}{r} 86 \\ -\ 30 \\ \hline 56 \end{array}$$

Another way: Subtract 7 to change the second number to 20.

$$\begin{array}{r} 83 \\ -\ 27 \\ \hline \end{array} \quad \begin{array}{l} \text{(subtract 7)} \\ \text{(subtract 7)} \end{array} \quad \begin{array}{r} 76 \\ -\ 20 \\ \hline 56 \end{array}$$

83 − 27 = **56** both ways.

The answer is reasonable because it is close to the estimate of 50.

Check Your Understanding

Subtract. Estimate to check whether your answers are reasonable.

1. 518 − 62 **2.** 744 − 227 **3.** 435 − 152 **4.** 3,125 − 417

Check your answers in the Answer Key.

Counting-Up Subtraction

You can use **counting-up subtraction** to find the difference between two numbers by counting up from the smaller number to the larger number. There are many ways to count up. One way is to start by counting up to the nearest multiple of 10, then count by 10s and 100s.

Example

425 − 48 = **?**

Estimate: 48 is close to 50.

An estimated difference is 425 − 50 = 375.

Write the smaller number, 48.

As you count from 48 to 425, circle each number that you count up.

	48	
+	(2)	Count up to the nearest 10.
	50	
+	(50)	Count up to the nearest 100.
	100	
+	(300)	Count up to the largest possible hundred.
	400	
+	(25)	Count up to the larger number
	425	

Add the numbers you circled: 2 + 50 + 300 + 25 = 377

You counted up by 377.

425 − 48 = **377**

The answer is reasonable because it is close to the estimate of 375.

Another way is to count up to close-but-easier numbers.

Example

127 − 74 = **?**

Estimate: 127 is close to 125, and 74 is close to 75.

An estimated difference is 125 − 75 = 50.

Write the smaller number.

	74	
+	(1)	Count up to 75, a close-but-easier number.
	75	
+	(25)	Count up to 100, another close-but-easier number.
	100	
+	(25)	Count up to 125, another close-but-easier number.
	125	
+	(2)	Count up to the larger number.
	127	

Add the numbers you circled: 1 + 25 + 25 + 2 = 53

You counted up 53.

127 − 74 = **53**

This answer is reasonable because it is close to the estimate of 50.

Check Your Understanding

Use counting-up subtraction to find the differences. Estimate to check whether your answers are reasonable.

1. 124 − 89 **2.** 1,000 − 899 **3.** 2,022 − 1,892

Check your answers in the Answer Key.

U.S. Traditional Subtraction

You can subtract numbers using **U.S. traditional subtraction.** Start at the right. Subtract column by column. Make any necessary trades as you go. Use an estimate to check whether the answer is reasonable.

Sometimes you need to make a trade in only one step.

Sometimes you need to make trades in two steps.

Example

572 – 385 = **?**

Estimate: You can round 572 to 600 and 385 to 400.

An estimated difference is 600 – 400 = 200.

Step 1 Start with the 1s. Since 2 < 5, you need to regroup. Trade 1 ten for 10 ones. Subtract the ones.

```
     6  12
  5  7̸  2̸
– 3  8  5
---------
        7
```

12 ones – 5 ones = 7 ones

Step 2 Go to the 10s. Since 6 < 8, you need to regroup. Trade 1 hundred for 10 tens. Subtract the tens.

```
    16
  4  6̸  12
  5̸  7̸  2̸
– 3  8  5
---------
     8  7
```

16 tens – 8 tens = 8 tens

Step 3 Subtract the hundreds.

```
    16
  4  6̸  12
  5̸  7̸  2̸
– 3  8  5
---------
  1  8  7
```

4 hundreds – 3 hundreds = 1 hundred

572 – 385 = **187**

This answer is reasonable because it is close to the estimate of 200.

When the minuend has a zero as one of its digits, sometimes two trades are made in one step. Often after making the double trade, no more trades are needed.

Note In a subtraction problem, the **minuend** is the number from which another number is subtracted. In the example, the minuend is 904.

Example

904 − 385 = **?**

Estimate: You can round 904 to 900 and 385 to 400.

An estimated difference is 900 − 400 = 500.

Step 1 Compare the digits in the ones place. You need to trade 1 ten for 10 ones, but there are no tens. So two trades are needed.

First trade 1 hundred for 10 tens. Then trade 1 ten for 10 ones. Subtract the ones.

```
      9
   8 1̶0̶ 14
   9̶  0̶  4̶
 −  3  8  5
 ----------
          9
```

14 ones − 5 ones = 9 ones

Step 2 Compare the digits in the tens place. No trade is needed. Subtract the tens.

```
      9
   8 1̶0̶ 14
   9̶  0̶  4̶
 −  3  8  5
 ----------
       1  9
```

9 tens − 8 tens = 1 ten

Step 3 Subtract the hundreds.

```
      9
   8 1̶0̶ 14
   9̶  0̶  4̶
 −  3  8  5
 ----------
    5  1  9
```

8 hundreds − 3 hundreds = 5 hundreds

904 − 385 = **519**

This answer is reasonable because it is close to the estimate of 500.

Note Another way to trade is to think of 904 as 90 tens and 4 ones. By trading 1 ten for 10 ones, the result is 89 tens and 14 ones, which is marked in the problem this way:

```
   8  9 14
   9̶  0̶  4̶
 −  3  8  5
 ----------
    5  1  9
```

Check Your Understanding

Subtract. Estimate to check whether your answers are reasonable.

1. 75 − 37 **2.** 853 − 471 **3.** 651 − 285 **4.** 704 − 447 **5.** 7,345 − 3,066

Check your answers in the Answer Key.

Multiplying by Powers of 10

You can use what you know about place value and exponential notation to multiply by powers of 10.

Note Numbers like 10, 100, and 1,000 are called *powers of 10.* They are numbers that can be written as products of 10s. For example, 1,000 = 10 * 10 * 10.

Multiplying A Whole Number by a Power of 10

In the base-10 place-value system, each digit in one place is worth 10 times as much as the place to its right. This attribute of the place-value system helps you multiply by powers of 10.

Example

1,000 * 100 = **?**

1,000s thousands	100s hundreds	10s tens	1s ones
1	0	0	0

Identify the place value of one of the factors. — 1 is in the thousands place

Write the second factor as a product of 10s. — 100 = 10 * 10

Write a new number sentence for 1,000 * 100 = ? — 1,000 * 10 * 10 = ?

100

Each time a number is multiplied by 10, the digits shift one place to the left. So the 1 in the first factor (1,000) is written two places to the left in the 100 thousands place, and 2 zeros are attached to show the place shift. — 100,000

2 zeros are attached to show that the 1 was moved two places to the left.

So, 1,000 * 100 = **100,000.**

Because of the base-10 place-value system, you can use patterns with zeros to multiply whole numbers by powers of 10.

Example

What patterns do you notice?

8 * 10 = 80	8 is multiplied by 10 one time. There is one zero in the product.
8 * 100 = 800	100 = 10 * 10, so 8 is multiplied by 10 two times. There are two zeros in the product.
8 * 1,000 = 8,000	1,000 = 10 * 10 * 10, so 8 is multiplied by 10 three times. There are three zeros in the product.

Each time 8 is multiplied by 10, the 8 shifts one place to the left and one zero is attached to the product.

A Shortcut for Multiplying A Whole Number by a Power of 10

You can use exponential notation to make a shortcut for multiplying whole numbers by powers of 10. The exponent tells how many places the digits of the number are shifted to the left and how many zeros to attach to show that the place value has shifted.

Example

1,000 ∗ 45 = **?**

Step 1 Write the power of 10 in exponential notation. $1{,}000 = 10 * 10 * 10 = 10^3$

Step 2 Write a new number sentence. $10^3 * 45 = ?$

Step 3 Write 45 and then attach 3 zeros. $10^3 * 45 = 45000$

The exponent tells that 45 is being multiplied by 10 three times. Each time a number is multiplied by 10, the original digits shift one place to the left. So each digit in 45 moves three places to the left. Three zeros are attached to show that the place value has shifted.

Step 4 Insert any needed commas. $10^3 * 45 = 45{,}000$

So, 1,000 ∗ 45 = **45,000.**

You can combine the steps to do this type of problem in your head.

Examples

137 ∗ 100 = **?**

Steps 1 & 2 $137 * 10^2 = ?$ ($100 = 10^2$)

Steps 3 & 4 $137 * 10^2 = 13700$ or 13,700

So, 137 ∗ 100 = **13,700.**

10,000 ∗ 14 = **?**

Steps 1 & 2 $10^4 * 14 = ?$ ($10{,}000 = 10^4$)

Steps 3 & 4 $10^4 * 14 = 140000$ or 140,000

So, 10,000 ∗ 14 = **140,000.**

Check Your Understanding

Multiply.

1. 42 ∗ 1,000 = ? **2.** 100 ∗ 3,457 = ? **3.** 20 ∗ 1,000 = ? **4.** 10,000 ∗ 32 = ?

Check your answers in the Answer Key.

Extended Multiplication Facts

You are using **extended multiplication facts** when you combine basic multiplication facts with multiplying by powers of 10.

One way to find answers to extended multiplication facts is to use basic multiplication facts with multiples of 10.

Example

8 * 70 = **?**

Think: 8 [7s] = 56

Then 8 [70s] is 10 times as much.

8 * 70 = 10 * 56 = 560

So, 8 * 70 = **560.**

You can use exponential notation to find the products of extended multiplication facts, especially when the numbers are large.

Example

3,000 * 200 = **?**

Step 1 Write the problem in exponential notation. $(3 * 10^3) * (2 * 10^2) = ?$

Step 2 Reorder the factors. Remember that the order and grouping of factors does not change the product. $(3 * 2) * (10^3 * 10^2) = ?$

- Collect the basic facts together.
- Collect the powers of 10 together.

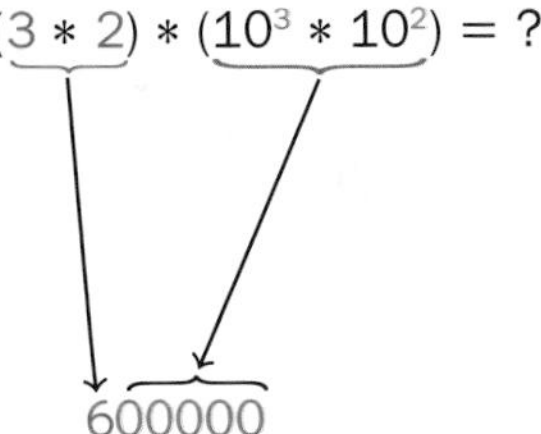

Step 3 Solve the basic fact: 3 * 2 = 6.

Step 4 Use patterns with powers of 10 to show how the place value shifts by attaching zeros to the product of the basic fact.

Multiplying by 10^3 means attach 3 zeros.

Multiplying by 10^2 means attach 2 zeros.
Attach a total of 5 zeros.

Step 5 Write the number sentence with the product and insert commas as needed. 3,000 * 200 = **600,000**

You can combine steps from the previous page to solve this type of problem in your head.

Examples

60,000 ∗ 5,000 = **?**

Steps 1 & 2 $(6 * 10^4) * (5 * 10^3) = ?$

$(6 * 5) * (10^4 * 10^3) = ?$

$10 * 10 * 10 * 10 * 10 * 10 * 10 = 10^7$

Steps 3, 4, & 5 $(30) * (10^7) = 300000000$ or 300,000,000

60,000 ∗ 5,000 = **300,000,000**

12,000 ∗ 200 = **?**

Steps 1 & 2 $(12 * 10^3) * (2 * 10^2) = ?$

$(12 * 2) * (10^3 * 10^2) = ?$

$10 * 10 * 10 * 10 * 10 = 10^5$

Steps 3, 4, & 5 $(24) * (10^5) = 2400000$ or 2,400,000

12,000 ∗ 200 = **2,400,000**

Check Your Understanding

Solve these problems using mental math.

1. 9 ∗ 500 = ? **2.** 4,000 ∗ 60 = ? **3.** 240 ∗ 20,000 = ? **4.** 8,000 ∗ 500

Check your answers in the Answer Key.

Area Models for Multiplication

An **area model** is a visual representation for multiplication problems in which the length and width of a rectangle represent the factors and the area of the rectangle represents the product. Area models can be used to represent several different multiplication methods, including partial-products multiplication and U.S. traditional multiplication.

This is an area model for 3 ∗ 6.

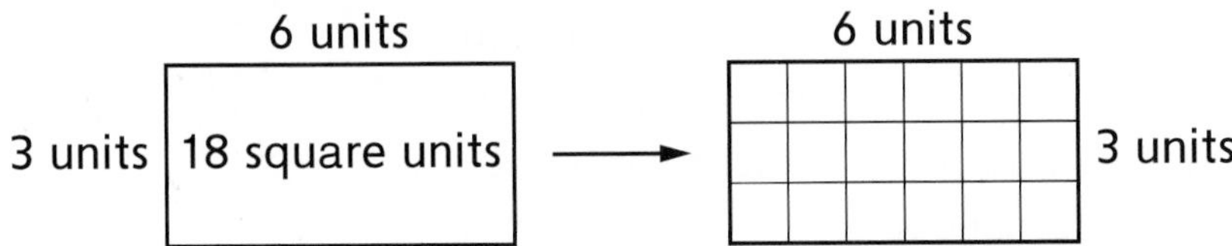

The factors are 3 and 6, the same as the side lengths. The product is 18. The width is 3 units and the length is 6 units. The area is 18 square units.

Area models can be divided into sections to find the area of a whole rectangle. The areas of the smaller rectangles can be added to find the area of the whole rectangle.

Example

Draw an area model for the problem: 4 ∗ 236 = **?**

Divide the rectangle into smaller rectangles that correspond to the place value of each digit in one of the factors.

Since 236 = 200 + 30 + 6, divide the large rectangle into three smaller rectangles with lengths of 200, 30, and 6, as in the sketch below.

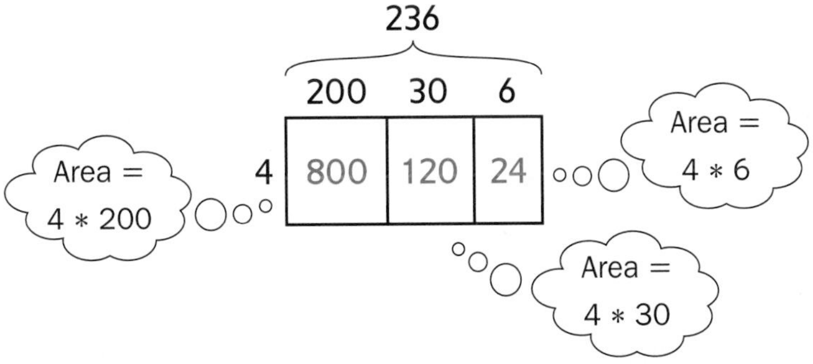

Write a number sentence to find the total area of the large rectangle. Remember that the sum of the areas of the smaller rectangles equals the area of the large rectangle.

The area of the large rectangle is 800 + 120 + 24 = 944.

So, 4 ∗ 236 = **944.**

Multiplication Methods

Partial-Products Multiplication

When you multiply using **partial-products multiplication,** the value of each digit in one factor is multiplied by the value of each digit in the other factor. The final product is the sum of these **partial products.**

When you use partial-products multiplication, factors are broken up by place value. This results in partial products that are either basic multiplication facts or extended multiplication facts that can be calculated mentally. You must keep track of the place value of each digit in the partial products.

Note The symbols × and ∗ are both used to indicate multiplication.

Example

4 ∗ 236 = **?**

Estimate: You can round 236 to 200.

Since 236 was rounded down, the product will be greater than 4 ∗ 200 = 800.

Think of 236 as 200 + 30 + 6.

Multiply each part of 236 by 4.

		2	3	6
	∗			4
4 ∗ 200 →		8	0	0
4 ∗ 30 →		1	2	0
4 ∗ 6 →			2	4
		9	4	4

Add the partial products.

4 ∗ 236 = **944**

The answer is reasonable because it is "in the hundreds" like the estimate of 800 and it is more than 800.

Compare the area model for multiplication on page 99 to using partial products. How are they alike? How are they different?

Note If you can estimate and multiply to find the partial products using mental math, then you do not need to write the steps shown in green.

You can draw an area model to help identify partial products.

Example

43 ∗ 26 = **?**

Estimate: 43 rounds to 40, and 26 is close to 25. An estimate is 40 ∗ 25 = 1,000.

Draw an area model. Draw a rectangle that is 43 by 26 and divide it into four smaller rectangles by breaking up the factors using place value.
The whole rectangle is 26 units wide and 43 units long.
Think about the width of 26 as 20 + 6.
Think about the length of 43 as 40 + 3.
Find the area of each of the smaller rectangles and you have found the partial products. Add the partial products together to find the answer.

43 (40, 3); 26 (20, 6)

	40	3
20	800	60
6	240	18

Think about the 40-by-20 rectangle. (800 shaded)

Think about the 40-by-6 rectangle. (240 shaded)

Think about the 3-by-20 rectangle. (60 shaded)

Think about the 3-by-6 rectangle. (18 shaded)

			2	6
	∗		4	3
40 ∗ 20 →		8	0	0
40 ∗ 6 →		2	4	0
3 ∗ 20 →			6	0
3 ∗ 6 →			1	8
	1,	1	1	8

43 ∗ 26 = **1,118**

The answer is reasonable because it is close to the estimate of 1,000.

Check Your Understanding

Multiply using partial products. Estimate to check whether your answers are reasonable.

1. 265 ∗ 3 **2.** 42 ∗ 67 **3.** 40 ∗ 58 **4.** 83 ∗ 54 **5.** 372 ∗ 50

Check your answers in the Answer Key.

U.S. Traditional Multiplication

When you use **U.S. traditional multiplication,** you add products as you go, rather than recording each partial product separately. Use an estimate to check whether the answer is reasonable.

Example

5 * 629 = **?**

Estimate: You can round 629 to 600.

Since 629 was rounded down, the product will be greater than 5 * 600 = 3,000.

Step 1 Multiply the ones.

5 * 9 ones = 45 ones = 4 tens + 5 ones
Write 5 in the 1s place below the line.
Write 4 above the 2 in the 10s place.

```
    4
  6 2 9
*     5
-------
      5
```

Step 2 Multiply the tens.

5 * 2 tens = 10 tens
Remember the 4 tens from Step 1.
10 tens + 4 tens = 14 tens in all.
14 tens = 1 hundred + 4 tens
Write 4 in the 10s place below the line.
Write 1 above the 6 in the 100s place.

```
  1 4
  6 2 9
*     5
-------
    4 5
```

Step 3 Multiply the hundreds.

5 * 6 hundreds = 30 hundreds
Remember the 1 hundred from Step 2.
30 hundreds + 1 hundred = 31 hundreds in all
31 hundreds = 3 thousands + 1 hundred
Write 1 in the 100s place below the line.
Write 3 in the 1,000s place below the line.

```
  1
  6 2 9
*     5
-------
3 1 4 5
```

5 * 629 = **3,145**

The answer is reasonable because it is close to and greater than the estimate of 3,000.

Did You Know?

Writing the numbers above the factors in U.S. traditional multiplication is sometimes called "carrying." The numbers are called "carry marks" or "carries."

When using U.S. traditional multiplication to multiply multidigit numbers by multidigit numbers, break up one of the numbers and record partial products on separate lines.

Example

73 ∗ 26 = **?**

Estimate: Round 73 to 70. Round 26 to 30. An estimated product is 70 ∗ 30 = 2,100.

The steps: | **Record the steps:** | **Think about the area model:**

Step 1 Multiply 73 by the 6 in 26 as if the problem were 6 ∗ 73.

Check: The product of 438 for Step 1 should be equal to the sum of the partial products involving 6: 6 ∗ 70 = 420 and 6 ∗ 3 = 18. 420 + 18 = 438

```
    1
    7 3
*   2 6
-------
  4 3 8
```

6 ∗ 73 corresponds to the rectangle that is 73 by 6.

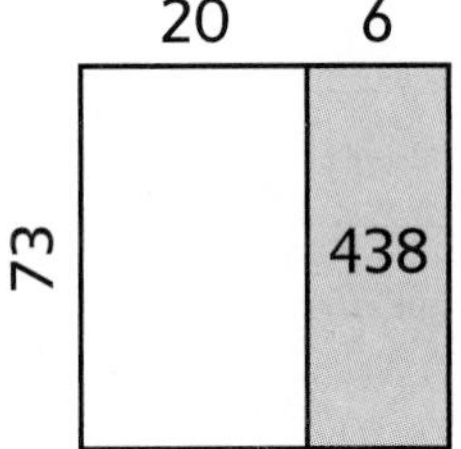

Step 2 *Think:* What is the value of the 2 in 26? It's 20, so multiply 73 by the 2 tens in 26 as if the problem were 20 ∗ 73.

Check: The product of 1,460 for Step 2 should be equal to the sum of the partial products involving 20: 20 ∗ 3 = 60 and 20 ∗ 70 = 1,400. 60 + 1,400 = 1,460

```
      1
      7 3
*     2 6
---------
    4 3 8
  1 4 6 0
```

20 ∗ 73 corresponds to the rectangle that is 73 by 20.

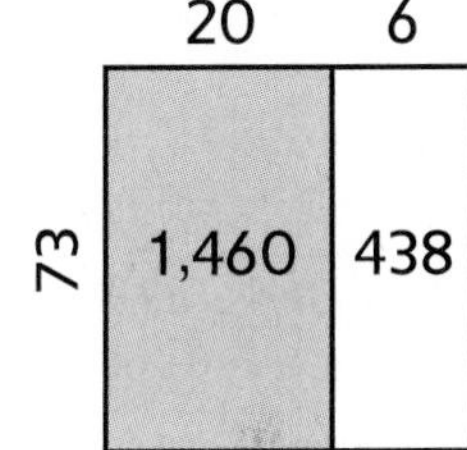

Step 3 Add the two partial products to get the final answer.

```
      1
      7 3
*     2 6
---------
    4 3 8
+ 1 4 6 0
---------
  1 8 9 8
```

Add the partial products to find the total area of the large rectangle.

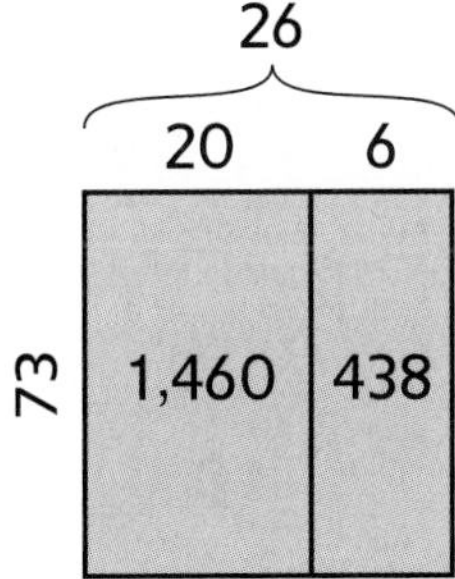

73 ∗ 26 = **1,898**

The answer is reasonable because it is close to the estimate of 2,100.

Check Your Understanding

Multiply. Estimate to check whether your answers are reasonable.

1. 38 ∗ 368 = ?

2. 829 ∗ 52 = ?

Check your answers in the Answer Key.

Lattice Multiplication

Lattice multiplication has been used for hundreds of years. It is based on placing answers to basic multiplication facts in each box and then adding along the diagonals. The box with cells and diagonals is called a **lattice.**

Lattice multiplication works because each diagonal is the same as a place-value column.

Example

6 * 815 = **?**

Estimate: 815 is close to 800. An estimated product is 6 * 800 = 4,800.

Write 815 above the lattice.
Write 6 on the right side of the lattice.

Multiply 6 * 5. Then multiply 6 * 1. Then multiply 6 * 8.
Write the answers as shown.

Add the numbers along each diagonal, starting at the right.
Read the answer. 6 * 815 = **4,890**

The answer is reasonable. It is close to the estimate of 4,800.

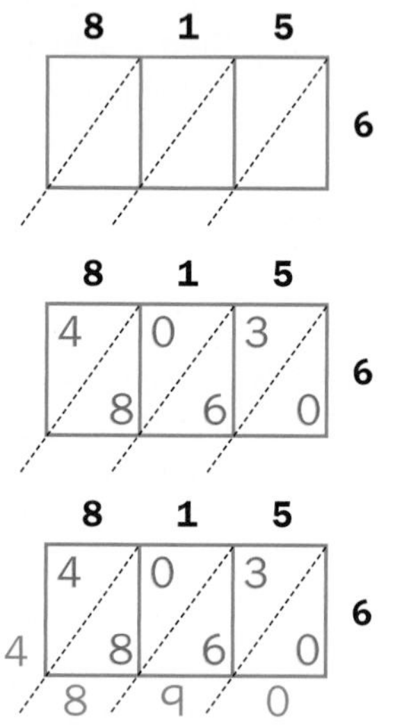

Example

42 * 37 = **?**

Estimate: 42 rounds to 40, and 37 rounds to 40. An estimate is 40 * 40 = 1,600.

Write 37 above the lattice. Write 42 on the right side of the lattice.
Multiply 4 * 7. Then multiply 4 * 3.
Multiply 2 * 7. Then multiply 2 * 3.

Write the answers as shown.

Add the numbers along each diagonal, starting at the right.
When the numbers along a diagonal add up to 10 or more:

- record the ones digit in the sum.
- add the tens digit to the sum along the next diagonal above.

Read the answer. 42 * 37 = **1,554**

The answer is reasonable because it is close to the estimate of 1,600.

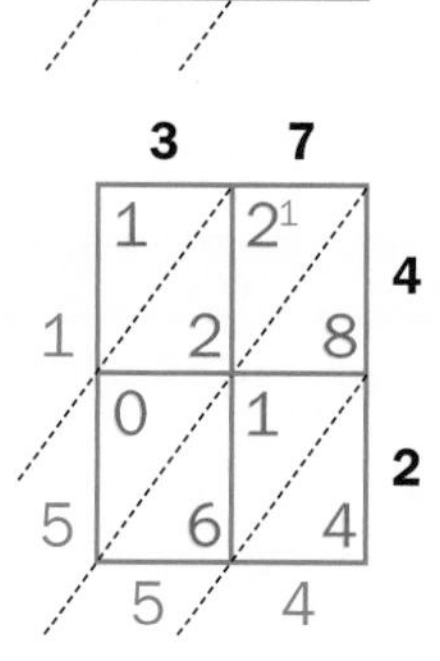

Dividing by Powers of 10

To divide by powers of 10 you can use what you know about place value and the relationship between multiplication and division.

Example

40,000 / 10 = **?**

Step 1 Write 40,000 / 10 as a missing factor multiplication number sentence.

40,000 = ? * 10

Step 2 What is the missing factor? What number times 10 is equal to 40,000?

Each digit in one place is worth 10 times as much as in the place to its right. So the missing factor must be 4,000, because 4,000 * 10 = 40,000.

Step 3 Use the multiplication number sentence to answer the division problem.

Since 4,000 * 10 = 40,000, 40,000 / 10 = 4,000.

So, 40,000 / 10 = **4,000.**

Note Sets of related multiplication and division facts are called **fact families.** When you rewrite a division problem as a missing factor multiplication problem, it is like using a fact family to solve a division problem. For example, you can solve 35 / 5 by thinking ? * 5 = 35. **7** * 5 = 35, so 35 / 5 equals 7.

Example

3,000 / 100 = **?**

Write 3,000 / 100 as a missing factor multiplication number sentence.

3,000 = ? * 100

What number times 100 is equal to 3,000?

Each digit in one place is worth 10 times as much as in the place to its right. So the missing factor must be 30, because 30 * 10 * 10 = 3,000.

Use the multiplication number sentence to answer the division problem.

Since 30 * 100 = 3,000, 3,000 / 100 = 30.

So, 3,000 / 100 = **30.**

Check Your Understanding

Solve these problems using mental math.

1. 900 / 10 = ? **2.** 12,000 / 100 = ? **3.** 200,000 / 10,000 = ? **4.** 1,000 / 10 = ?

Check your answers in the Answer Key.

Extended Division Facts

You are using **extended division facts** when you combine basic division facts and multiples of 10.

One way to find answers to extended division facts is to relate the problem to basic facts and to multiplication with powers of 10.

Example

240 / 4 = **?**

Think: 24 / 4 = 6

Then 240 / 4 is 10 times as much.

240 / 4 = 10 * 6 = 60

So, 240 / 4 = **60.**

You can also use the relationship between multiplication and division to solve extended division facts, especially when the numbers are large.

Example

45,000 / 900 = **?**

Write 45,000 / 900 as a missing factor multiplication number sentence.

45,000 = ? * 900

What is the related division fact?

45 / 9 = 5

Try 5 as the missing factor for your multiplication number sentence.

5 * 900 = 4,500, and that's too small.

You want 45,000. So you need a number that is 10 times as much as 5.

Try 5 * 10, or 50, as your missing factor.

50 * 900 = 45,000

This works.

Fill in the missing factor for the multiplication number sentences. Write a related division number sentence to show the answer.

Since 50 * 900 = 45,000,

45,000 / 900 = 50.

45,000 / 900 = **50**

Check Your Understanding

Solve these problems using mental math.

1. 27,000 / 900 = ? **2.** 4,800 / 60 = ? **3.** 4,000 / 200 = ? **4.** 3,000 / 60 = ?

Check your answers in the Answer Key.

Mental Math with Division

You can often divide mentally by breaking the **dividend** into parts that are multiples of the **divisor.** Divide each part by the divisor and add the partial quotients to find the answer.

Example

96 / 6 = **?**

Step 1 Find a multiple of the divisor that is less than the divisor and easy to use in mental math.

Try 60. It is a multiple of 6 that is easy to use in mental math for this problem.

Step 2 What do you need to add to the multiple to total the dividend? Is that number also an easy multiple of 6? If yes, rename the dividend as an addition number sentence.

NOTE: If the remaining part is not an easy multiple of the divisor, keep breaking it up.

You need to add 36 to 60 to get 96.
36 is also an easy multiple of 6.
60 + 36 = 96

Step 3 Divide each part of the dividend by the divisor.

60 / 6 = 10 and 36 / 6 = 6

Step 4 Find the sum of the partial quotients. The sum is the answer to the division problem.

10 + 6 = 16, so 96 / 6 = **16.**

Sometimes you have to break up the dividend into more parts. In some cases, there is a small number left that cannot be broken up into any more multiples of the divisor. The leftover number is the remainder.

Example

78 / 5 = **?**

What is an easy multiple of 5 that is less than 78?

Try 50. It is an easy multiple of 5.

Write an addition number sentence that renames 78 using 50 as an addend.

50 + 28 = 78

Is the new part an easy multiple of 5? No. Break up the 28 by using an easy multiple of 5. Write a new addition number sentence that renames 78.

28 can be broken up into 25 and 3.

$78 = 50 + \underbrace{25 + 3}_{28}$

Are you finished? Yes, all of the addends are easy multiples of 5, and the 3 left over is less than 5.

Divide the parts by 5. Figure out the answer by adding the partial quotients and writing the remainder. The remainder is the part less than 5 that is left over.

50 / 5 = 10, 25 / 5 = 5, and R3
10 + 5 = 15, and there is 3 left over.

Check: 5 * 15 = 75, then add the remainder, and the total is 78.

78 / 5 → **15 R3**

Division Methods

Different symbols may be used to indicate division. For example, "94 divided by 6" may be written as $94 \div 6$, $6\overline{)94}$, 94 / 6, and $\frac{94}{6}$.

- The number being divided is the **dividend.**
- The number that divides the dividend is the **divisor.**
- The answer to a division problem is the **quotient.**
- When numbers cannot be divided evenly, the answer includes a quotient and a **remainder.**

Four ways to show "123 divided by 4"	
$123 \div 4 \rightarrow$ 30 R3	123 / 4 $\rightarrow$ 30 R3
$\begin{array}{r} 30 \text{ R3} \\ 4\overline{)123} \end{array}$	$\frac{123}{4} \rightarrow$ 30 R3
123 is the dividend. 4 is the divisor. 30 is the quotient. 3 is the remainder.	

Partial-Quotients Division

Partial-quotients division takes several steps to find the quotient. At each step, you find a partial answer (called a **partial quotient**). Then add the partial answers to find the quotient.

Study the example below. To find the number of 6s in 1,010 first find partial quotients and then add them. Record the partial quotients in a column to the right of the original problem.

Example

1,010 / 6 = **?**

Estimate: 1,010 is close to 1,200. The quotient will be less than 1,200 / 6 = 200.

	Write partial quotients in this column.	
$6\overline{)1{,}010}$	↓	*Think:* How many 6s are in 1,010? At least 100.
− 600	100	The first partial quotient is 100. 100 ∗ 6 = 600
410		Subtract 600 from 1,010. At least 50 [6s] are left in 410.
− 300	50	The second partial quotient is 50. 50 ∗ 6 = 300
110		Subtract. At least 10 [6s] are left in 110.
− 60	10	The third partial quotient is 10. 10 ∗ 6 = 60
50		Subtract. At least 8 [6s] are left in 50.
− 48	8	The fourth partial quotient is 8. 8 ∗ 6 = 48
2 ↑ Remainder	168 ↑ Quotient	Subtract. Add the partial quotients.

The answer is **168 R2.** Record the answer as $\begin{array}{r} 168 \text{ R2} \\ 6\overline{)1{,}010} \end{array}$ or write 1,010 / 6 → 168 R2.

The answer is reasonable because it is close to and less than the estimate of 200.

Partial-quotients division works the same whether you divide by a 2-digit or a 1-digit divisor.

Examples

463 / 21 = **?**

Estimate: 463 / 21 is close to 460 / 20. An estimated quotient is 460 / 20 = 23.

	Write partial quotients in this column.	
21)463	↓	*Think:* How many 21s are in 463? At least 10.
− 210	10	The first partial quotient is 10. 10 ∗ 21 = 210
253		Subtract 210 from 463. At least 10 [21s] are left in 253.
− 210	10	The second partial quotient is 10. 10 ∗ 21 = 210
43		Subtract 210 from 253. At least 2 [21s] are left in 43.
− 42	2	The third partial quotient is 2. 2 ∗ 21 = 42
1	22	Subtract. Add the partial quotients.
↑	↑	
Remainder	Quotient	

The answer is **22 R1.** Record the answer as 21)463 with 22 R1 written above, or write 463 / 21 → 22 R1.

The answer is reasonable because 22 R1 is close to the estimate of 23.

It often helps to write down some easy facts for the divisor.

Example

Divide 600 by 22.

Estimate: 22 is close to 20. An estimated quotient is 600 / 20 = 30.

22)600		
− 440	20	(20 [22s] in 600)
160		
− 110	5	(5 [22s] in 160)
50		
− 44	2	(2 [22s] in 50)
6	27	

Some facts for 22 (to help find partial quotients):

1 ∗ 22 = 22
2 ∗ 22 = 44
5 ∗ 22 = 110
10 ∗ 22 = 220
20 ∗ 22 = 440

The answer is **27 R6.** Record the answer as 22)600 with 27 R6 written above, or write 600 / 22 → 27 R6.

The answer is reasonable because 27 R6 is close to the estimate of 30.

There are different ways to find partial quotients when you use partial-quotients division. Study the examples below. The answer in each example box is the same for each way.

Example

381 / 4 = **?**

Estimate: 381 is close to 400. An estimated quotient is 400 / 4 = 100.

One way:

4)381	
− 200	50
181	
− 120	30
61	
− 40	10
21	
− 20	5
1	95

A second way:

4)381	
− 200	50
181	
− 160	40
21	
− 20	5
1	95

A third way:

4)381	
− 360	90
21	
− 20	5
1	95

The answer, **95 R1,** is reasonable because it is close to the estimate of 100.

Example

779 / 24 = **?**

Estimate: 779 is close to 750, and 24 is close to 25. An estimate is 750 / 25 = 30.

24)779	
− 240	10
539	
− 240	10
299	
− 240	10
59	
− 48	2
11	32

24)779	
− 720	30
59	
− 48	2
11	32

Which of the two ways is more efficient? How can you choose your partial quotients to make the division efficient?

The answer, **32 R11,** is reasonable because it is close to the estimate of 30.

Area Models for Division

An **area model** is a visual representation for division problems in which the area of the rectangle represents the **dividend,** the length represents the **divisor,** and the width represents the **quotient.**

Example

Draw an area model for 24 / 6.

The dividend is 24, so the area model must have an area of 24 square units.

The divisor is 6, so the area model must have a length of 6.

The answer to the division problem, the quotient, is the width.

Think about what division means. 24 / 6 is another way of asking, "How many groups of 6 are in 24?"

You can show groups of 6 by making rows in a rectangular array. It takes 4 rows of 6 to make 24, so there are 4 [6s] in 24.

This means that 24 / 6 = 4.

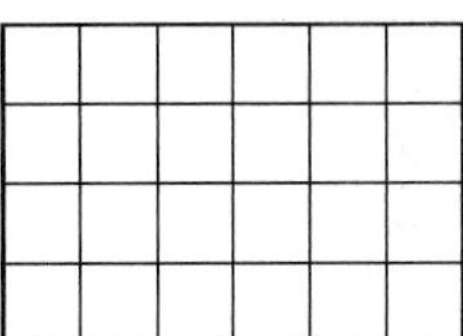

A rectangular array showing that 4 groups of 6 make 24

To make an area model for division, draw a rectangle with an area of 24 square units, a length of 6 units, and a width of 4 units. Label the area, length, and width.

Check: Is the area of a 6-by-4 rectangle equal to 24 square units?

Yes. 6 ∗ 4 = 24

6 units

4 units | 24 square units

An area model showing 24 / 6 = 4

Dividing area models into smaller sections can help you make sense of the steps of partial-quotients division.

Example

154 / 11 = **?**

Sketch a rectangle to represent the problem. Think of the dividend, 154, as the total area. Think of the divisor, 11, as the length of one side.

Divide the rectangle into smaller rectangles that correspond to each partial quotient.

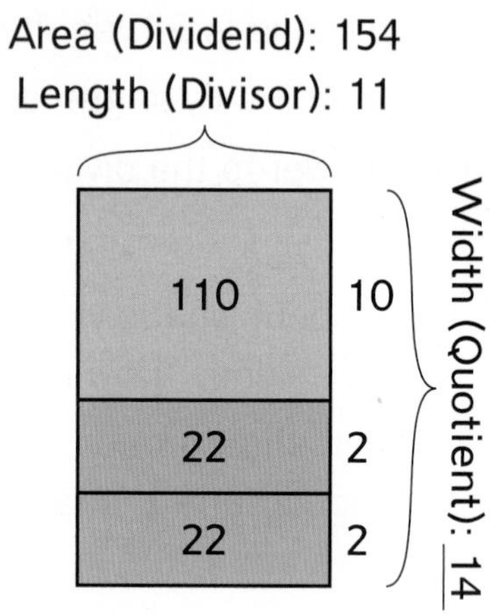

Think: How many 11s are in 154? There are at least 10 [11s] in 154, so use 10 as the first partial quotient. This is like 10 rows of 11. 10 ∗ 11 = 110, so the area of the first section is 110 square units.

The total area is 154 square units, and the first section covers 110 square units. 44 square units are left.

Think: How many 11s are in 44? There are at least 2 [11s] in 44, so use 2 as the next partial quotient. This is like 2 rows of 11. 2 ∗ 11 = 22, so the area of this section is 22 square units.

There were 44 square units left, and the second section covers 22 square units. 22 square units are left.

Think: How many 11s are in 22? There are 2 [11s] in 22, so use 2 as the next partial quotient. This is like 2 more rows of 11. 2 ∗ 11 = 22, so the area of this section is 22 square units.

The area of the three sections together is 110 + 22 + 22 = 154 square units, which matches the dividend.

Write a number sentence to find the quotient. Remember that the sum of the widths of the smaller rectangles equals the width of the large rectangle.

The width of the large rectangle is

10 + 2 + 2 = 14.

So, 154 / 11 = **14.**

Area (Dividend): 154
Length (Divisor): 11

Area	Width
110	10
22	2
22	2

Width (Quotient): 14

Check Your Understanding

Divide using any method. Estimate to check whether your answers are reasonable.

1. $4\overline{)71}$ **2.** 735 / 25 **3.** 342 ÷ 4 **4.** $32\overline{)674}$

Check your answers in the Answer Key.

Interpreting Remainders

Some numbers cannot be divided evenly. When this happens, the answer includes a quotient and a **remainder.** The way you represent the quotient and remainder depends on the problem situation. There are several ways to think about remainders:

- Ignore the remainder. Use the quotient as the answer.
- Round the quotient up to the next whole number.
- Rewrite the remainder as a fraction or decimal. Use this fraction or decimal as part of the answer.

Example

Suppose 3 students share 14 counters equally. How many counters will each student get?

Person 1 Person 2 Person 3 Remainder

3)14
− 12 | 4
2 4
↑ ↑
Remainder Quotient

Ignore the remainder. You cannot split the remaining counters among 3 students.

Answer: Each student will get 4 counters, with 2 counters left over.

Example

Suppose 14 photos are placed in a photo album. How many pages are needed if 3 photos can fit on a page?

Page 1 Page 2 Page 3 Page 4 Page 5

14 / 3 → 4 R2

Round the quotient up to the next whole number. A fifth page is needed to include all the photos. The album will have 4 pages filled and another page only partially filled.

Answer: 5 pages are needed.

To rewrite a remainder as a fraction, make the remainder the **numerator** of the fraction and the divisor the **denominator** of the fraction.

Examples

Suppose 3 friends share 14 sheets of yellow construction paper for an art project. How much yellow construction paper does each friend get?

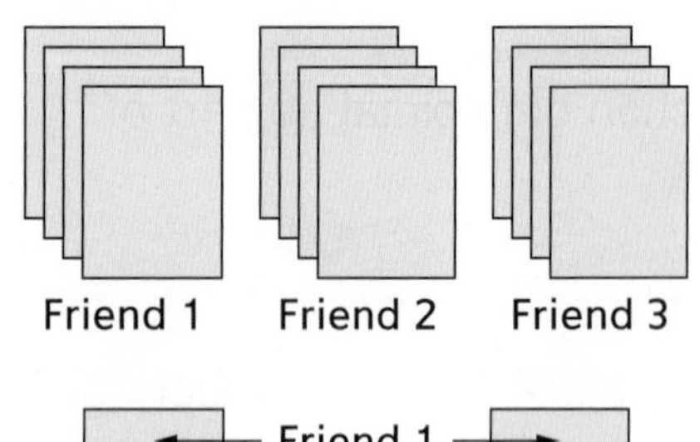

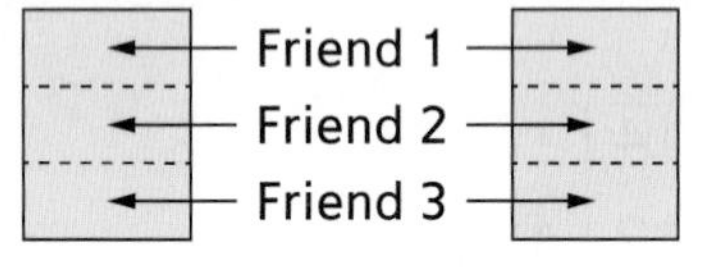

14 / 3 → 4 R2

The answer, 4 R2, shows that each friend receives 4 sheets of yellow construction paper, and 2 sheets are left over. Each of the remaining 2 sheets can be divided into thirds. Each friend receives 2 of these $\frac{1}{3}$-sheets, or $\frac{2}{3}$ of a sheet of yellow construction paper.

The remainder (2) is rewritten as a fraction $\left(\frac{2}{3}\right)$. Use this as part of the answer.

Answer: Each friend will get $4\frac{2}{3}$ sheets of yellow construction paper.

Check Your Understanding

Solve each number story and decide what to do with the remainder. Explain what you did and why.

1. At the school picnic, there are 10 watermelons that need to be divided among 3 stations. How many watermelons will each station get?
2. 6 people can fit at a picnic table. If there are 38 people at the picnic, how many tables are needed for everyone to have a seat?
3. Raffle tickets cost \$2.00 each. Abner has \$15.00. How many raffle tickets can he buy?

Check your answers in the Answer Key.

Decimals

Both decimals and fractions are used to write numbers that are between whole numbers. While fractions are written using fraction notation, decimals are written using the base-10 place-value system just like whole numbers. That means that you can compute with decimals in similar ways that you compute with whole numbers.

Example

The amount shaded is a number between 1 and 2.

Fractions that show the amount shaded: $1\frac{1}{2}$, $\frac{3}{2}$.

These fractions are read "one and one-half" and "three-halves."

1 Whole | $\frac{1}{2}$ | $\frac{1}{2}$

To write a decimal for the amount shaded, first shade the same amount using tenths, hundredths, thousandths, and so forth.

Example

Show $1\frac{1}{2}$ shaded as tenths.

The fraction for the amount shaded is $1\frac{5}{10}$ and is read "one and five-tenths."

1 Whole

You can write a decimal for the amount shaded. Use a dot, called a **decimal point,** to separate the whole from the parts.

decimal point

13.205

whole number | part of a whole

Example

To write the decimal for $1\frac{5}{10}$:

- Write 1 to the left of the decimal point to show 1 whole.
- Write 5 to the right of the decimal point to show 5 *tenths*. The first digit after the decimal point *always* names tenths.

So the decimal is written 1.5.

This decimal is read "one and five-tenths," just like the fraction name.

Note Both decimals and fractions are used to name part of a whole thing or part of a collection.

Decimals and fractions are also used to make more precise measurements than can be made using only whole numbers.

Fractional parts of a dollar are almost always written as decimals.

Writing Decimals

Many fractions have denominators of 10, 100, 1,000, and so on. It is easy to write the decimal names for fractions like these.

Examples

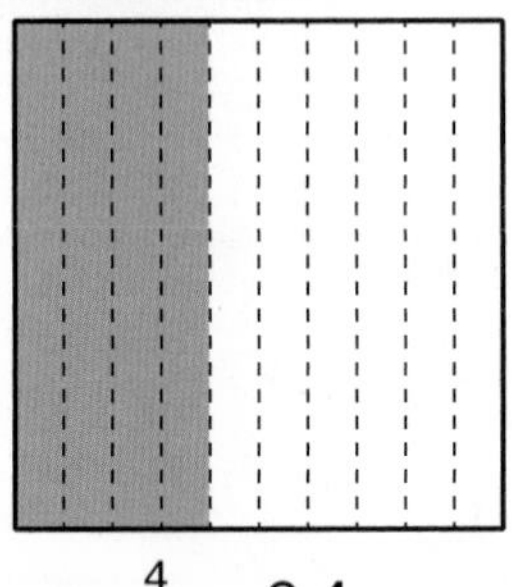

$\frac{4}{10} = 0.4$

This square is divided into 10 equal parts. Each part is $\frac{1}{10}$ of the square. When one digit is written to the right of the decimal point, it names *tenths*. So the decimal name for $\frac{1}{10}$ is 0.1.

Since $\frac{4}{10}$ of the square above is shaded, the decimal name is:

0.4

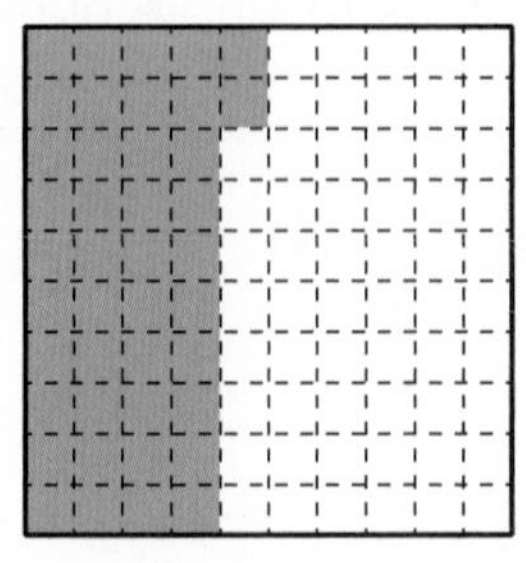

$\frac{42}{100} = 0.42$

This square is divided into 100 equal parts. Each part is $\frac{1}{100}$ of the square. When two digits are written to the right of the decimal point, they name *hundredths*. So the decimal name for $\frac{1}{100}$ is 0.01.

Since $\frac{42}{100}$ of the square above is shaded, the decimal name is:

0.42

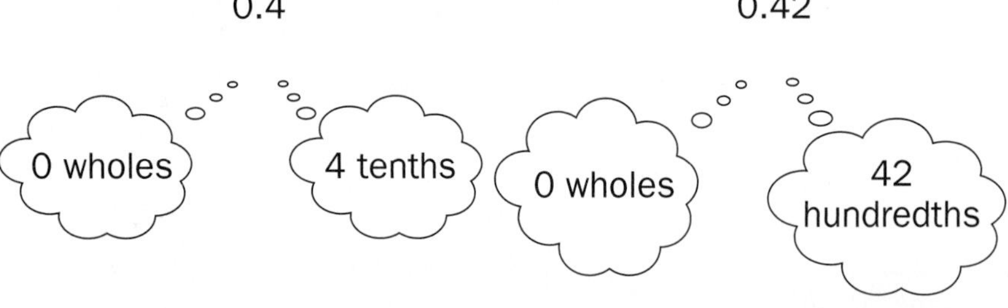

Did You Know?

Decimals were invented by the Dutch scientist Simon Stevin in 1585. But there is no single, worldwide form for writing decimals. People in the United States, Australia, and most Asian countries write the decimal 3.25. In the United Kingdom, the decimal is written 3·25, and in some parts of Europe and South America, the decimal is written 3,25.

The picture to the right shows 1 whole and 57 thousandths shaded. The fraction is written $1\frac{57}{1{,}000}$. Use three digits to the right of the decimal point to name *thousandths*. For this fraction, a 0 is needed in the tenths place as a placeholder. The decimal is written: 1.057.

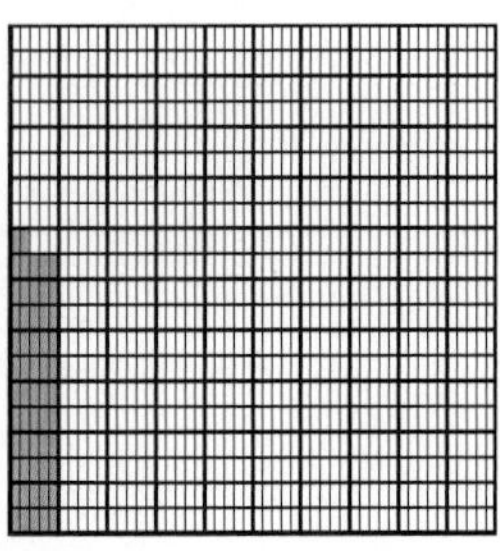

$1\frac{57}{1{,}000} = 1.057$

Reading Decimals

You can use place value to help you read decimals aloud.

- Non-zero digits to the left of the decimal point are read just like a whole number.
- Follow the whole number name with the word *and* to locate the decimal point.
- Then read digits to the right of the decimal point just like whole numbers followed by:
  - *tenths* if there is exactly one digit to the right of the decimal point
  - *hundredths* if there are exactly two digits to the right of the decimal point
  - *thousandths* if there are exactly three digits to the right of the decimal point

Examples

104.6 The 104 to the left of the decimal point is read as "one hundred four," followed by "and" to show the location of the decimal point. The 6 uses exactly one digit to the right of the decimal point. It is read "six *tenths*."

104.6 is read "one hundred four and six tenths."

0.03 The zero to the left of the decimal point means there are no wholes, so you don't say anything for the whole number name. The 03 uses exactly two digits to the right of the decimal point, so it is read "three *hundredths*."

0.03 is read "three hundredths."

1.804 The 1 to the left of the decimal point is read as "one," followed by "and" to show the location of the decimal point. The 804 uses exactly three digits to the right of the decimal point. It is read "eight hundred four *thousandths*."

1.804 is read "one and eight hundred four thousandths."

Check Your Understanding

For each picture, write a decimal and the words you would use to read the decimal.

1. 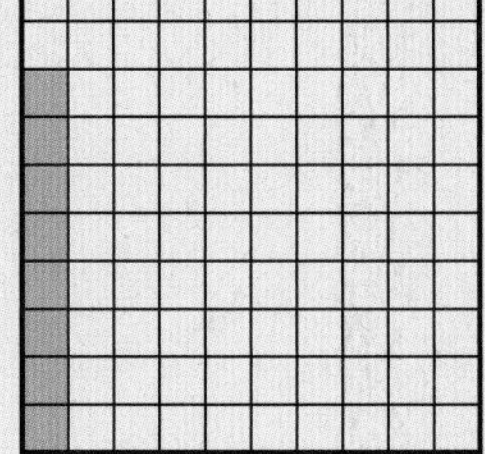

2. 

Write the words you would use to read each decimal. Then write each decimal as a fraction.

3. 0.70 **4.** 4.506

5. 24.68 **6.** 0.014

Check your answers in the Answer Key.

Extending Place Value to Decimals

The base-10 system works the same way for decimals as it does for whole numbers.

Examples

1,000s thousands	100s hundreds	10s tens	1s ones	.	0.1s tenths	0.01s hundredths	0.001s thousandths
		4	7	.	8	0	5
			4	.	3	6	0

In the number 47.805,

8 is in the **tenths** place; its value is 8 tenths, or $\frac{8}{10}$, or 0.8.

0 is in the **hundredths** place; its value is 0.

5 is in the **thousandths** place; its value is 5 thousandths, or $\frac{5}{1{,}000}$, or 0.005.

In the number 4.360,

3 is in the **tenths** place; its value is 3 tenths, or $\frac{3}{10}$, or 0.3.

6 is in the **hundredths** place; its value is 6 hundredths, or $\frac{6}{100}$, or 0.06.

0 is in the **thousandths** place; its value is 0.

Decimals can also be written in **expanded form.** For the examples above,

$$47.805 = (4 * 10) + (7 * 1) + \left(8 * \frac{1}{10}\right) + \left(0 * \frac{1}{100}\right) + \left(5 * \frac{1}{1{,}000}\right)$$

$$= (4 * 10^1) + (7 * 10^0) + \left(8 * \frac{1}{10^1}\right) + \left(0 * \frac{1}{10^2}\right) + \left(5 * \frac{1}{10^3}\right)$$

$$4.360 = (4 * 1) + \left(3 * \frac{1}{10}\right) + \left(6 * \frac{1}{100}\right) + \left(0 * \frac{1}{1{,}000}\right)$$

$$= (4 * 10^0) + \left(3 * \frac{1}{10^1}\right) + \left(6 * \frac{1}{10^2}\right) + \left(0 * \frac{1}{10^3}\right)$$

Right to Left in the Place-Value Chart

Study the place-value chart below. Look at the numbers that name the places. A digit in one place represents **10 times as much as it represents in the place to its right.**

∗ 10 ∗ 10 ∗ 10 ∗ 10 ∗ 10 ∗ 10

1,000s thousands	100s hundreds	10s tens	1s ones	.	0.1s tenths	0.01s hundredths	0.001s thousandths

1 [1,000] = 10 [100s]

1 [100] = 10 [10s]

1 [10] = 10 [1s]

$1\,[1] = 10\left[\frac{1}{10}\text{s}\right]$

$1\left[\frac{1}{10}\right] = 10\left[\frac{1}{100}\text{s}\right]$

$1\left[\frac{1}{100}\right] = 10\left[\frac{1}{1{,}000}\text{s}\right]$

Left to Right in the Place-Value Chart

Study the place-value chart below. Look at the numbers that name the places. As you move from left to right across the chart, a digit in one place represents $\frac{1}{10}$ **of what it represents in the place to its left.** Finding $\frac{1}{10}$ of a number is the same as dividing that number by 10.

$\frac{1}{10}$ of → $\frac{1}{10}$ of → $\frac{1}{10}$ of → $\frac{1}{10}$ of → $\frac{1}{10}$ of → $\frac{1}{10}$ of →

1,000s thousands	100s hundreds	10s tens	1s ones	.	0.1s tenths	0.01s hundredths	0.001s thousandths

$1\ [100] = \frac{1}{10}$ of 1,000

$1\ [10] = \frac{1}{10}$ of 100

$1\ [1] = \frac{1}{10}$ of 10

$1\left[\frac{1}{10}\right] = \frac{1}{10}$ of 1

$1\left[\frac{1}{100}\right] = \frac{1}{10}$ of $\frac{1}{10}$

$1\left[\frac{1}{1{,}000}\right] = \frac{1}{10}$ of $\frac{1}{100}$

Examples

Write a number that is $\frac{1}{10}$ of 0.4.

Since 4 is in the tenths place, move the 4 to the hundredths place to make it $\frac{1}{10}$ of its original value.

Answer: 0.04

Write a number that is $\frac{1}{10}$ of 6.

Since 6 is in the ones place, move the 6 to the tenths place to make it $\frac{1}{10}$ of its original value.

Answer: 0.6

Write a number that is worth 10 times as much as 0.007.

Since 7 is in the thousandths place, move it to the hundredths place to make it 10 times its original value.

Answer: 0.07

Write a number that is worth 10 times as much as 0.8.

Since 8 is in the tenths place, move it to the ones place to make it 10 times its original value.

Answer: 8

The grids for decimals show the same base-10 relationships between places as the place-value charts.

1 Whole

Each column is 1 tenth.

Since 10 columns make 1 whole square, 1 whole is 10 times as much as 1 tenth, *and* 1 tenth is $\frac{1}{10}$ of a whole.

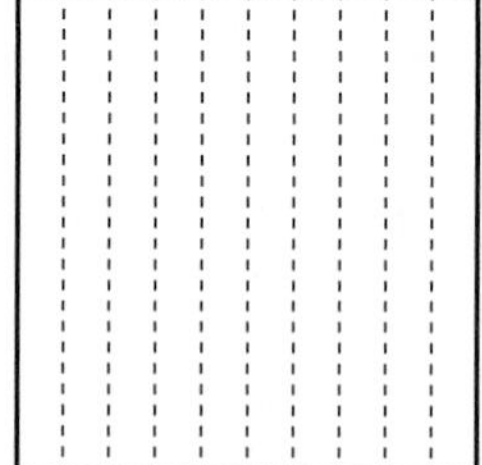

Each little square is 1 hundredth.

Since 10 hundredth squares make 1 tenth, 1 tenth is 10 times as much as 1 hundredth, *and* 1 hundredth is $\frac{1}{10}$ of a tenth.

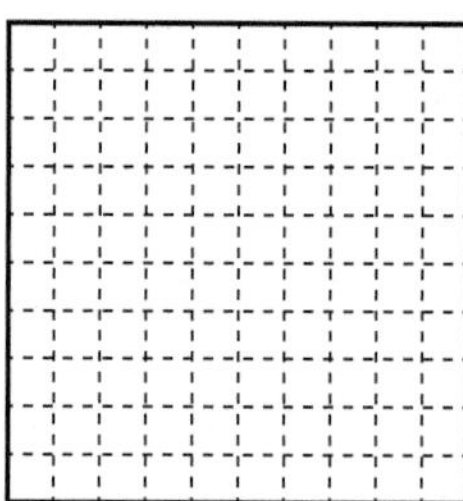

Each tiny rectangle is 1 thousandth.

Since 10 thousandth rectangles make 1 hundredth, 1 hundredth is 10 times as much as 1 thousandth, *and* 1 thousandth is $\frac{1}{10}$ of a hundredth.

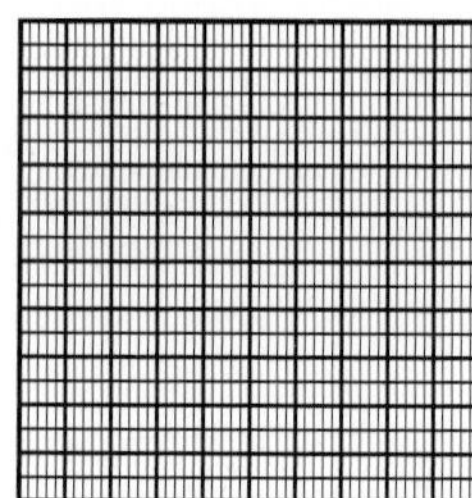

Check Your Understanding

1. What is the value of the digit 2 in each of these numbers?

a. 20,006.8

b. 0.02

c. 34.502

2. Write a number that is $\frac{1}{10}$ the value of 0.03.

3. Write a number that is 10 times the value of 0.09.

Check your answers in the Answer Key.

Comparing Decimals

One way to compare decimals is to shade in decimal grids. If you don't have grids, you can sketch the grids as shown here.

Grid	Name	Sketch of Grid
	This square is one whole.	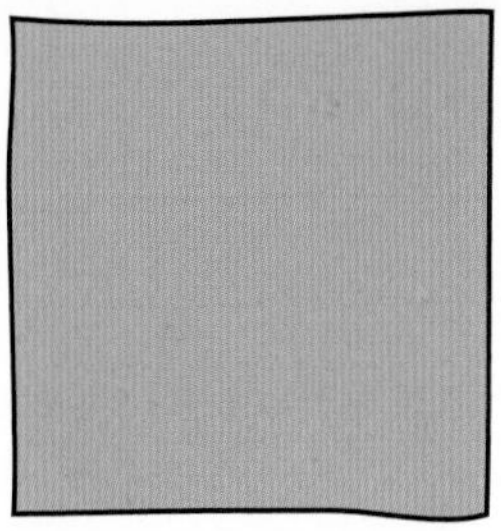
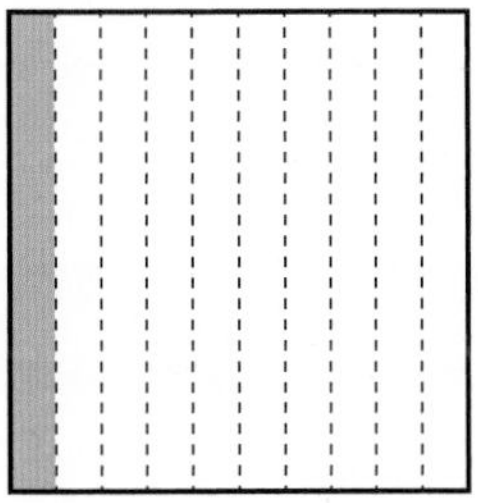	The shaded column is one-tenth, or 0.1.	
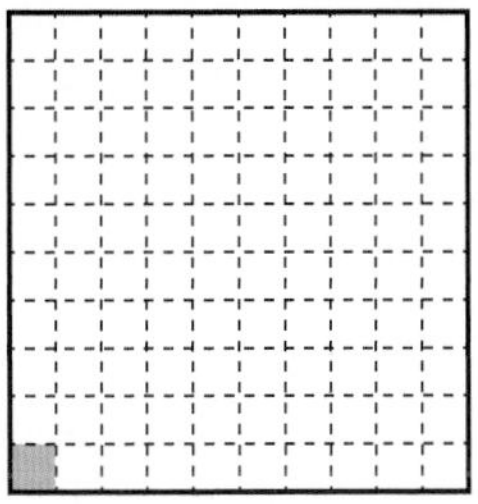	The little shaded square is one-hundredth, or 0.01.	
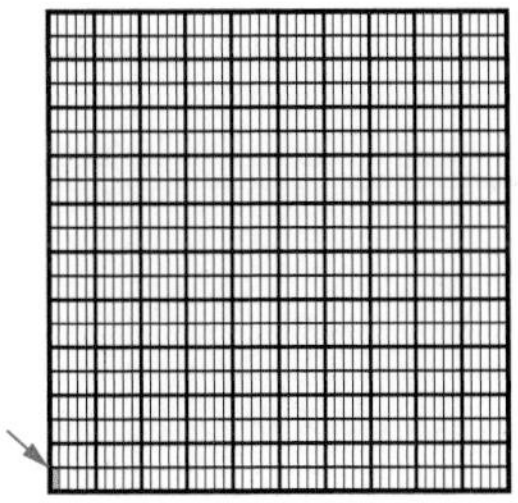	The tiny rectangle is one-thousandth, or 0.001.	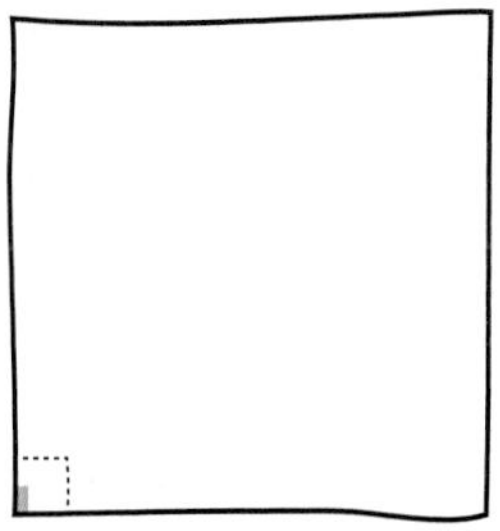

Example

Compare 1.3 and 1.16.

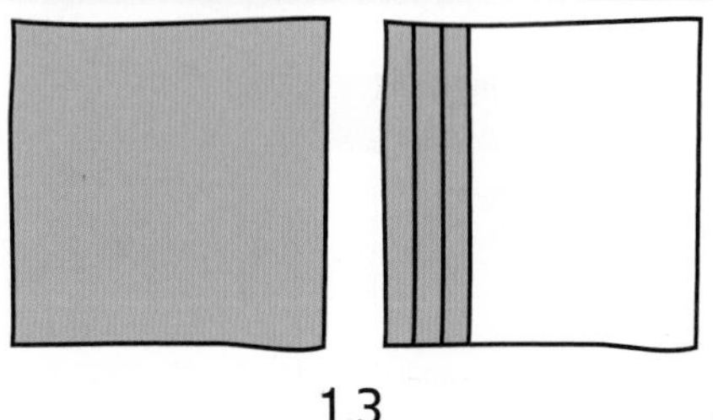

1.3

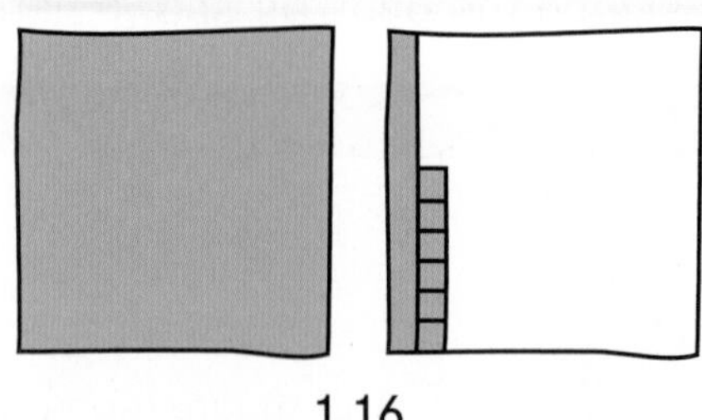

1.16

The sketches show that both numbers are more than one whole. When you compare the decimal parts, you see that 3 tenths covers more than 1 tenth and 6 hundredths (or 16 hundredths).

So, 1.3 is more than 1.16. $1.3 > 1.16$

Another way to compare decimals is to attach 0s to the end of the fractional part of each decimal without changing the value.

Example

$0.3 = 0.30$

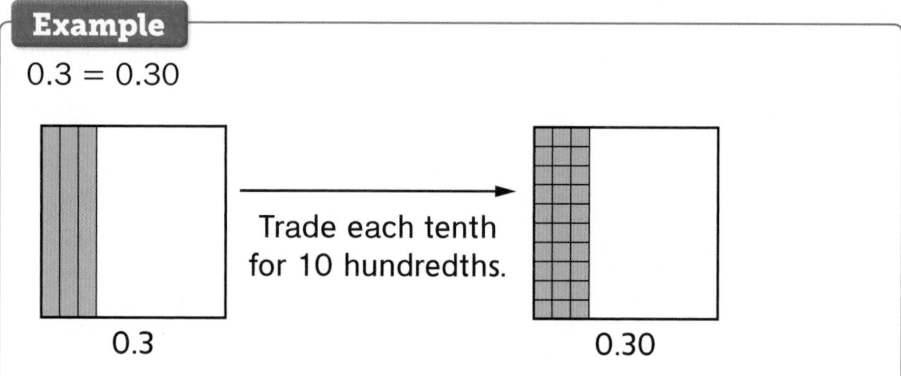

Did You Know?

Imagine two points that are 100 meters apart. Sound would travel that distance in about 0.3 second. A beam of light would travel that distance in about 0.0000003 second (3 ten-millionths of a second). Light travels about 1 million times as fast as sound.

Attaching 0s without changing the value of the number makes comparing decimals easier.

Examples

Compare 0.3 and 0.06.

You can attach a 0 to 0.3 without changing its value by writing 0.30.
0.3 covers the same amount as 0.30.

Now compare 0.30 and 0.06.
30 hundredths is more than 6 hundredths.

$0.30 > 0.06$, so $0.3 > 0.06$.

Compare 1.172 and 1.4.

You can attach 0s to 1.4 without changing its value by writing 1.400.
1.4 covers the same amount as 1.400.

Now compare 1.172 and 1.400.
Both show one whole and a part of a whole.
172 thousandths is less than 400 thousandths.

$1.172 < 1.400$, so $1.172 < 1.4$.

A decimal place-value chart is another way to compare decimals.

Example

Compare 3.14 and 3.029.

1s ones	.	0.1s tenths	0.01s hundredths	0.001s thousandths
3	.	1	4	
3	.	0	2	9

The ones digits are the same. They are both worth 3 wholes.

The tenths digits are the not the same.

The 1 is worth 1 tenth or 0.1, and the 0 is worth 0 tenths. The 1 in the tenths place is worth more than the 0 in the tenths place.

So, 3.14 is greater than 3.029. 3.14 > 3.029

Example

Compare 27.14 and 27.105.

10s tens	1s ones	.	0.1s tenths	0.01s hundredths	0.001s thousandths
2	7	.	1	4	
2	7	.	1	0	5

The tens digits are the same. They are both worth 2 tens.

The ones digits are the same. They are both worth 7 ones.

The tenths digits are the same. They are both worth 1 tenth.

The hundredths digits are not the same.

The 4 is worth 4 hundredths, or 0.04, and the 0 is worth 0 hundredths. The 4 in the hundredths place is worth more than the 0 in the hundredths place.

So, 27.14 is greater than 27.105. 27.14 > 27.105

Check Your Understanding

1. Compare 0.43 and 0.152 by sketching a picture of shaded grids.
2. Compare 2.8 and 2.102 by attaching zeros.
3. Compare 1.402 and 1.41 by using a place-value chart.
4. Using the digits 3, 5, and 8, what is the largest decimal less than 1 that you can write?

Check your answers in the Answer Key.

Rounding Decimals

Rounding is a way to make sense of decimals and to estimate calculations. Rounding decimals is similar to rounding whole numbers.

Rounding Decimals Using Grids

Tenths, hundredths, and thousandths grids can be used to round decimals.

Example

Sales tax on a purchase was calculated to be $0.249. How many cents is the sales tax?

Cents are hundredths of a dollar, so to find the sales tax in cents, round $0.249 to the nearest hundredth.

Shade a thousandths grid to show 0.249.

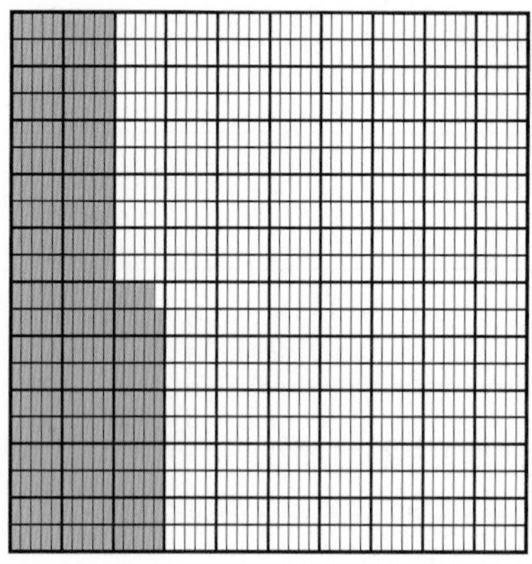

Ask: Between what two hundredths is 0.249?

It is between 0.24 and 0.25, because the picture shows more than 24 shaded hundredths squares, but fewer than 25 shaded hundredths squares.

Ask: Is 0.249 more or less than halfway between 0.24 and 0.25?

Most of the 25th hundredth square is filled. So, 0.249 is more than halfway to 0.25.

Round 0.249 to the nearest hundredth and write an answer to the problem.

0.249 is closer to 0.25 than 0.24, so 0.249 rounds up to 0.25.

The tax should be $0.25, or 25 cents.

Rounding Decimals Using Number Lines

Number lines are useful for rounding decimals, just as they are for rounding whole numbers.

- Identify the two numbers you can round to.
- Sketch a number line that shows the distance between the two numbers, mark and label the location of the halfway point, and mark and label the approximate location of the number to be rounded.
- Decide whether to round up to the higher number or down to the lower number.

Examples

Mr. Martin told his students to do only the problems on a homework page that have estimated answers less than 10.

Do students need to do the first problem? Problem 1: 22.37 − 10.8 = ?

One way to estimate the answer is to round each decimal to the nearest whole number, and then mentally calculate with the rounded numbers.

Round 22.37 to the nearest whole:

Ask: Between what two whole numbers is 22.37? 22 and 23

Show the higher and lower numbers on a number line.

Ask: What is halfway from 22 to 23? 22.5 Mark that point.

Ask: Is 37 hundredths more or less than 5 tenths? Since 37 hundredths is 3 tenths 7 hundredths, it is less than 5 tenths. Mark 22.37 at a location that is less than halfway from 22 to 23.

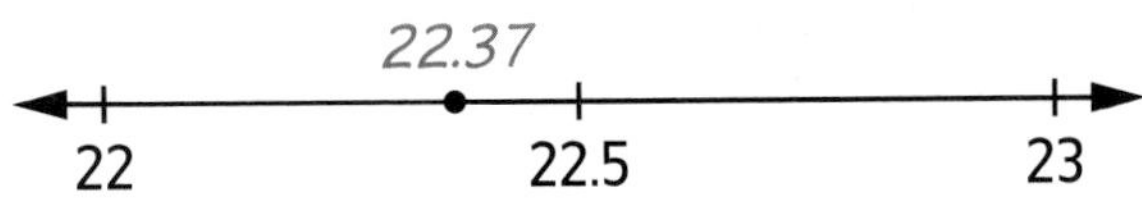

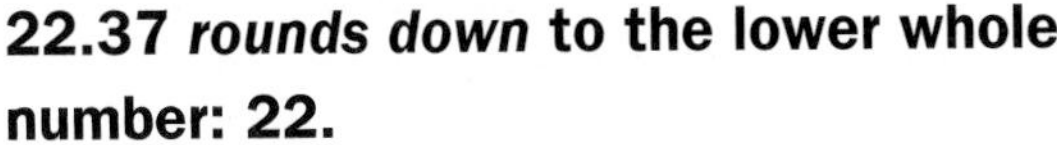

22.37 *rounds down* to the lower whole number: 22.

Round 10.8 to the nearest whole:

Ask: Between what two whole numbers is 10.8? 10 and 11

Show the higher and lower numbers on a number line.

Ask: What is halfway from 10 to 11? 10.5 Mark that point.

Ask: Is 8 tenths more or less than 5 tenths? 8 tenths is more than 5 tenths. Mark 10.8 at a location that is more than halfway from 10 to 11.

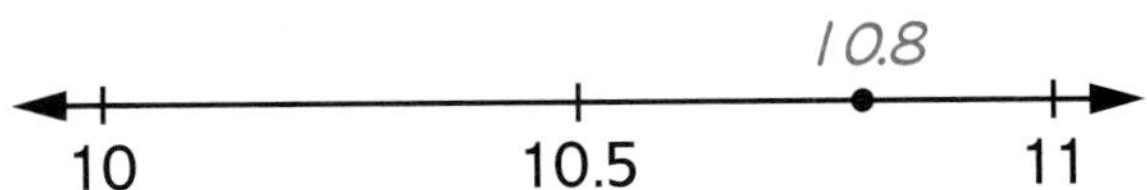

10.8 *rounds up* to the higher whole number: 11.

Use the rounded numbers to estimate the difference between 22.37 and 10.8.

Estimate: 22 − 11 = 11

The students do not need to do Problem 1 because the estimated answer is more than 10.

Rounding Decimals Using a Shortcut

You can use a shortcut for rounding decimals:

Step 1 Underline the digit in the place you are rounding to.

Step 2 Identify the two multiples of that place you can round to.

Step 3 Look at the digit to the right of the underlined digit.

If that digit is 5 or more, round up to the higher number. If that digit is less than 5, round down to the lower number.

Example

Before taking their items to the cashier, Mr. and Mrs. Wilson estimated the total cost of their items in different ways.

$5.39 $0.43 $1.09

Mr. Wilson rounded to the nearest tenth, or the nearest dime.

$5.$\underline{3}$9	3 is in the tenths place. Round to either $5.30 or $5.40. 9 is the digit to the right of the tenths place. 9 is greater than 5. So, $5.39 rounds up to $5.40.
$0.$\underline{4}$3	4 is in the tenths place. Round to either $0.40 or $0.50. 3 is the digit to the right of the tenths place. 3 is less than 5. $0.43 rounds down to $0.40.
$1.$\underline{0}$9	0 is in the tenths place. Round to either $1.00 or $1.10. 9 is the digit to the right of the tenths place. 9 is greater than 5. $1.09 rounds up to $1.10.

Estimate: $5.40 + $0.40 + $1.10 = $6.90

Mrs. Wilson rounded to the nearest whole number, or the nearest dollar.

$$\underline{5}$.39	5 is in the ones place. Round to either $5.00 or $6.00. 3 is the digit to the right of the ones place. 3 is less than 5. $5.39 rounds down to $5.00.
$$\underline{0}$.43	0 is in the ones place. Round to either $0.00 or $1.00. 4 is the digit to the right of the ones place. 4 is less than 5. $0.43 rounds down to $0.00.
$$\underline{1}$.09	1 is in the ones place. Round to either $1.00 or $2.00. 0 is the digit to the right of the ones place. 0 is less than 5. $1.09 rounds down to $1.00.

Estimate: $5.00 + $0.00 + $1.00 = $6.00

The cashier found a total of $6.91. Both estimates indicate that the cashier's total is reasonable. Mr. Wilson's estimate is closer to the exact answer because he rounded to the nearest tenth, which is more **precise** than rounding to the nearest whole dollar. However, Mrs. Wilson's estimate is easier to calculate mentally.

Rounding Halfway Numbers

When the number being rounded is exactly halfway between the two numbers you can round to, use the real-world or problem situation to decide whether to round up or down.

For example, to estimate the sum of 3.67 + 2.5, you can round the addends to the nearest whole number. Round 3.67 to 4, because 3.67 is closer to 4 than 3. Since 2.5 is exactly halfway between 2 and 3, you can round up or down. It makes sense to round 2.5 down to 2, since you rounded up 3.67. A reasonable estimate for 3.67 + 2.5 is 4 + 2 = 6.

When there is not any information in the real-world or problem situation to help you decide whether to round up or down, the "rounding up" rule usually applies. That means you round up to the higher number.

Check Your Understanding

Estimate these sums by rounding to the place indicated.

1. Estimate the sum twice. First round each addend to the nearest tenth and then to the nearest whole number (one): 5.41 + 2.07 = ?

2. Estimate this sum by rounding to the nearest whole number (one): 13.771 + 21.26 = ?

3. Estimate this sum twice. First round each addend to the nearest dollar (one) and then to the nearest dime (tenth): \$3.62 + \$1.09 = ?

Check your answers in the Answer Key.

Estimating with Decimals

You can use **close-but-easier numbers** to estimate when you see nearby numbers that are easy to calculate mentally.

Example

Estimate: 23.2 ∗ 0.43 = **?**

One way:

0.43 is close to a half, or 0.5, and 23.2 is close to 24.
24 ∗ 0.5 → half of 24 is 12.

So, 23.2 ∗ 0.43 is about 12.

Another way:

0.43 is close to a half, or 0.5, and 23.2 is close to 20.
20 ∗ 0.5 → half of 20 is 10.

So, 23.2 ∗ 0.43 is about 10.

Another way:

Use an easy mental math problem.

25 ∗ 4 = 100 is easy.
Since 23.2 is close to 25 and 0.43 is close to 0.4, you can think about 25 ∗ 0.4 = ?
Since 25 ∗ 4 = 100, 25 ∗ 0.4 (or 25 ∗ 4 tenths) would be 100 tenths, or 10.

So, 23.2 ∗ 0.43 is about 10.

10 and 12 are both reasonable estimates for 23.2 ∗ 0.43.

When all the numbers in the problem have whole number parts, you can use the whole number parts to estimate answers. You can also use close-but-easier numbers.

Example

Estimate: 28.91 / 6.7 = **?**

One way: Divide using the whole number parts. Ignore the remainder.

28 / 6 → 4 R4

So, 28.91 / 6.7 is about 4.

Another way: Divide using close-but-easier numbers.

28 / 7 is a basic fact that uses close-but-easier numbers. 28 / 7 = 4

So, 28.91 / 6.7 is about 4.

Another way: Divide using different close-but-easier numbers.

30 / 6 is a basic fact that uses close-but-easier numbers. 30 / 6 = 5

So, 28.91 / 6.7 is about 5.

4 and 5 are both reasonable estimates for 28.91 / 6.7.

Check Your Understanding

Estimate an answer for each of the following. Show how you estimated.

1. 15.41 − 12.96 = ? **2.** 19.91 ∗ 2.103 = ? **3.** 29.334 / 14.2 = ?

Check your answers in the Answer Key.

Using Grids to Add and Subtract Decimals

There are many ways to add and subtract decimals. Grids can be helpful to figure out sums and differences for many decimal problems.

Example

0.6 + 0.63 = **?**

Think about each addend as a different color: 0.6 + 0.63 = ?

Shade each decimal part on a grid, filling columns as you go.

Count the total number of shaded columns.

Each column represents one-tenth. There are 6 red columns and 6 blue columns. That makes 12 tenths in all.

Trade 10 tenths for 1 whole. 12 tenths is 1 whole and 2 tenths.

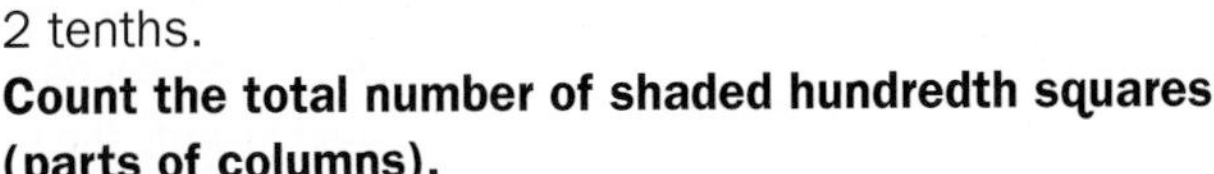

Count the total number of shaded hundredth squares (parts of columns).

Remember, each little square represents one-hundredth. Add the remaining 3 small blue squares, which represent 3 hundredths.

Write the sum as a decimal.

1 whole, 2 tenths, and 3 hundredths is the same as 1.23.

So, 0.6 + 0.63 = **1.23.**

Example

0.48 − 0.3 = **?**

Shade the starting number on a grid: 0.48

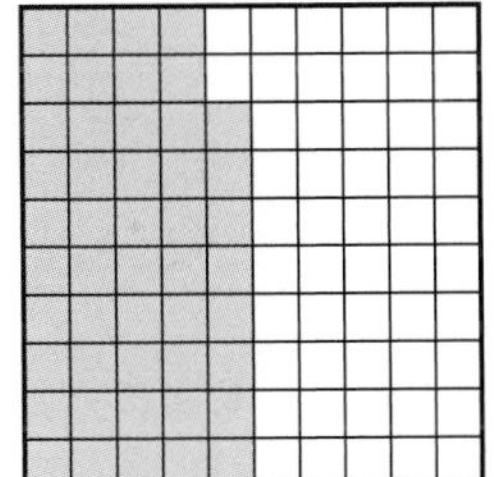

Show the subtracted amount: 0.3

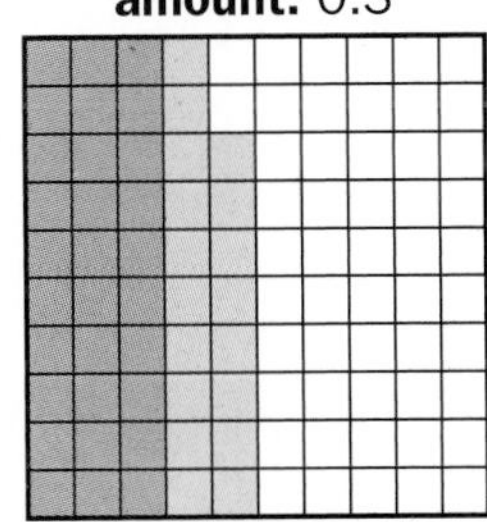

Figure out how much remains.

The darker gray shows the amount taken away. The amount remaining is light gray. There is one column, or 1 tenth, that is light gray, and there are 8 small squares, or 8 hundredths, that are light gray.

That means 0.18 remains.

0.48 − 0.3 = **0.18**

More Methods for Adding Decimals

Most paper-and-pencil methods for adding and subtracting whole numbers also work for decimals. To add and subtract whole numbers and decimals, you must line up the place values of the numbers correctly. Estimate to check whether your answer is reasonable.

Examples

4.56 + 7.9 = **?**

Estimate: You can round 4.56 to 4.6 and 7.9 to 8. An estimate is 4.6 + 8 = 12.6.

Partial-Sums Addition:

		1s		0.1s	0.01s
		4	.	5	6
		+ 7	.	9	0
Add the ones.	4 + 7 →	11	.	0	0
Add tenths.	0.5 + 0.9 →	1	.	4	0
Add the hundredths.	0.06 + 0.00 →	0	.	0	6
Add the partial sums.		12	.	4	6

Note If you can find the partial sums in your head, then you do not need to write the steps shown in green.

Column Addition:

	1s		0.1s	0.01s
	4	.	5	6
	+ 7	.	9	0
Add the numbers in each column.	11	.	14	6
Trade 14 tenths for 1 and 4 tenths.				
Move the 1 into the ones column.	12	.	4	6

U.S. Traditional Addition:

Step 1 Add the 0.01s:

```
   4 . 5 6
+  7 . 9 0
----------
         6
```

6 hundredths + 0 hundredths = 6 hundredths

Step 2 Add the 0.1s:

```
   1
   4 . 5 6
+  7 . 9 0
----------
       4 6
```

5 tenths + 9 tenths = 14 tenths
14 tenths = 1 whole + 4 tenths

Step 3 Add the 1s: 1 + 4 + 7 = 12.

```
    1
    4 . 5 6
+   7 . 9 0
-----------
  1 2 . 4 6
```

12 ones = 1 ten and 2 ones

Remember to include the decimal point in the answer.

4.56 + 7.9 = **12.46** using all three methods.

The answer of 12.46 makes sense, because it is close to the estimate of 12.6.

More Methods for Subtracting Decimals

Examples

9.4 − 4.85 = **?**

Estimate: 9.4 rounds to 9. You can round 4.85 to 5.

An estimated difference is 9 − 5 = 4.

Trade-First Subtraction:

	1s	0.1s	0.01s
	9 .	4	0
−	4 .	8	5

Write a 0 in the hundredths place of 9.4. Line up the places in both numbers. Then look at the 0.01s place.

0 < 5, so you need to make a trade.

	1s	0.1s	0.01s
		3	10
	9 .	~~4~~	~~0~~
−	4 .	8	5

Trade 1 tenth for 10 hundredths. Mark the problem to show the trade. Now look at the 0.1s place. 3 < 8, so you need to make a trade.

	1s	0.1s	0.01s
		13	
	8	~~3~~	10
	~~9~~ .	~~4~~	~~0~~
−	4 .	8	5
	4 .	5	5

Trade 1 one for 10 tenths. Now subtract in each column.

9.4 − 4.85 = **4.55**

Counting-Up Subtraction:

Write the smaller number, 4.85.

As you count from 4.85 to 9.4, circle each number that you count up.

4.85	
+ (0.15)	Count up to the nearest whole.
5.00	
+ (4.00)	Count up to the largest possible whole.
9.00	
+ (0.40)	Count up to the larger number.
9.40	

Add the numbers you circled: 0.15 + 4.00 + 0.40 = 4.55

You counted up by 4.55.

9.40 − 4.85 = **4.55** using either method.

The answer 4.55 makes sense because it is close to the estimate of 4.

Example

9.4 − 4.85 = **?**

Estimate: 9.4 rounds to 9. You can round 4.85 to 5.

An estimated difference is 9 − 5 = 4.

U.S. Traditional Subtraction:

Step 1 Write the problem. Write a 0 in the hundredths place of 9.4. This lines up all the places without changing the value of the starting number.

$$\begin{array}{r} 9.40 \\ -\ 4.85 \\ \hline \end{array}$$

Step 2 Start with the hundredths place. Since 0 < 5, you need to regroup. Trade 1 tenth for 10 hundredths. Subtract the hundredths.

$$\begin{array}{r} 9.\overset{3}{\cancel{4}}\overset{10}{\cancel{0}} \\ -\ 4.85 \\ \hline 5 \end{array}$$

10 hundredths − 5 hundredths = 5 hundredths

Step 3 Go to the tenths place. Since 3 < 8, you need to regroup. Trade 1 whole for 10 tenths. Subtract the tenths.

$$\begin{array}{r} \overset{8}{\cancel{9}}.\overset{13}{\cancel{4}}\overset{10}{\cancel{0}} \\ -\ 4.85 \\ \hline 55 \end{array}$$

13 tenths − 8 tenths = 5 tenths

Step 4 Go to the ones place. Subtract 8 − 4 = 4.

$$\begin{array}{r} \overset{8}{\cancel{9}}.\overset{13}{\cancel{4}}\overset{10}{\cancel{0}} \\ -\ 4.85 \\ \hline 4.55 \end{array}$$

8 ones − 4 ones = 4 ones

9.4 − 4.85 = **4.55**

The answer 4.55 makes sense because it is close to the estimate of 4.

Check Your Understanding

Add or subtract using any method. Estimate to check whether your answers are reasonable.

1. 0.62 + 0.4 = ?

2. 13.18 + 1.4 = ?

3. 2.4 − 2.377 = ?

Check your answers in the Answer Key.

Multiplying Decimals by Powers of 10

You can use patterns you already know to multiply decimals by powers of 10 greater than 1. One way is to use **partial-products multiplication.**

Example

Solve 1,000 ∗ 45.6 using partial-products multiplication.

Step 1 Solve the problem as if there were no decimal point.

			1	0	0	0
	∗			4	5	6
400 ∗ 1,000 →	4	0	0	0	0	0
50 ∗ 1,000 →		5	0	0	0	0
6 ∗ 1,000 →			6	0	0	0
	4	5	6	0	0	0

Note Some powers of 10 greater than one:

$10^1 = 10$

$10^2 = 10 * 10 = 100$

$10^3 = 10 * 10 * 10 = 1{,}000$

$10^4 = 10 * 10 * 10 * 10 = 10{,}000$

Step 2 Estimate the answer to 1,000 ∗ 45.6 and place the decimal point where it belongs.

1,000 ∗ 45 = 45,000, so 1,000 ∗ 45.6 must be close to 45,000.

So the answer to 1,000 ∗ 45.6 is **45,600.**

Another method for multiplying a number by a power of 10 greater than 1 is to move the decimal point. Think of this as an efficient shortcut.

Example

1,000 ∗ 45.6 = **?**

Write the power of 10 in exponential notation.

$1{,}000 = 10 * 10 * 10 = 10^3$

Move the decimal point in the other factor. Use the exponent to find how many places to move the decimal point.

4 5 . 6 0 0 .

The exponent tells the number of times a number is multiplied by 10. Each time a number is multiplied by 10, the original digits shift one place to the left, which moves the decimal point one place to the right.

The decimal point moves 3 places to the right because 45.6 is multiplied by 10 three times. This makes each digit 10^3 times as large.

So, 1,000 ∗ 45.6 = 45600 or **45,600.**

Check Your Understanding

Multiply. Estimate to check whether your answers are reasonable.

1. 100 ∗ 4.56 **2.** 0.28 ∗ 10,000 **3.** 1,000 ∗ $4.50 **4.** 1.04 ∗ 10

Check your answers in the Answer Key.

Multiplying Decimals

You can use the same procedures for multiplying decimals that you use for whole numbers. The main difference is that with decimals you have to decide where to place the decimal point in the product. To multiply decimals:

- Estimate the product.
- Multiply as if the factors were whole numbers.
- Use your estimate to place the decimal point in the answer.

Note You can find estimates by using close-but-easier numbers or by rounding.

Example

15.2 ∗ 3.6 = **?**

Step 1 Make an estimate.

Round 15.2 to 15 and 3.6 to 4.

An estimated product is 15 ∗ 4 = 60.

Step 2 Multiply as you would with whole numbers. One way to do this is to use partial-products multiplication. Ignore the decimal points.

		1	5	2
	∗		3	6
30 ∗ 100 →	3	0	0	0
30 ∗ 50 →	1	5	0	0
30 ∗ 2 →			6	0
6 ∗ 100 →		6	0	0
6 ∗ 50 →		3	0	0
6 ∗ 2 →			1	2
	5	4	7	2

Note If you can find the partial products in your head, then you do not need to write the steps shown in green.

Step 3 Use your estimate to place the decimal point in the answer. Your estimate was 60. To have a product that is closest to 60, the decimal point should be placed between the 4 and 7 in 5472.

So, 15.2 ∗ 3.6 = **54.72.**

Example

3.27 ∗ 0.8 = **?**

Step 1 Make an estimate.

The factor 0.8 is almost 1. An estimated product is 3.27 ∗ 1 = 3.27. The answer will be less than 3.27, since the factor 0.8 was rounded up.

Step 2 Multiply as you would with whole numbers. One way to do this is to use U.S. traditional multiplication. Ignore the decimal points.

```
   2 5
   3 2 7
∗      8
--------
 2 6 1 6     327 ∗ 8 = 2616
```

Step 3 Use your estimate to place the decimal point in the answer. The estimate was less than 3.27, so the decimal point should be placed between the 2 and 6 in 2616.

So, 3.27 ∗ 0.8 = **2.616.**

Sometimes when you multiply a decimal less than 1 by another decimal less than 1, making an estimate to place the decimal point is difficult. Another way to multiply decimals is to use powers of 10 to place the decimal point.

Example

0.2 ∗ 0.041 = **?**

Multiply both factors by a power of 10 to make them whole numbers. Keep track of both powers of 10, or the total number of places the decimal point shifted in each factor.

$0.2 * 10^1 = 2$

$0.041 * 10^3 = 41$

Multiply the whole numbers.

$2 * 41 = 82$

Undo the multiplication by powers of 10 by dividing by powers of 10. Divide the product by the powers of 10 used to change the factors to whole numbers, or shift the decimal point to the left the total number of places it shifted when changing the factors to whole numbers.

$82 / (10^1 * 10^3) = 0.0082$

This is the same as (10) ∗ (10 ∗ 10 ∗ 10).

So, 0.2 ∗ 0.041 = **0.0082.**

Check Your Understanding

Multiply. Estimate to check whether your answers are reasonable.

1. 1.7 ∗ 5.7 **2.** 2.33 ∗ 8.4 **3.** 0.61 ∗ 4.04 **4.** 0.3 ∗ 0.021

Check your answers in the Answer Key.

Dividing Decimals by Powers of 10

Here is one way to divide a decimal by a power of 10 greater than 1.

Example

45.6 / 1,000 = **?**

Write the power of 10 in exponential notation.

$1{,}000 = 10 * 10 * 10 = 10^3$

Move the decimal point in the dividend. Use the exponent to find how many places to move the decimal point.

Sometimes you need to insert 0s to show the correct place value.

0 . 0 4 5 . 6

The decimal point moves 3 places to the left because 45.6 is divided by 10 three times. This makes each digit worth $\frac{1}{10^3}$ as much as it was worth before.

Dividing by 10 is the same as multiplying by $\frac{1}{10}$. Each time a number is multiplied by $\frac{1}{10}$, the original digits shift one place to the right, which moves the decimal point one place to the left.

So, 45.6 / 1,000 = **0.0456.**

Note Some powers of 10 greater than one:

$10^1 = 10$

$10^2 = 100$

$10^3 = 1{,}000$

$10^4 = 10{,}000$

$10^5 = 100{,}000$

$10^6 = 1{,}000{,}000$

Examples

350 / 100 = ?

$100 = 10^2$

Move the decimal 2 places to the left.

3 . 5 0 .

350 / 100 = 3.50

350 / 10,000 = ?

$10{,}000 = 10^4$

Move the decimal 4 places to the left.

0 . 0 3 5 0 .

350 / 10,000 = 0.0350

$290.50 / 1,000 = ?

$1{,}000 = 10^3$

Move the decimal 3 places to the left.

0 . 2 9 0 . 5 0

$290.50 / 1,000 = $0.29 (rounded to the nearest cent)

Note When the dividend (the number you are dividing) does not have a decimal point, you must locate the decimal point before moving it. Note that in two of the examples above, 350 is written with a decimal point after the 0 (350 = 350.).

Check Your Understanding

Divide.

1. 56.7 / 10

2. 0.47 / 100

3. $290 / 1,000

4. 60 / 10,000

Check your answers in the Answer Key.

Dividing Decimals by Whole Numbers

You can use the same procedures for dividing decimals that you use for whole numbers. The main difference is that with decimals you have to decide where to place the decimal point in the quotient. Here is one way to divide decimals.

- Estimate the quotient.
- Divide as if the divisor and dividend were whole numbers.
- Use your estimate to place the decimal point in the answer.

Note You can estimate by using close-but-easier numbers or by rounding.

Example

97.24 / 26 = **?**

Step 1 Make an estimate.

You can round 97.24 to 100. 26 is close to 25.
An estimated quotient is 100 / 25 = 4.

Step 2 Divide as you would with whole numbers. Ignore the decimal point. You can use partial-quotients division.

26)9724	
− 7800	300
1924	
− 1040	40
884	
− 780	30
104	
− 104	4
0	374

9724 / 26 = 374

Step 3 Decide where to place the decimal point in the answer. The estimate was 4. To have a quotient that is close to 4, the decimal point should be placed between the 3 and 7 in 374.

So, 97.24 / 26 = **3.74.**

Check Your Understanding

Divide. Estimate to decide where to place the decimal point.

1. 148.8 / 6 **2.** 25.32 / 12 **3.** 45.5 / 35

Check your answers in the Answer Key.

Division of Decimals: Rounding Quotients

Sometimes the quotient for division of decimals does not come out evenly. When you divide as if the divisor and dividend are whole numbers, there may be a remainder. When this happens, one approach is to round the quotient before placing the decimal point.

Note The symbol ≈ means *is about equal to.*

Example

Four teachers will share 7.5 pounds of granola to take on a field trip with their classes. How many pounds of granola should each teacher take?

7.5 / 4 = **?**

Step 1 Make an estimate.

You can round 7.5 pounds to 8 pounds.
An estimated quotient is 8 / 4 = 2 pounds.

Step 2 Divide, ignoring the decimal point. Round up if the fraction remainder is greater than or equal to a half.

75 / 4 =?

```
 4)75  |
 - 40  | 10
 ----
   35  |
 - 32  |  8
 ----  -----
    3    18
```

The quotient is $18\frac{3}{4}$.

$\frac{3}{4}$ is greater than $\frac{1}{2}$. The quotient rounds to 19.

Step 3 Use your estimate to place the decimal point.

75 / 4 is about 19.
Since the estimate of the quotient is 2, place the decimal point between the 1 and the 9.

7.5 / 4 ≈ **1.9**

Each class gets about 1.9 pounds of granola for the field trip.

This answer is reasonable since the original estimate was 2 pounds.

Example

9.1 / 6 = **?**

Step 1 Make an estimate.

Round 9.1 to 9.
9 / 6 → 1 R3
An estimated quotient is a little more than 1.

Step 2 Divide, ignoring the decimal point. Round up if the fraction remainder is greater than or equal to a half.

91 / 6 = ?

$$\begin{array}{r|l} 6\overline{)91} & \\ -\ 60 & 10 \\ \hline 31 & \\ -\ 30 & 5 \\ \hline 1 & 15 \end{array}$$

The quotient is $15\frac{1}{6}$.

$\frac{1}{6} < \frac{1}{2}$, so round $15\frac{1}{6}$ down to 15.

Step 3 Use your estimate to place the decimal point.

91 / 6 is about 15.
Since the estimate for the quotient is more than 1, place the decimal point between the 1 and the 5.

9.1 / 6 is about 1.5.

9.1 / 6 ≈ **1.5**

The quotient is reasonable since the original estimate was that the answer should be a little more than 1.

Dividing Decimals by Decimals

When division problems involve decimal divisors, sometimes estimating the quotient is difficult. For these types of problems, you can find an equivalent division problem that is easier to solve.

Note Fractions can be used to show division problems. The fraction $\frac{a}{b}$ is another way of saying *a* divided by *b*, $a \div b$, or a / b.

Just as an equivalent fraction expresses the same value with a different numerator and denominator, an **equivalent division problem** produces the same quotient using a different dividend and divisor. You can find an equivalent problem by making the same change to the dividend and divisor.

Example

2.84 / 0.004 = ?

Write the division problem as a fraction. $2.84 / 0.004 = \frac{2.84}{0.004}$

Find an equivalent fraction with no decimals. $\frac{2.84 * \mathbf{1{,}000}}{0.004 * \mathbf{1{,}000}} = \frac{2{,}840}{4}$

Write the equivalent fraction as a division problem. $\frac{2{,}840}{4} = 2{,}840 / 4$

Since the two fractions $\frac{2.84}{0.004}$ and $\frac{2{,}840}{4}$ are equivalent, the answer to 2.84 / 0.004 is the same as the answer to 2,840 / 4.

So, 2.84 / 0.004 = **710.**

To summarize, you can use these steps to divide a decimal by a decimal.

Step 1: Estimate.

You can estimate using the original problem or an equivalent problem.

Step 2: Write an equivalent problem with a whole-number divisor.

Use the multiplication rule to find an equivalent problem with no decimals. This makes the same change to the dividend and the divisor.

Step 3: Solve the equivalent problem using partial-quotients division or another method.

The answer to the equivalent problem is the same as the answer to the original problem.

Example

0.78 / 0.013 = **?**

Step 1 Estimate.

One way to estimate is to use an equivalent problem.

Step 2 Find an equivalent problem with a whole number divisor.

0.78 ∗ **1,000** = 780
0.013 ∗ **1,000** = 13

Equivalent problem: 780 / 13

Estimate (Step 1): 13 is close to 10. An estimated quotient is 780 / 10 = 78.

Step 3 Solve the equivalent problem.

780 / 13 = 60

Since $\frac{0.78}{0.013} = \frac{780}{13}$,

0.78 / 0.013 = **60.**

The answer is reasonable because it is close to the estimate of 78.

If the quotient does not come out evenly, remember that you can use the remainder to decide whether to round up or round down.

Example

0.73 / 0.03 = **?**

Step 1 Estimate.

69 / 3 is easy to do mentally and is close to 73 / 3.
An estimated quotient is $\frac{0.69 * \mathbf{100}}{0.03 * \mathbf{100}} = \frac{69}{3}$, or 23.

Step 2 Write an equivalent problem with a whole number divisor.

0.73 ∗ **100** = 73 and 0.03 ∗ **100** = 3
The equivalent problem is 73 / 3 = ?

Step 3 Solve the equivalent problem.

73 / 3 → 24 R1, or 73 / 3 = $24\frac{1}{3}$
Since $\frac{1}{3}$ is less than half, the quotient rounds down to 24.
0.73 / 0.03 is about 24

0.73 / 0.03 ≈ **24**

This quotient is reasonable because it is close to the original estimate of 23.

Check Your Understanding

Divide. Estimate to check whether your answers are reasonable.

1. 8.01 / 0.3 **2.** 3.9 / 0.013 **3.** 135 / 0.45

Check your answers in the Answer Key.

Column Division with Decimal Quotients

Column division can be used to find quotients that have a decimal point. In the example below, think of sharing $15 equally among 4 people.

Example

$4\overline{)15} =$ **?**

1. Set up the problem. Draw a line to separate the digits in the dividend. Work left to right. Think of the 1 in the tens column as 1 $10 bill.

4)	1	5

2. The 1 $10 bill cannot be shared by 4 people. So trade it for 10 $1 bills. Think of the 5 in the ones column as 5 $1 bills. That makes 10 + 5, or 15 $1 bills in all.

4)	~~1~~	~~5~~
		15

3. If 4 people share 15 $1 bills, each person gets 3 $1 bills. There are 3 $1 bills left over.

		3
4)	~~1~~	~~5~~
		15
		−12
		3

4. Draw a line and make decimal points to show amounts less than $1. Write 0 after the decimal point in the dividend to show there are 0 dimes. Then trade 3 $1 bills for 30 dimes.

		3 .	
4)	~~1~~	~~5~~ .	~~0~~
		15	30
		−12	
		~~3~~	

5. If 4 people share 30 dimes, each person gets 7 dimes. There are 2 dimes left over. Draw another line and another 0 in the dividend to show pennies.

		3 .	7	
4)	~~1~~	~~5~~ .	~~0~~	0
		15	30	
		−12	−28	
		~~3~~	2	

6. Trade 2 dimes for 20 pennies.

		3 .	7	
4)	~~1~~	~~5~~ .	~~0~~	~~0~~
		15	30	20
		−12	−28	
		~~3~~	~~2~~	

7. If 4 people share 20 pennies, each person gets 5 pennies.

		3 .	7	5
4)	~~1~~	~~5~~ .	~~0~~	~~0~~
		15	30	20
		−12	−28	−20
		~~3~~	~~2~~	0

The column division shows that 15 / 4 = **3.75.**

This means that $15 shared equally among 4 people is $3.75 each.

Space Travel

When we travel on Earth, we typically measure distances in miles or kilometers. In space, where the distances are much greater, astronomers use units of measure based on the fastest thing in the universe: light.

Traveling at the Speed of Light

A common unit of measure is the light-year. One light-year is the distance light travels in a year. Light travels 186,000 miles every second. In one year, light can travel nearly 6 trillion ($5.9 * 10^{12}$) miles.

To get a sense of the speed of light and the vastness of the universe, imagine that you are exploring the universe in a spaceship that travels at the speed of light. The bright star in this picture is 8 light-years away. That means that its light traveled for 8 years before reaching Earth, and it would take your spaceship 8 years to reach the star.

First, try out your ship close to home by orbiting Earth. At the speed of light, you can make about 7 laps around Earth every second!

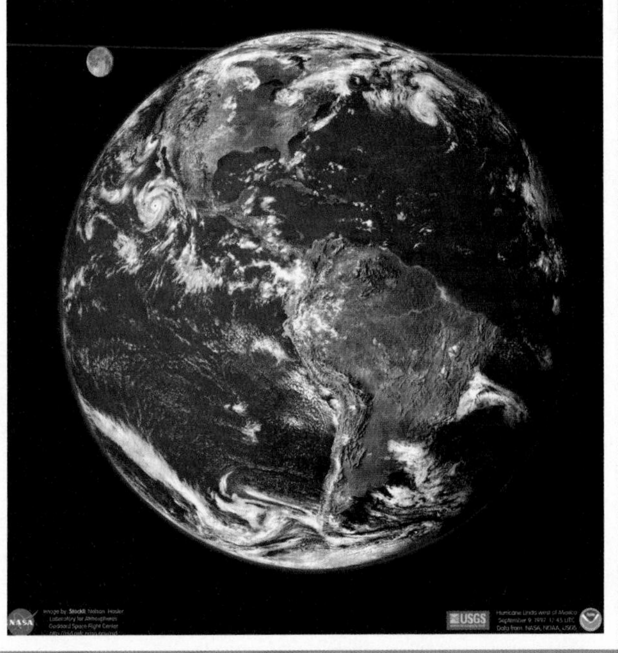

Next, take a trip to the moon. The moon is 240,000 miles away, so your spaceship will get you there in about 1.3 seconds.

Our Solar System

Earth and other planets revolve around the Sun in our solar system. Your next few trips in your imaginary spaceship will take you to other places in our solar system.

Traveling at light speed, a trip from Earth to the Sun would take a little more than 8 minutes. The distance from the Sun to Earth is 93 million miles. Astronomers call that distance 1 astronomical unit (AU). This unit helps them calculate and compare distances to other celestial bodies, such as planets, moons, and stars. Note that this picture is not to scale. The distances between the stars and planets are so great that it is almost impossible to show all of the planets in relation to the Sun in the same picture with the correct scale.

Now travel from the Sun to Saturn, the sixth planet from the Sun. Saturn is nearly 10 AU from the Sun, so your light-speed trip will take about 10 times as long as the trip from Earth to the Sun, or about 80 minutes.

This illustration shows your final destination in our solar system, Pluto, and one of its moons, Charon. Pluto is a dwarf planet that is about 40 AU, or around 3.6 billion ($3.6 * 10^9$) miles, from the Sun. So a trip in your spaceship from the Sun to Pluto will take a little more than 5 hours.

Stars

Once you leave our solar system on your imaginary journey, you can visit the stars. The nearest star, Proxima Centauri, is about 4 light-years away. This may seem like a short distance when measured in light-years, but it is actually a long way: about $3 * 10^5$ AU, or more than 20 trillion ($2 * 10^{13}$) miles.

You could use your light-speed ship to visit some of the more well-known stars, like Sirius, the brightest star seen in the Northern Hemisphere. In this picture, Sirius is the bright star close to the horizon. Your journey from Earth to Sirius would take about 8 years. This may seem like a long time, but if you made the same trip in the U.S. Space Shuttle traveling at about 18,000 miles per hour, it would take nearly 320,000 years, or 40,000 times as long.

Another famous star you could visit is Polaris, or the North Star. Polaris is the last star in the handle of the Little Dipper. At light speed, a trip to Polaris would take more than 300 years, or over 75 times as long as your trip to Proxima Centauri.

Galaxies

The Sun, Sirius, Polaris, and all the other stars we can see without a telescope are part of the Milky Way galaxy. The Milky Way contains at least 100 billion ($1 * 10^{11}$) stars. Scientists believe it is one of about 200 billion ($2 * 10^{11}$) galaxies in the entire universe.

The Milky Way is a spiral galaxy like the Whirlpool galaxy in this photograph.

From Earth, the Milky Way looks like this. Our home is about $\frac{2}{3}$ of the way from the center of the galaxy to the outside edge. In your light-speed ship, it will take about 30,000 years to reach the center of our galaxy.

The Small Magellanic Cloud is one of the galaxies closest to the Milky Way. It is named after the explorer Ferdinand Magellan, who observed and noted this galaxy during his voyage around the world in the 1500s. A trip to this galaxy in your light-speed ship would take about 210,000 years.

(tl)NASA, ESA, S. Beckwith (STScI), and The Hubble Heritage Team (STScI/AURA); (cr)Khlongwangchao/iStock/Getty Images Plus/Getty Images; (br)NASA, ESA, and the Hubble Heritage Team (STScI/AURA) - ESA/Hubble Collaboration

The Andromeda galaxy is the closest spiral galaxy to ours. At the speed of light, it will take you about 2.5 million ($2.5 * 10^6$) years to get there. The Andromeda galaxy, the Milky Way, and more than 30 other galaxies are part of a cluster of galaxies known as the Local Group.

The Local Group is one of many clusters that make up the Local Supercluster, which itself is only a small part of the greater Laniakea Supercluster. There are around 100,000 galaxies in the Laniakea Supercluster. The picture below shows a small string of galaxies in the Virgo Cluster, the largest cluster in the Local Supercluster. Traveling at the speed of light, you could reach the Virgo Cluster in around 60 million ($6 * 10^7$) years. To travel across the entire Laniakea Supercluster, it would take over 8 times as long, or about 500 million ($5 * 10^8$) years.

It is a long way across the Laniakea Supercluster, but that's only a tiny fraction of the distance to the most distant galaxies. Astronomers think that the faint objects in this photograph are galaxies that are 13 billion ($1.3 * 10^{10}$) light-years away. That means the light we are now seeing from these distant stars started its journey to Earth about 13 billion years ago.

(t)NASA/JPL-Caltech; (cr)Stocktrek Images/Getty Images; (bl)Robert Williams and the Hubble Deep Field Team (STScI) and NASA

Back to Reality

Unfortunately, there are no light-speed spaceships to take us to other planets, stars, and galaxies. But technological achievements have enabled us to look deep into space and time to see an expanding universe.

The Hubble Space Telescope, first launched in 1990, orbits Earth at an altitude of 353 miles once every 97 minutes. The Hubble can track and photograph planets, stars, galaxies, and other celestial bodies. It then transmits the photographs to Earth by radio.

The International Space Station is operated by the United States, Russia, and several other countries. The space station is a zero-gravity science laboratory that orbits Earth at more than 17,000 miles per hour at an altitude of 120 to 300 miles. Normally a crew of 6 astronauts lives and works in the space station for several months. A trip to the space station for the crew can take as long as 2 days or as little as 6 hours.

Frequently crew members need to leave the space station and go on "spacewalks" to conduct science experiments and station maintenance.

What new destinations in outer space do you think humans will be able to visit in your lifetime?

(tl)NASA/JSC; (tr)Purestock/SuperStock; (bl)NASA

Number and Operations—Fractions

Fractions

Fractions were invented thousands of years ago to name numbers that are between whole numbers. People likely used these in-between numbers for making more precise measurements. Today most rulers and other measuring tools have marks to name numbers that are between whole-unit measures. Learning how to read these in-between marks is an important part of learning to use these tools. Here are some examples of measurements that use fractions: $1\frac{1}{4}$ inches, $\frac{2}{3}$ cup, $\frac{3}{4}$ hour, $\frac{9}{10}$ kilometer, and $13\frac{1}{2}$ pounds.

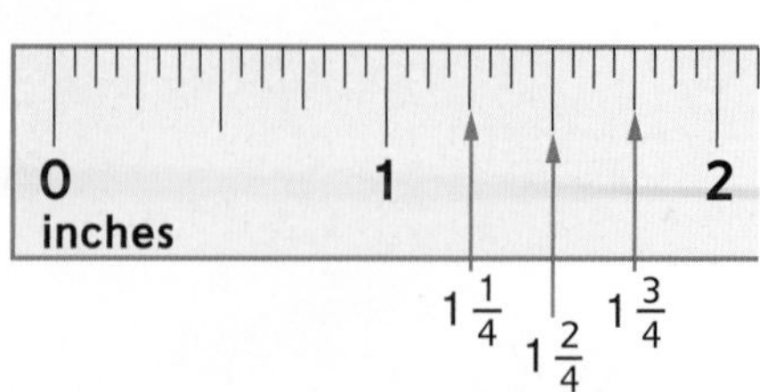

The $\frac{1}{4}$-inch marks between 1 and 2 inches are labeled.

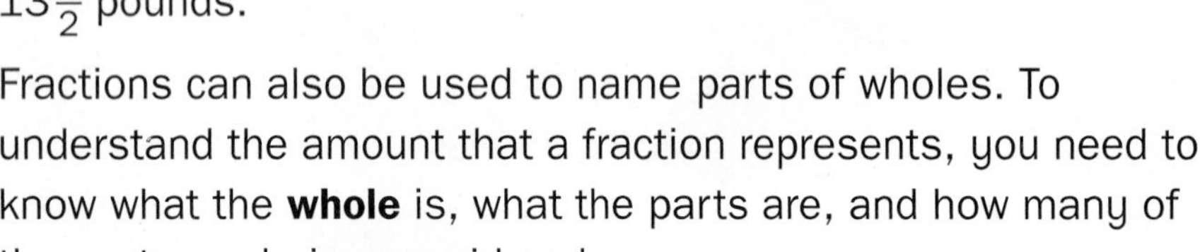

Fractions can also be used to name parts of wholes. To understand the amount that a fraction represents, you need to know what the **whole** is, what the parts are, and how many of the parts are being considered.

The whole might be one single thing, like a stick of butter. The whole might be a distance, like an inch or a mile. The whole might also be a collection of things, like a box of crayons. The **whole** is sometimes called *one*, *one whole*, or a *unit*.

The box of six muffins is the whole. One muffin is one-sixth of the whole box.

A fraction communicates an amount by describing how the parts are related to the whole. This means that a fraction like $\frac{1}{2}$ can communicate different amounts depending on how the whole is defined. For example, half an inch is much smaller than half a mile. Similarly, half a box of crayons could be many crayons or just a few crayons, depending on the number of crayons in the whole box.

Fractions can also be used to show division, ratios, rates, scales, and other relationships.

There are many more crayons in half a box of 64 crayons than in half a box of 16.

Reading and Writing Fractions

Fractions can be written $\frac{a}{b}$, where *a* and *b* are two whole numbers (*b* cannot be 0) separated by a fraction bar. For example, in the fraction $\frac{2}{3}$, 2 and 3 are both whole numbers. Fractions can also be written in different ways. For example, two-thirds can be written as 2-thirds or 2 out of 3.

In fractions that name parts of wholes, the numerator and denominator work together to make one number that describes the amount of the whole that the fraction represents. The **denominator** determines the size of each part of the whole by telling how many equal parts it takes to make the whole. The denominator cannot be 0. The **numerator** describes the number of equal-size parts that are being considered. When reading a fraction, say the numerator first. Then say the size of the equal parts represented by the denominator.

numerator ⟶ 2
denominator ⟶ 3
$\frac{2}{3}$ two-thirds

Example

Write a fraction that names the shaded part of the hexagon.

The hexagon is divided into 6 equal parts, so each part is 1 sixth of the hexagon. Since 3 of the parts are shaded, 1 sixth + 1 sixth + 1 sixth, or 3 sixths, of the hexagon is shaded.

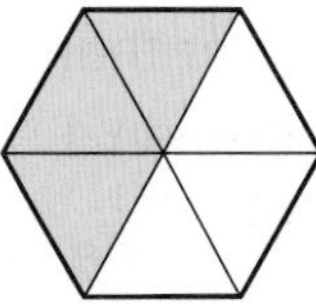

The whole hexagon is divided into 6 equal parts, so the denominator is 6.

Three of the parts are shaded, so the numerator is 3.

Write the fraction $\frac{3}{6}$.

Example

Read $\frac{3}{4}$ and sketch a picture to show what $\frac{3}{4}$ means.

3 ⟵ The *numerator* 3 tells the number of *shaded* parts.
4 ⟵ The *denominator* 4 tells the number of equal parts in the *whole*. Since there are 4 equal parts, the size of each part is a fourth of the whole.

Say, "three-fourths."

You can draw a circle as the whole. Partition the whole into 4 equal parts, then shade 3 of the parts.

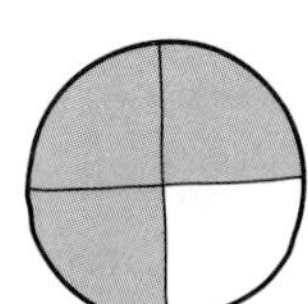

Fractions Equal To and Greater Than One

Some fractions are numbers that are equal to one or greater than one. Just as in other fractions, the denominator tells how many equal parts it takes to make the whole. The numerator describes how many parts are being considered.

If the numerator is equal to the denominator, the amount represented is equal to one whole, or 1.

Examples

$\frac{4}{4}$ means 4 out of 4 equal parts. That's all of the parts of 1 whole.

$\frac{4}{4} = 1$ whole

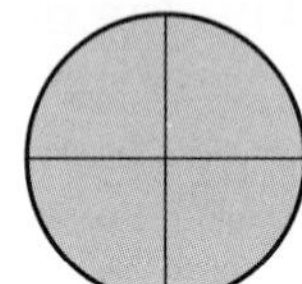

$\frac{6}{6}$ means 6 out of 6 equal parts. That's all of the parts of 1 whole.

$\frac{6}{6} = 1$ whole

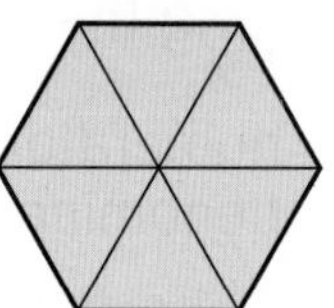

If the numerator is larger than the denominator, the amount represented is greater than one whole. For example, the fraction $\frac{7}{2}$ is greater than 1.

Fractions greater than one can also be written as **mixed numbers.** A mixed number has a whole-number part and a fraction part. In the mixed number $3\frac{1}{2}$, the whole-number part is 3 and the fraction part is $\frac{1}{2}$. A mixed number is equal to the sum of the whole-number part and the fraction part: $3\frac{1}{2} = 3 + \frac{1}{2}$.

Note Fractions greater than one are sometimes called *improper fractions,* although there is nothing wrong with them. It is often easier to use fractions written with the numerator greater than the denominator than to use mixed numbers when using number sentences to solve problems.

Example

What does $\frac{13}{5}$ mean? Write $\frac{13}{5}$ as a mixed number.

The numerator is greater than the denominator. The fraction is a number greater than one and is read, "thirteen-fifths." So this fraction represents an amount that is greater than one whole.

$\frac{13}{5}$

13 ⟵ The *numerator* 13 tells the number of parts.

5 ⟵ The *denominator* 5 tells the number of equal parts in the whole. The size of each part is a fifth of the whole.

You know $\frac{5}{5} = 1$ whole.

You can represent $\frac{13}{5}$ as two whole circles and three-fifths of another circle.

$\frac{13}{5}$ can be written as $2\frac{3}{5}$.

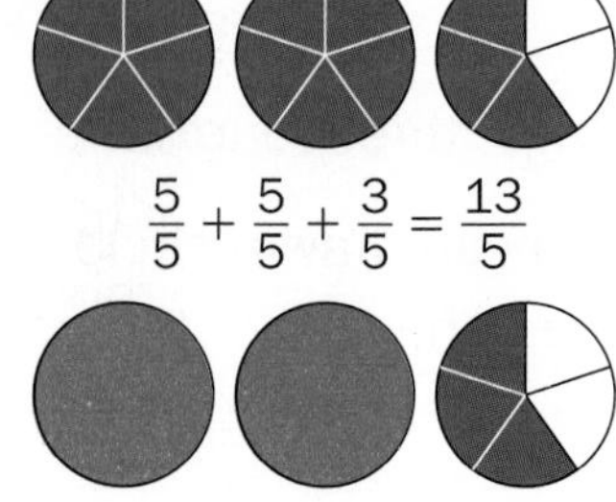

$\frac{5}{5} + \frac{5}{5} + \frac{3}{5} = \frac{13}{5}$

$2\frac{3}{5}$

Meanings of Fractions

Fractions can have different meanings.

Parts of Regions	Fractions can be used to name part of a whole region: $\frac{5}{8}$ of the circle is shaded.	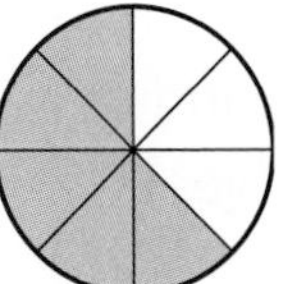
Parts of Collections	Fractions can be used to mean part of a whole collection of objects: $\frac{3}{10}$ of this collection of coins is quarters.	

Points on a Number Line Fractions can be used to mean distance from 0 on a number line. The whole is the distance from 0 to 1, and each whole can be divided into equal distances.

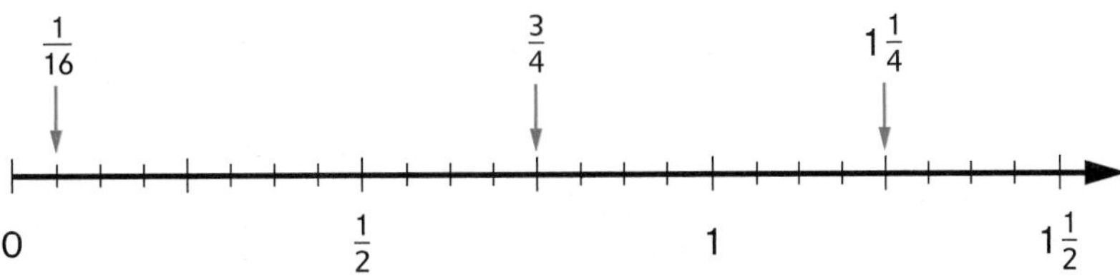

Division Fractions can be used to mean division. The numerator is divided by the denominator. The division problem 3 divided by 4 can be written in any of these ways: $3 \div 4$, $4\overline{)3}$, 3 / 4, and $\frac{3}{4}$.

Ratios and Rates Fractions can be used to compare quantities. For example, DuSable School won 6 games and lost 14 games last year. The total number of games played was $6 + 14 = 20$ games. The fraction $\frac{6}{20}$ compares the number of games won to the number of games played. This fraction compares quantities with the same unit (games).

Fractions can also be used to compare quantities with different units. For example, Bill's car can travel about 30 miles on 1 gallon of gasoline. The fraction $\frac{30 \text{ miles}}{1 \text{ gallon}}$ compares quantities with different units (miles and gallons). At this rate, Bill's car can travel about 150 miles on 5 gallons of gasoline.

$$\frac{30 \text{ miles}}{1 \text{ gallon}} = \frac{150 \text{ miles}}{5 \text{ gallons}}$$

Uses of Fractions

Fractions have many uses in everyday life.

Scale

Fractions can be used to compare the size of a drawing or model to the size of the actual object.

Maps often include a **scale.** You can use the scale to estimate real-world distances. The scale on the map shown here is 1:10,000. So every distance on the map is $\frac{1}{10,000}$ of the real-world distance. A 1-centimeter distance on the map stands for a real-world distance of 10,000 centimeters (100 meters).

Lake Place
Rt 150 E.
2nd St.
3rd St.
Rock Creek
Alexander Ave.
Rock Rd.
Mansfield Park
Scale 1:10,000
0 100 m
0 1 cm

Decimals

Fractions can be used to rename decimals.

This coin is called a quarter because it is $\frac{1}{4}$ of a dollar. $\frac{1}{4}$ is equivalent to $\frac{25}{100}$. $\frac{25}{100}$ of a dollar is the same as 0.25 or \$0.25.

1 quarter

1 dollar

Percents

Fractions can be used to rename percents. Percent means *per hundred*, or *out of a hundred*. So 1% (one percent) has the same meaning as the fraction $\frac{1}{100}$ and the decimal 0.01.

Saving 50% means you will save half the cost. That's because 50% means $\frac{50}{100}$, or $\frac{1}{2}$.

"In-Between" Measures

Fractions can name measures that are between whole-number measures.

Fractions can be used to name lengths measured with rulers, metersticks, yardsticks, and tape measures.

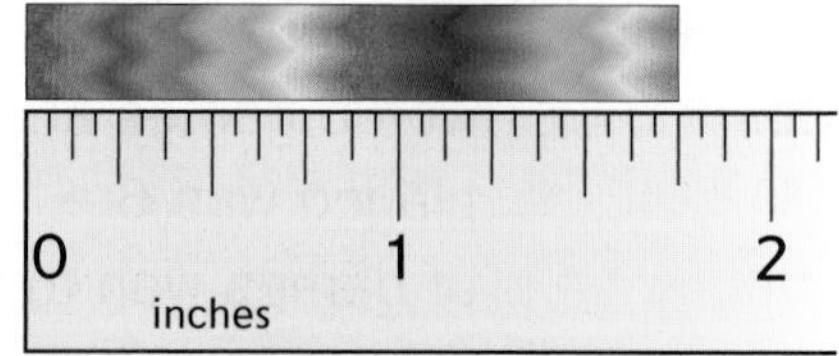

The ribbon measures $1\frac{3}{4}$ inches.

Fractions can be used to tell how far it is to a destination.

A scenic view is $\frac{3}{4}$ mile ahead.

Fractions can be used to name measures of volume.

The liquid in the measuring cup measures $\frac{1}{2}$ cup.

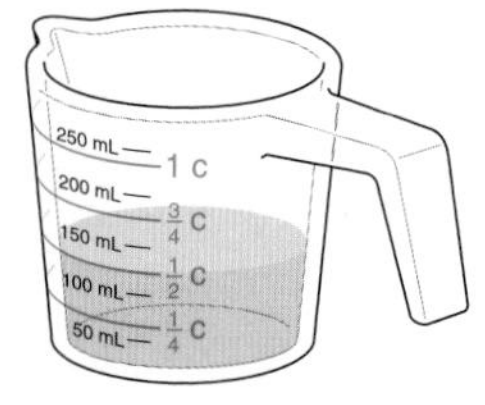

Fraction Circles

There are many ways to represent fractions. One way is with fraction circles.

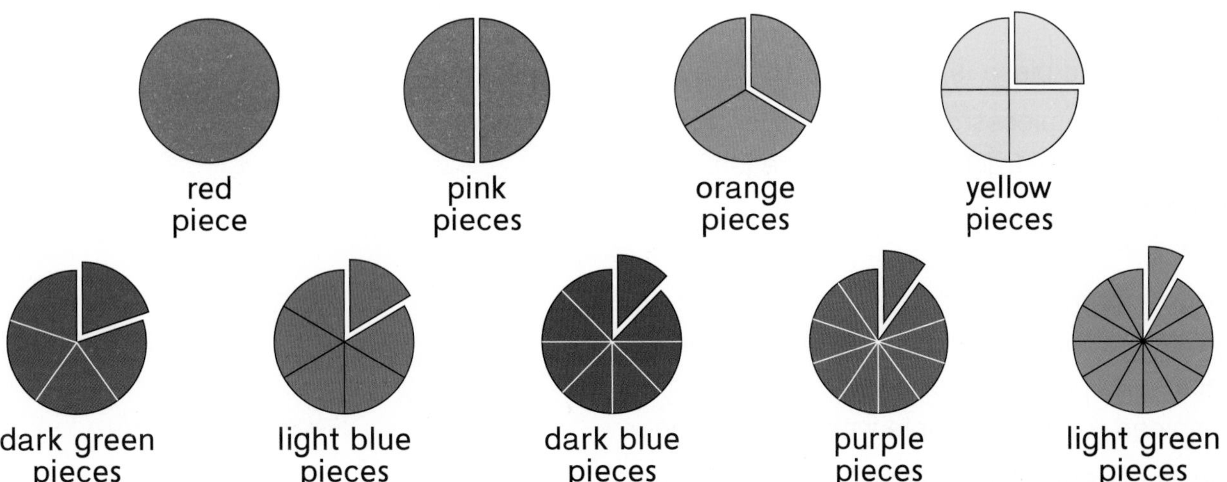

The fraction name for any of the fraction circle pieces depends on which piece is the whole.

Examples

The red fraction circle piece is the whole. Show $\frac{1}{3}$ (one-third) of the red piece. Explain how you know.

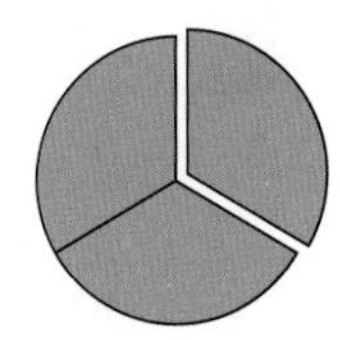

$\frac{1}{3}$ means one out of three equal parts. Three equal-size pieces cover the whole red piece.

Three orange pieces cover the whole red piece, so each orange piece is a third.

One orange piece is $\frac{1}{3}$ (one-third) of the red piece.

The yellow fraction circle piece is the whole. Which fraction circle piece is $\frac{1}{3}$ (one-third) of the whole? Explain how you know.

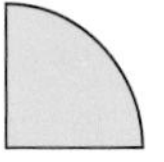

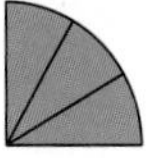

Find three equal-size pieces that cover the whole yellow piece.

Three light green pieces cover the whole yellow piece, so each light green piece is a third.

One light green piece is $\frac{1}{3}$ (one-third) of the yellow piece.

The fraction circle piece that represents $\frac{1}{3}$ is *not* the same when the wholes are different sizes. One-third of the red piece is *greater than* one-third of the yellow piece because the red piece is *larger than* the yellow piece.

Representing Numbers Greater Than One

Fraction circle pieces can be used to show fractions greater than one.

Example

The red circle piece is the whole. Use fraction circle pieces to show $\frac{4}{3}$.

Three orange pieces completely cover a red piece. So each orange fraction circle piece is one-third of the whole, or $\frac{1}{3}$. Four orange pieces show $\frac{1}{3} + \frac{1}{3} + \frac{1}{3} + \frac{1}{3} = \frac{4}{3}$.

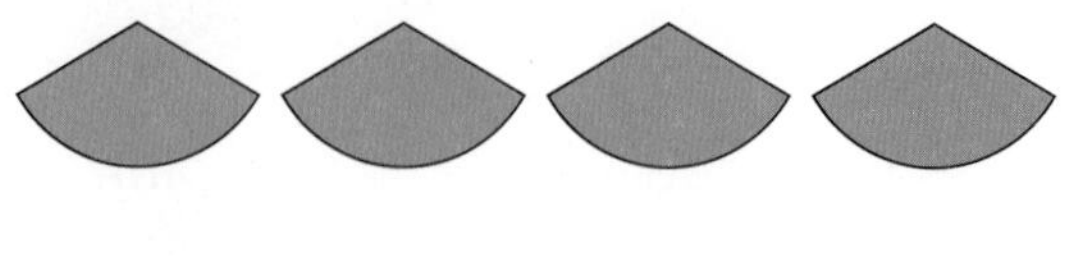

The four orange pieces can also be used to make as many wholes as possible. Since 3 orange pieces completely cover a red piece, $\frac{3}{3}$ can be combined to show 1 whole with $\frac{1}{3}$ remaining.

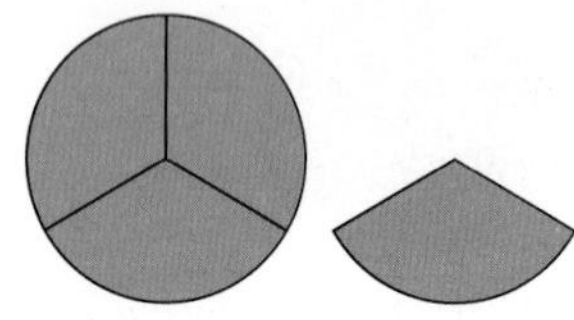

One whole and one-third can be named using a mixed number: $1\frac{1}{3}$.

Example

The red circle piece is the whole. How can you name the amount shown?

One way:

8 equal-size pieces make 1 whole, so each dark blue piece is an eighth of a whole, or $\frac{1}{8}$. In all, there are 21 dark blue pieces. So the amount shown is $\frac{21}{8}$.

Another way:

The complete fraction circle shows 1 whole, and there are 13 more dark blue pieces. That means there is another $\frac{13}{8}$ of a whole.

The amount shown can be written as $1\frac{13}{8}$.

Another way:

Put 8 more pieces together to make another circle.

The 2 complete fraction circles show 2 wholes. There are 5 more dark blue pieces, or another $\frac{5}{8}$ of a whole.

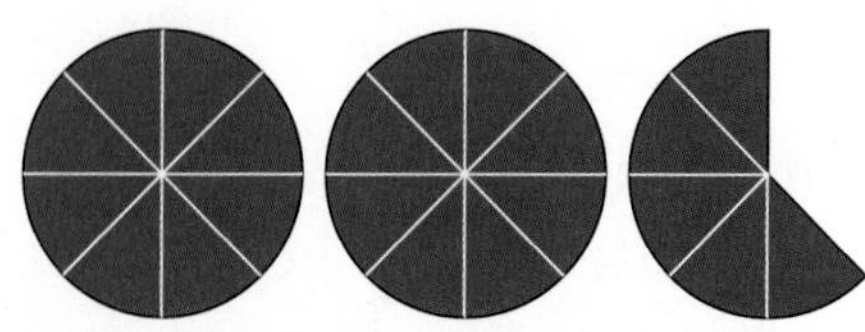

In all, there are 2 wholes and $\frac{5}{8}$ of a whole. Two and five-eighths can be written as $2\frac{5}{8}$.

$\frac{21}{8}$, $1\frac{13}{8}$, and $2\frac{5}{8}$ are all names for the amount shown by the fraction circle pieces.

$\frac{21}{8} = 1\frac{13}{8} = 2\frac{5}{8}$

Fraction Strips

In previous grades, you used fraction strips to represent fractions. A fraction strip is a long rectangle that represents the **whole.** It can be folded into equal-size parts. You can show amounts less than one by folding the fraction strip.

This fraction strip is divided into 3 equal parts, or thirds. $\frac{3}{3} = 1$ whole

$\frac{1}{3}$	$\frac{1}{3}$	$\frac{1}{3}$

This fraction strip is folded to represent $\frac{2}{3}$.

$\frac{1}{3}$	$\frac{1}{3}$

You can draw your own fraction strips to represent fractions. You can show amounts less than one whole by shading equal parts.

Example

Name the amount shaded.

There are three equal parts in the whole. So each part is one-third, or $\frac{1}{3}$.

Two of the equal parts are shaded.

The amount shaded is $\frac{1}{3} + \frac{1}{3} = \frac{2}{3}$.

You can show amounts between whole numbers by shading whole fraction strips together with strips that have been divided into equal-size pieces.

Example

Name the amount shaded.

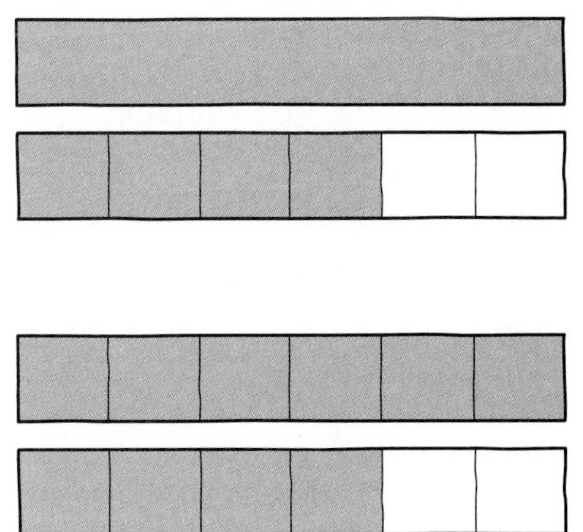

The amount shaded is more than one but less than two whole strips. So the name will be 1 and some fraction.

Since four-sixths of the second fraction strip is shaded, the total amount shaded is 1 and four-sixths strips, or $1\frac{4}{6}$.

You can find another name for the amount shaded by partitioning the whole into sixths.

The whole is $\frac{6}{6}$. The remaining fractional part is $\frac{4}{6}$.

In all, 6 sixths and 4 sixths are shaded. $\frac{6}{6} + \frac{4}{6} = \frac{10}{6}$

$1\frac{4}{6}$ and $\frac{10}{6}$ are both names for the amount shaded.

Fractions on Number Lines

On number lines, the **whole** is the distance from 0 to 1. The parts are equal-size distances, or **intervals,** within the whole. A location, or a **point,** on a number line is labeled to name a distance from 0. Points on number lines show the length of the distance from 0 to the point.

On the number line below, use your finger to trace a distance that starts at 0 and ends at 1. The distance your finger traveled is *one whole*, *a whole*, or *one*. 1 is labeled at the end of the distance to show the length of the distance from 0 to 1.

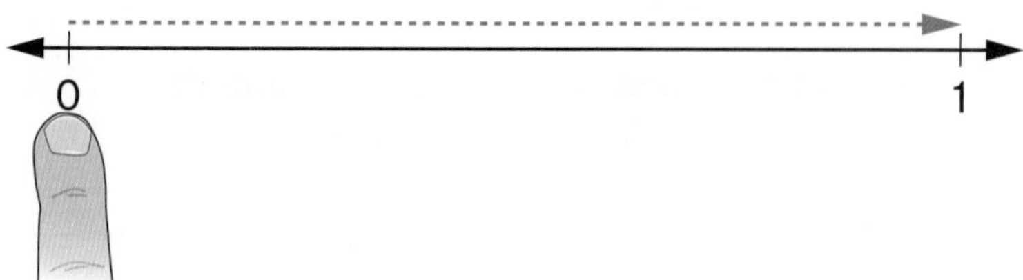

Naming Points between 0 and 1 on a Number Line

You can name a point between 0 and 1 on a number line by using the number of equal intervals to determine the numerator and the denominator of the fraction.

Example

What is a name for point *A*?

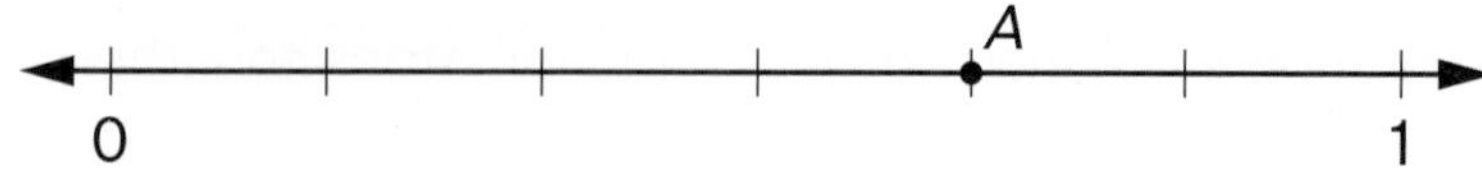

Estimate the location:

The distance from 0 to point *A* is more than halfway to 1. So the name of the point will be a fraction greater than one-half.

Think about the denominator:

Count the intervals between 0 and 1. There are six equal intervals, so the length of each interval is one-sixth. The denominator is 6.

NOTE: Even though the number of tick marks between 0 and 1 is five, the number of intervals, or small distances, between 0 and 1 is six.

Think about the numerator:

Count the number of intervals as you use your finger to trace the distance from 0 to point *A*. Starting at 0, your finger passes **1** sixth, **2** sixths, and **3** sixths, and then stops at **4** sixths, or point *A*. The numerator is 4, since you counted 4 intervals from 0 to point *A*.

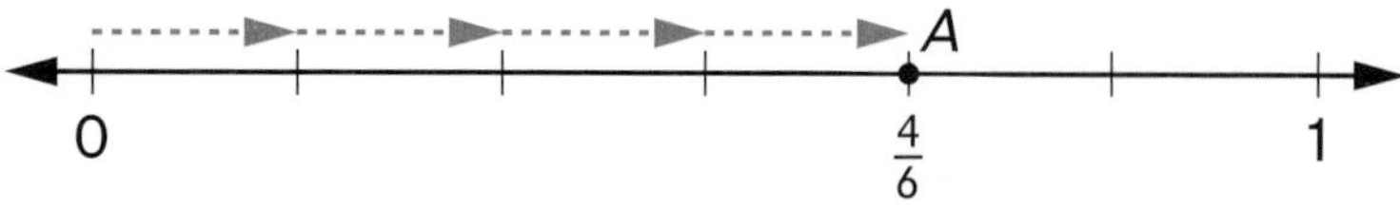

A name for point *A* is four-sixths, or $\frac{4}{6}$.

Naming Points Greater Than One

You can use mixed numbers or fractions greater than one to name points between whole numbers. Find which two whole numbers the point is between and then figure out the fractional part of the distance by examining the intervals between the whole numbers.

Example

What is a name for point B?

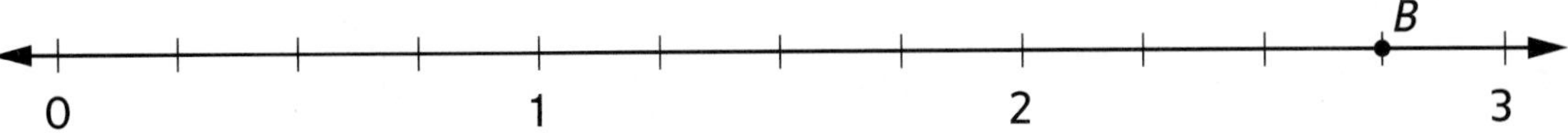

One way:

Think about whole numbers first:

The distance from 0 to B is more than 2 wholes but less than 3. Point B is closer to 3 than it is to 2. So the name of the point will be 2 and some fraction greater than one-half.

Think about the fractional part:

What is the denominator? There are four equal intervals between 2 and 3. So the small distances between 2 and 3 are fourths. The denominator is 4.

What is the numerator? Count the number of intervals you travel as you trace the distance from 2 to point B. The numerator is 3, since you travel 3 intervals.

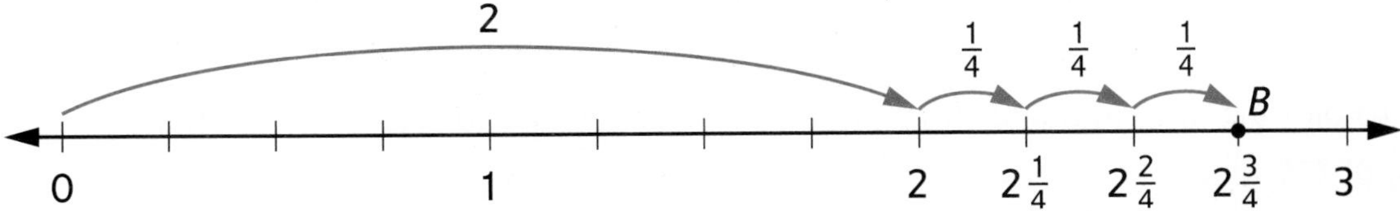

Since point B is 3 fourths beyond 2, a name for point B is two and three-fourths, or $2\frac{3}{4}$.

Another way:

Think about the unit fractions on the number line:

There are 4 intervals between any two consecutive whole numbers, so the small distances are fourths. There are 11 fourths between 0 and point B, so a name for point B is $\frac{11}{4}$.

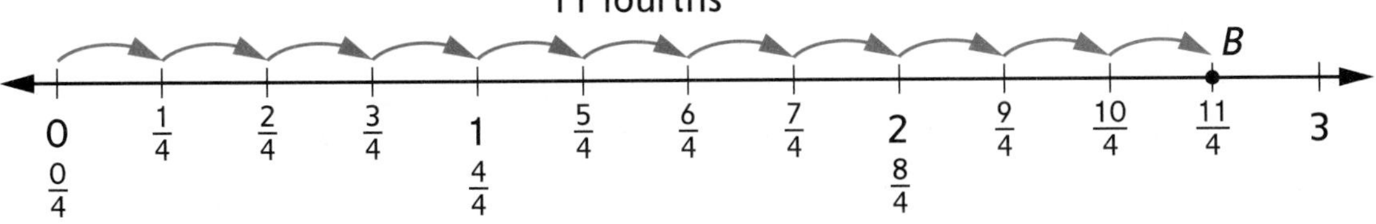

Both $\frac{11}{4}$ and $2\frac{3}{4}$ are names for point B.

Check Your Understanding

Write a fraction and a mixed number for the location of point D on the number line.

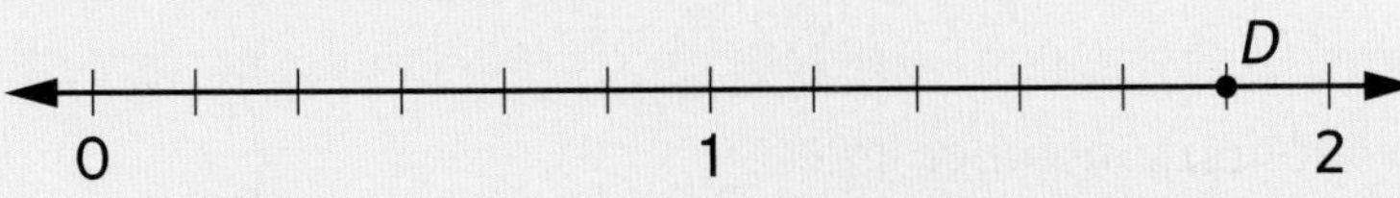

Check your answers in the Answer Key.

Plotting Points on a Number Line

To plot a point on a number line, first draw a number line and label it with whole numbers. Estimate the location of the point by using benchmarks such as 0, $\frac{1}{2}$, and 1. Then find the precise location of a point by making tick marks for equal-length intervals. Travel from 0 to the number you are locating. Place a dot at the point and label it.

Example

Plot $1\frac{1}{3}$ on a number line.

Sketch a number line. Because $1\frac{1}{3}$ is between 1 and 2, the number line should go from 0 to at least 2.

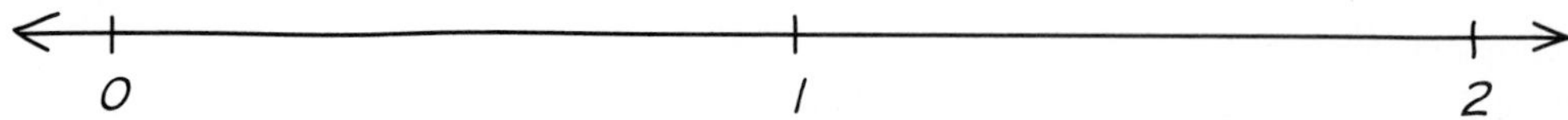

Estimate: Since the point is between 1 and 2, and because $\frac{1}{3}$ is less than halfway, the location of the point will be closer to 1 than it is to 2.

Show the intervals using tick marks. The fractional parts are thirds, so partition the distances between whole numbers into 3 equal parts.

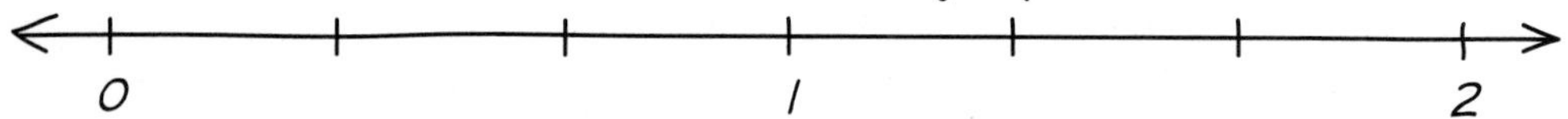

One way to plot $1\frac{1}{3}$ is to imagine traveling the whole distance to 1 and then continue another $\frac{1}{3}$ of the distance from 1 to 2. Label the endpoint $1\frac{1}{3}$.

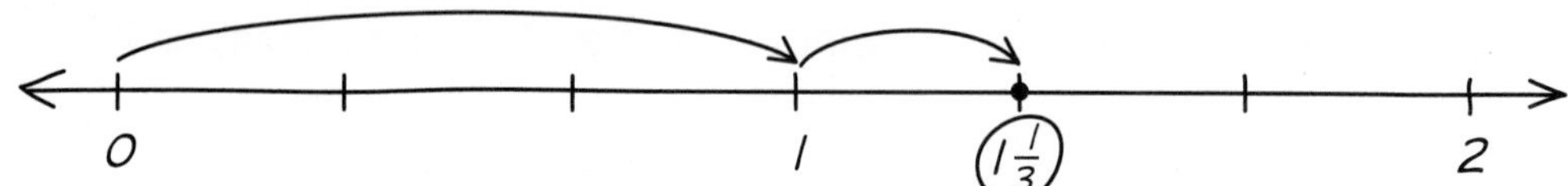

Another way to plot $1\frac{1}{3}$ is to think of $1\frac{1}{3}$ as $\frac{3}{3} + \frac{1}{3}$, or $\frac{4}{3}$. Start at 0 and count 4 thirds. Label the endpoint $\frac{4}{3}$.

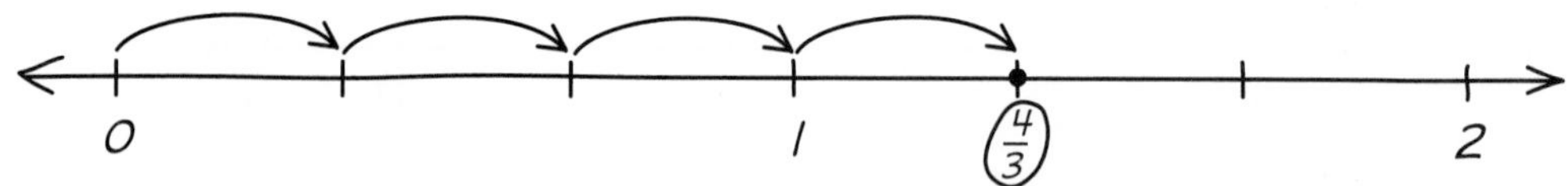

The location for both methods is reasonable because in each case the plotted point is closer to 1 than it is to 2.

The endpoints are the same for both methods because $1\frac{1}{3}$ is the same distance from 0 as $\frac{4}{3}$.

Check Your Understanding

Plot and label each of the following points on a number line.

1. $\frac{3}{5}$ **2.** $2\frac{2}{3}$ **3.** $\frac{5}{3}$

Check your answers in the Answer Key.

Fraction Number Lines Poster

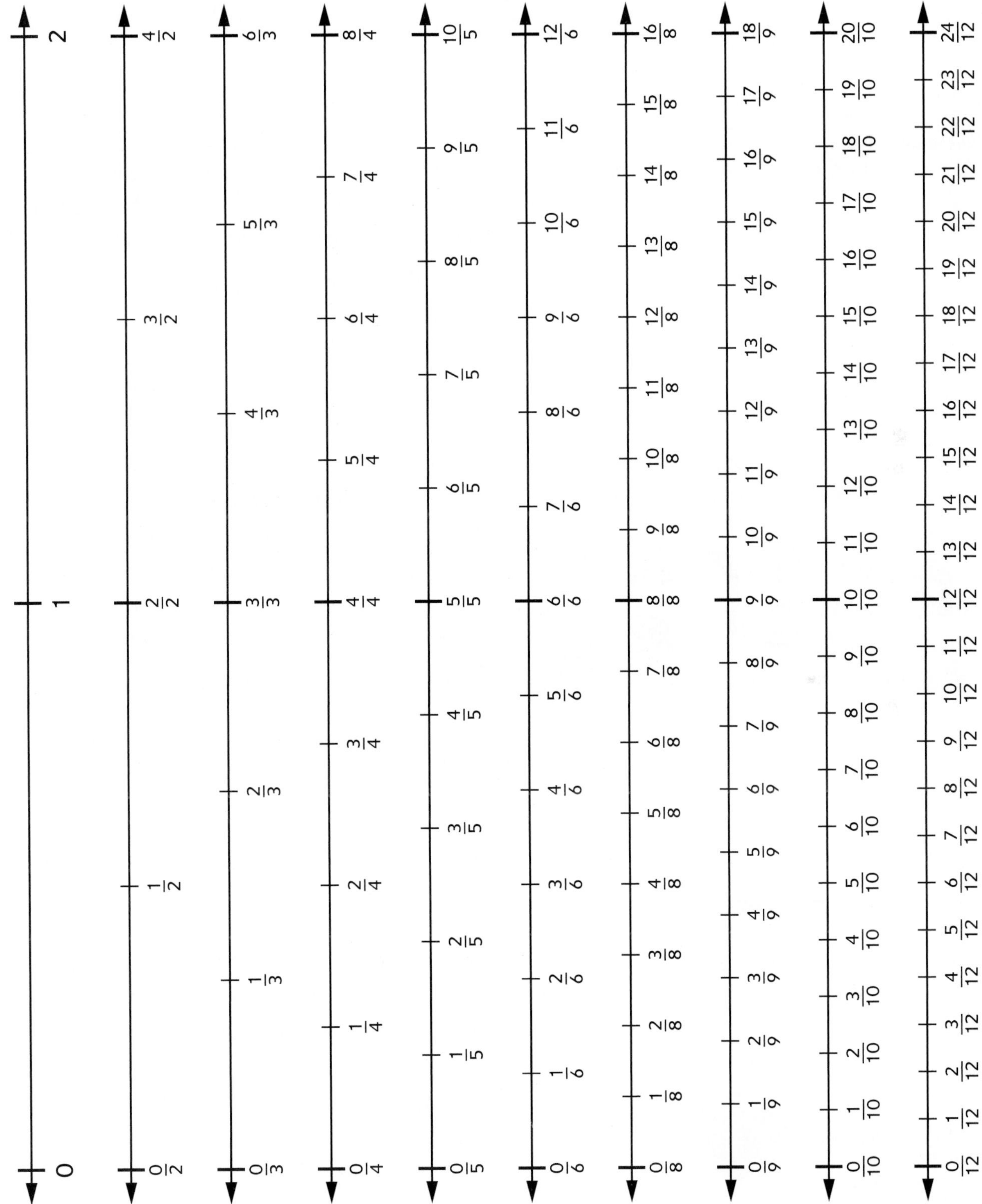

Fractions that Name Parts of Collections

When you have a collection of objects, you can name part of the collection with a fraction. You know what the **whole** is by the way the collection of objects is described. Each of the objects in the collection is a part.

Examples

What fraction of the marbles in the bag is purple?

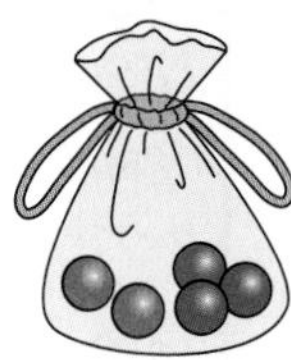

The whole is the collection of all the marbles in the bag. There are 5 marbles in the bag, so each marble is $\frac{1}{5}$ of the whole.

Two-fifths of the marbles in the bag are purple because 2 of 5 marbles in the bag are purple.

Two-fifths and $\frac{2}{5}$ are names for the fraction of marbles that are purple.

What fraction of the group of students at the table is wearing glasses?

The whole is the group of students at the table. Since there are 6 students at the table, each student is one-sixth of the set.

3 of the 6 students in the group are wearing glasses, so three-sixths of the students are wearing glasses. Three-sixths and $\frac{3}{6}$ are names for this fraction.

There are often many ways to name parts of a collection. Because 3 out of 6 is half the students, $\frac{1}{2}$ is another name for the fraction of students wearing glasses.

How many cartons of eggs are filled?

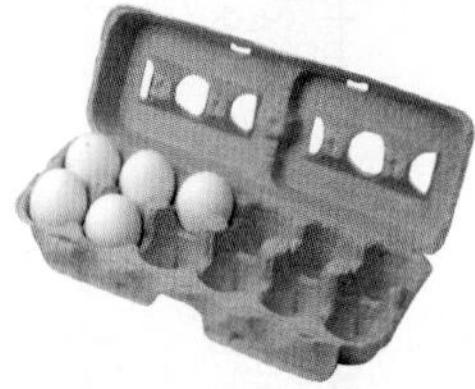

The whole is one carton of eggs. Since one carton holds 12 eggs, each egg is one-twelfth of a carton.

There is 1 whole carton and part of another carton of eggs filled. Since 5 eggs are in the partial carton, that carton is five-twelfths full.

One and five-twelfths and $1\frac{5}{12}$ are names for this amount.

Fractions as Division

Fair-share stories involve sharing something equally, or fairly, among a given number of people or shares. The following is an example of a fair-share number story:

Five loaves of bread are shared equally among 8 people. How much bread does each person get?

You can think of fair-share stories as division situations. Often the numbers being divided are whole numbers and the result is a fraction. For example, 5 loaves of bread ÷ 8 people = $\frac{5}{8}$ loaf per person.

You can model fair-share stories in different ways. You can use fraction circle pieces, fraction strips, number lines, or drawings to model and solve fair-share stories. When you solve problems like these, you learn that the fraction $\frac{a}{b}$ is another way of saying *a* divided by *b*.

Note **Fair-share** number stories can also be called **equal-sharing** number stories. In an equal-sharing number story, you know the number of shares and you need to find the number or amount in each equal share. For example, four boys share 24 marbles fairly, how many marbles does each boy get?

Example

Three bagels are equally shared among Rick, Sue, Kim, and Lee. What part of a whole bagel does each person get?

One way: Draw a picture that shows each bagel equally divided into 4 pieces, and assign 1 piece from each bagel to each person.

Each person gets $\frac{1}{4}$ of each of the 3 bagels, or $\frac{1}{4} + \frac{1}{4} + \frac{1}{4} = 3 * (\frac{1}{4}) = \frac{3}{4}$ bagel.

3 bagels divided into 4 equal parts results in each person getting 3 one-fourths of a bagel. Each person gets $\frac{3}{4}$ of a bagel.

So, $3 \div 4 = \frac{3}{4}$.

Another way: Draw a picture of the 3 bagels with 4 people below the bagels.

Show each of the first two bagels cut into 2 equal parts. Draw lines to show that each person gets $\frac{1}{2}$ of a bagel.

Show the last bagel cut into 4 equal parts. Draw lines to show that each person gets another $\frac{1}{4}$ of a bagel.

3 bagels divided equally among 4 people results in each person getting $\frac{1}{2}$ and $\frac{1}{4}$ of a bagel. $\frac{1}{2} + \frac{1}{4} = \frac{3}{4}$

So, $3 \div 4 = \frac{3}{4}$.

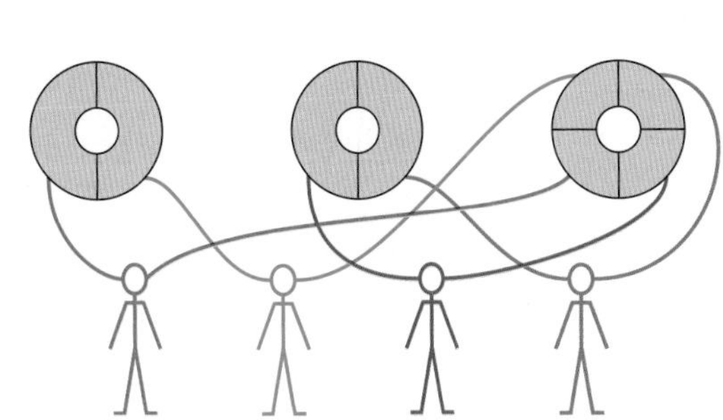

Both strategies show that $\frac{3}{4}$ is another way of saying 3 ÷ 4.

You can use the relationship between fractions and division to write different number models to solve a fair-share number story in more than one way.

Example

If 4 people share a 13-pound bag of rice equally, how many pounds of rice will each person get?

One way: Think of the problem as whole-number division. Solve $13 \div 4 =$ **?**

$4\overline{)13}$ → 3 R1

Each person gets 3 pounds. There is 1 pound left. If you divide the last 1 pound into 4 equal parts, each person will get another $\frac{1}{4}$ pound.

Each person will get $3\frac{1}{4}$ pounds of rice.

Another way: Think of the problem as a fraction. $13 \div 4$ is another way of saying $\frac{13}{4}$, so rename $\frac{13}{4}$ as a mixed number to solve the problem.

Since $\frac{4}{4}$ makes one whole, make as many wholes as you can.

$\frac{13}{4} = \frac{4}{4} + \frac{4}{4} + \frac{4}{4} + \frac{1}{4} = 3\frac{1}{4}$

Each person will get $3\frac{1}{4}$ pounds of rice.

The number model for the first way is $13 \div 4 \rightarrow 3$ R1, or $3\frac{1}{4}$.

The number model for the second way is $\frac{13}{4} = 3\frac{1}{4}$.

So, $13 \div 4 = \frac{13}{4} = \mathbf{3\frac{1}{4}}$.

Writing Division Number Stories Using Fractions

A fraction can represent many division number stories.

Examples

Write a division number story that is solved by finding the answer to $\frac{6}{4} = ?$.

6 is the dividend and 4 is the divisor.

6 apples are split into 4 equal servings. How many apples are in each serving?

Answer: $\frac{6}{4} = 1\frac{2}{4}$, so each serving is $1\frac{2}{4}$ apples.

Write a division number story that is solved by finding the answer to $3 \div 5 = ?$.

3 is the dividend and 5 is the divisor.

3 yards of ribbon will be shared equally among 5 students. How much ribbon will each student get?

Answer: Since $3 \div 5 = \frac{3}{5}$, each student will get $\frac{3}{5}$ yard of ribbon.

Equivalent Fractions

Fractions that name the same amount or the same distance from 0 on a number line are called **equivalent fractions.**

Example

The four circles below are the same size, but they are divided into different numbers of equal parts. The shaded areas are the same in each circle. These circles show different fractions that are equivalent to $\frac{1}{2}$.

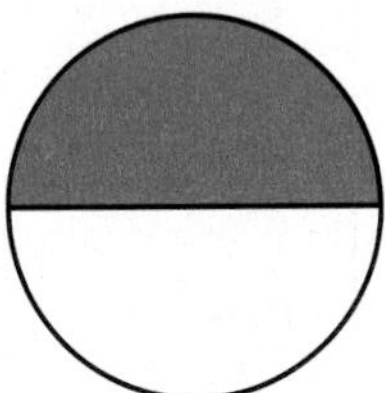

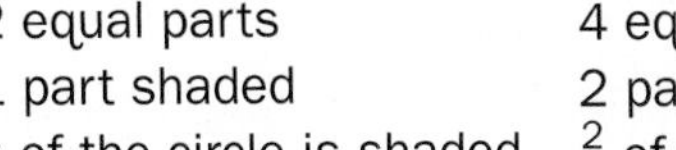

2 equal parts
1 part shaded
$\frac{1}{2}$ of the circle is shaded.

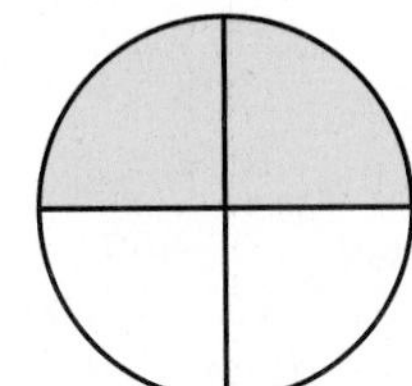

4 equal parts
2 parts shaded
$\frac{2}{4}$ of the circle is shaded.

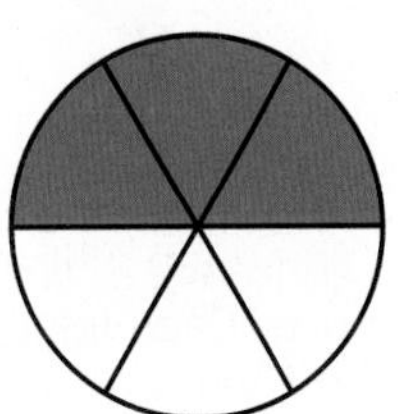

6 equal parts
3 parts shaded
$\frac{3}{6}$ of the circle is shaded.

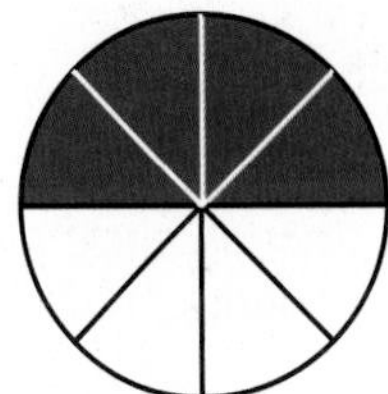

8 equal parts
4 parts shaded
$\frac{4}{8}$ of the circle is shaded.

The fractions $\frac{1}{2}$, $\frac{2}{4}$, $\frac{3}{6}$, and $\frac{4}{8}$ are all equivalent because they represent the same amount of the whole. They are different names for the part of the circle that is shaded.

You can write: $\frac{1}{2} = \frac{2}{4}$ $\frac{1}{2} = \frac{4}{8}$ $\frac{2}{4} = \frac{4}{8}$

$\frac{1}{2} = \frac{3}{6}$ $\frac{2}{4} = \frac{3}{6}$ $\frac{3}{6} = \frac{4}{8}$

Example

Ms. Klein picks up a total of 24 students on her bus route. She picks up 18 boys and 6 girls. Write fractions for the part of the students that are girls.

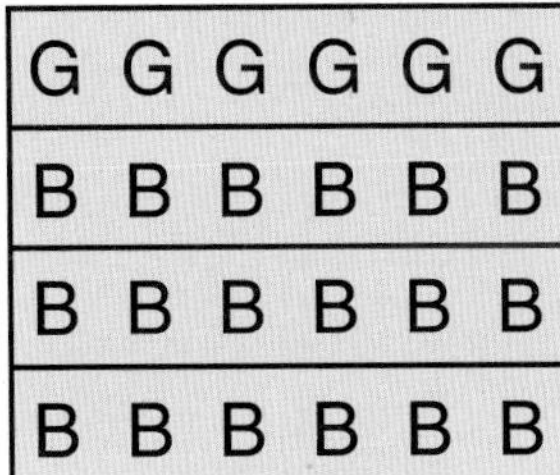

G G G G G G
B B B B B B
B B B B B B
B B B B B B

4 equal groups
Each group is $\frac{1}{4}$ of the total.
1 group of girls
$\frac{1}{4}$ of the students are girls.

G G	G G	G G
B B	B B	B B
B B	B B	B B
B B	B B	B B

12 equal groups
Each group is $\frac{1}{12}$ of the total.
3 groups of girls
$\frac{3}{12}$ of the students are girls.

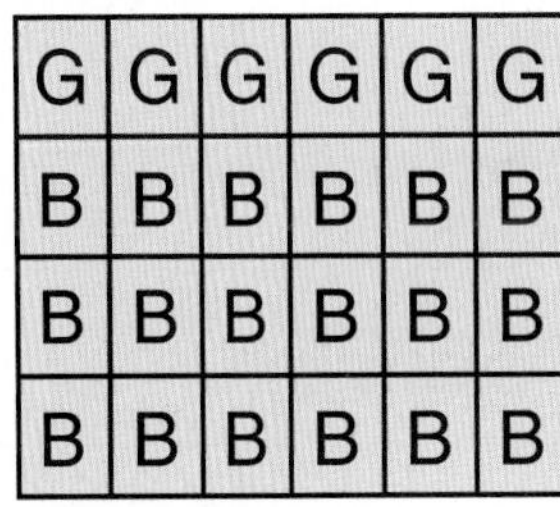

G	G	G	G	G	G
B	B	B	B	B	B
B	B	B	B	B	B
B	B	B	B	B	B

24 equal groups
Each group is $\frac{1}{24}$ of the total.
6 groups of girls
$\frac{6}{24}$ of the students are girls.

The fractions $\frac{1}{4}$, $\frac{3}{12}$, and $\frac{6}{24}$ are equivalent.

You can write $\frac{1}{4} = \frac{3}{12} = \frac{6}{24}$.

Finding Equivalent Fractions

You can use fraction tools to find equivalent fractions. Fraction circle pieces show equivalent fractions when they cover the same amount of space, or when they show equal areas.

Example

Name a fraction that is equivalent to $\frac{1}{3}$.

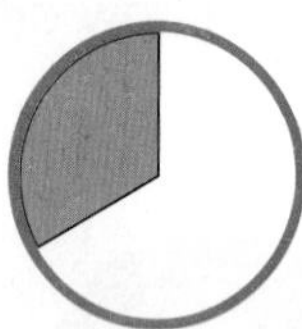

If the red circle is the whole, then the orange fraction circle piece represents $\frac{1}{3}$.

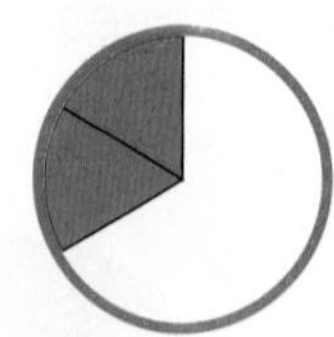

Two light blue pieces cover the same amount as one orange piece. Each light blue piece is $\frac{1}{6}$ of the whole, so two-sixths is the same as one-third.

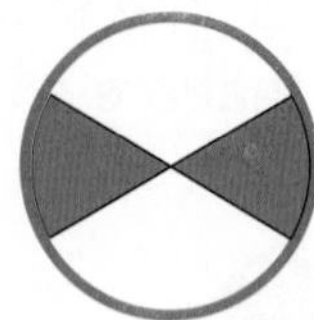

Two-sixths of this circle covers the same area even when the sixths are in different places.

$\frac{2}{6}$ is equivalent to $\frac{1}{3}$. Write $\frac{2}{6} = \frac{1}{3}$.

Number lines can show equivalent fractions when the wholes are the same length on each number line. Fractions are equivalent if they name points that are the same distance from 0.

Example

Name two fractions that are equivalent to $\frac{3}{2}$.

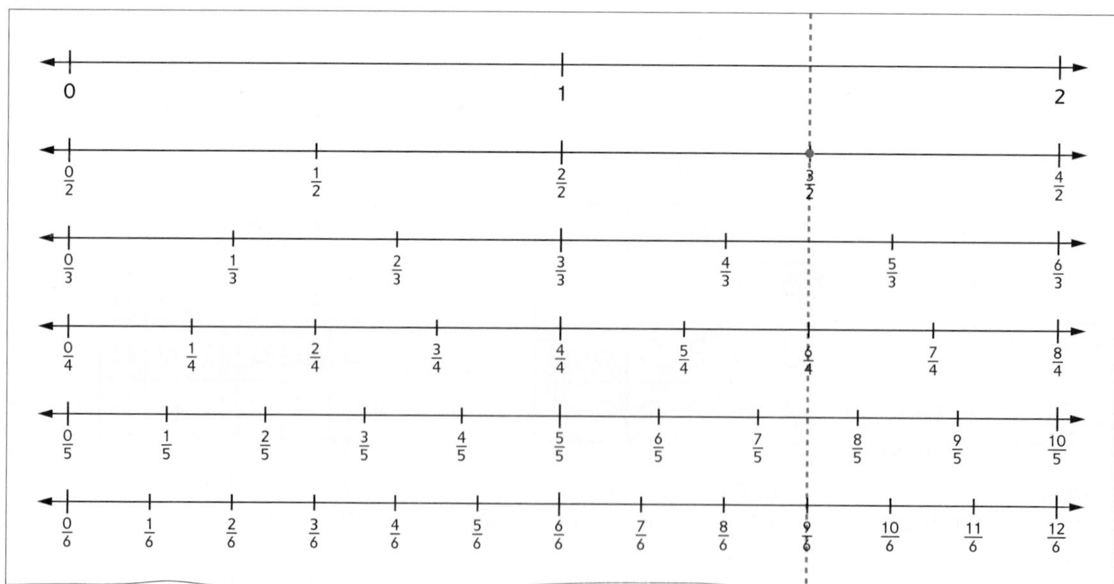

$\frac{3}{2}$ names the same distance from 0 as $\frac{6}{4}$ and $\frac{9}{6}$. This means that these are equivalent fractions. $\frac{3}{2} = \frac{6}{4} = \frac{9}{6}$

Equivalent Fractions on a Ruler

Rulers marked in inches usually have tick marks of different lengths. The longest tick marks on the ruler below show whole inches. The marks used to show half inches are shorter than the marks for whole inches; the marks for quarter inches are shorter than the marks for half inches; and so on. The shortest marks show sixteenths of an inch.

Every tick mark on this ruler can be named as a number of sixteenths. Some tick marks can also be named eighths, fourths, halves, and whole inches. The picture at the right shows the pattern of fraction names for parts of 1 inch.

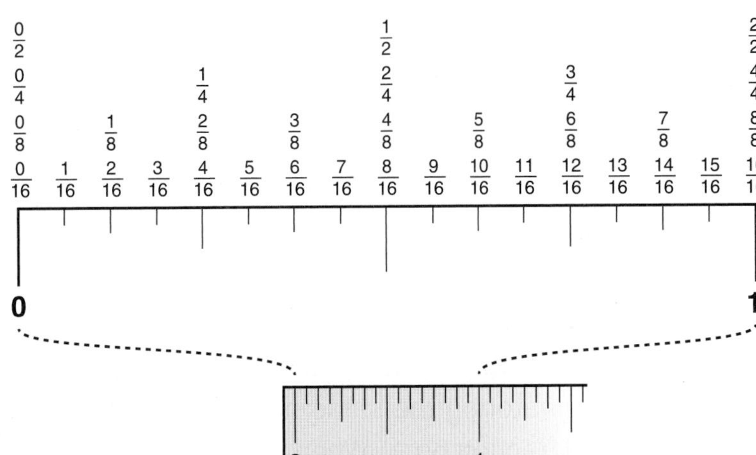

This pattern of naming tick marks continues past 1 inch, but mixed numbers are used to name the lengths.

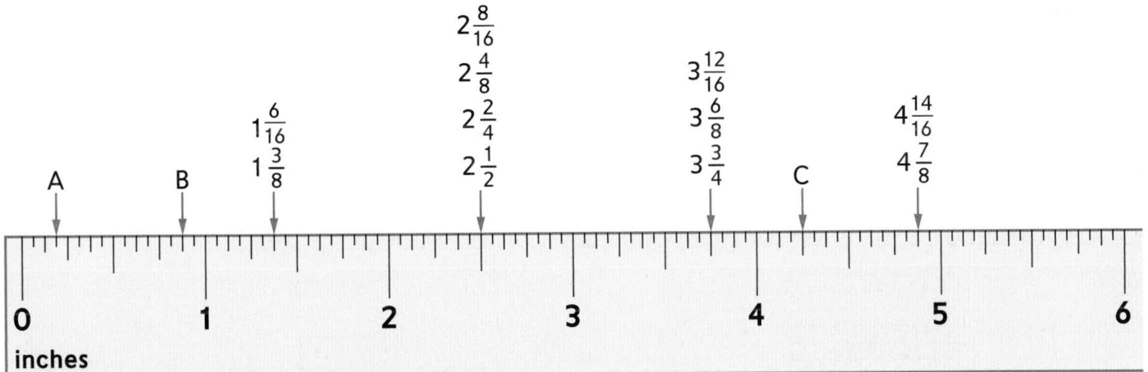

Check Your Understanding

1. Name a fraction or mixed number for each mark labeled A, B, and C on the ruler above.
2. What is the length of this nail?
   - **a.** in quarter inches
   - **b.** in eighths of an inch
   - **c.** in sixteenths of an inch

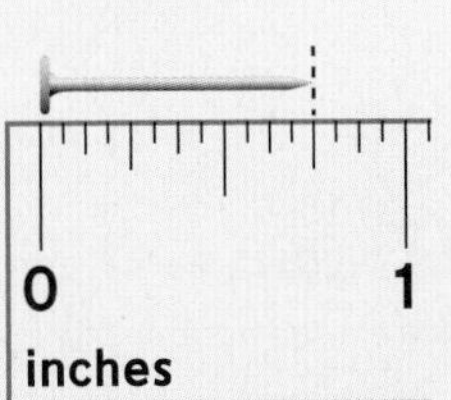

Check your answers in the Answer Key.

Equivalent Fraction Rules

Here are two rules you can use to find equivalent fractions.

The Multiplication Rule for Equivalent Fractions

To find a fraction that is equivalent to a given fraction, multiply the numerator and the denominator by the same number (not 0).

Example

Find fractions that are equivalent to $\frac{2}{5}$.
Draw a rectangle divided into 5 equal parts. Shade 2 parts.

Use horizontal lines to divide the rectangle into a greater number of equal parts. Notice that the exact same area of the rectangle is still shaded.

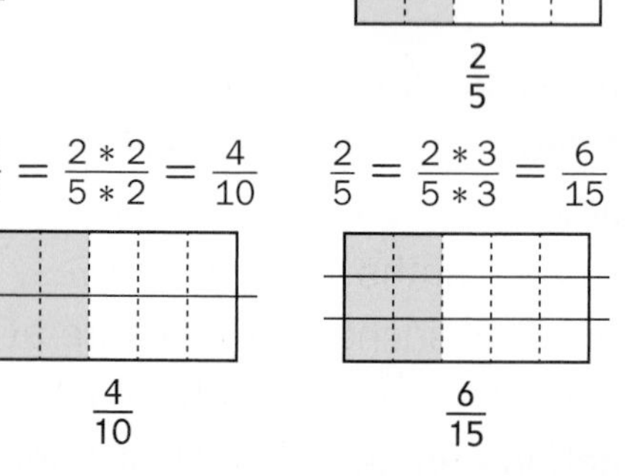

$\frac{2}{5} = \frac{4}{10} = \frac{6}{15}$

This rule works because when you multiply the numerator and the denominator by the same number, it is the same as multiplying the fraction by 1. When you multiply a fraction or any number by 1, the product equals the original number. For example, $6 * 1 = 6$, $192 * 1 = 192$, and $\frac{3}{4} * 1 = \frac{3}{4}$.

Note When the numerator and denominator are the same, the fraction is equal to 1. For example, $\frac{2}{2} = 1$; $\frac{5}{5} = 1$; $\frac{23}{23} = 1$.

Example

Find a fraction that is equivalent to $\frac{3}{4}$.

$\frac{3}{4} = \frac{3}{4} * 1$

Rewrite 1 as a fraction that is equal to 1, such as $\frac{5}{5}$. Then multiply $\frac{3}{4} * \frac{5}{5}$.

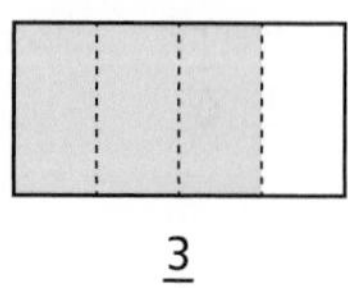

$\frac{3}{4}$

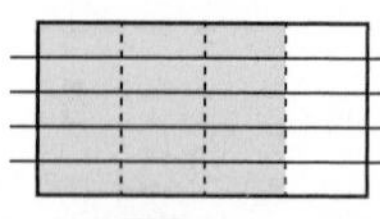

$\frac{3}{4} * \frac{5}{5} = \frac{3 * 5}{4 * 5} = \frac{15}{20}$

$\frac{5}{5} = 1$

$\frac{3}{4}$ is equivalent to $\frac{15}{20}$.

The Division Rule for Equivalent Fractions

If the numerator and the denominator of a fraction are both divided by the same number (not 0), the result is a fraction that is equivalent to the original fraction.

Example

Find an equivalent fraction for $\frac{15}{20}$.

Think: What number divides both 15 and 20 evenly?

Since 5 is a common factor of 15 and 20, divide the numerator and the denominator by 5.

$$\frac{15}{20} = \frac{15 \div 5}{20 \div 5} = \frac{3}{4}$$

Dividing both numbers in $\frac{15}{20}$ by 5 gives an equivalent fraction, $\frac{3}{4}$.

Notice that when you divide the numerator and the denominator by the same number, it's the same as dividing the fraction by 1. When you divide any number by 1, the quotient is the same as the dividend. For example, $6 \div 1 = 6$, $192 \div 1 = 192$, and $\frac{3}{4} \div 1 = \frac{3}{4}$. This rule uses this property of 1 to help you find equivalent fractions that have smaller numerators and denominators.

Examples

Find an equivalent fraction for $\frac{6}{15}$.

$\frac{3}{3} = 1$

$$\frac{6}{15} = \frac{6}{15} \div \boxed{\frac{3}{3}} = \frac{6 \div 3}{15 \div 3} = \frac{2}{5}$$

Find an equivalent fraction for $\frac{24}{60}$.

$$\frac{24}{60} = \frac{24 \div 4}{60 \div 4} = \frac{6}{15}$$

Check Your Understanding

1. Use the multiplication rule to find an equivalent fraction.

a. $\frac{3}{4}$ **b.** $\frac{3}{8}$ **c.** $\frac{2}{5}$ **d.** $\frac{4}{7}$ **e.** $\frac{8}{3}$ **f.** $\frac{11}{12}$

2. Use the division rule to find an equivalent fraction.

a. $\frac{8}{12}$ **b.** $\frac{10}{25}$ **c.** $\frac{24}{36}$ **d.** $\frac{45}{60}$ **e.** $\frac{36}{24}$ **f.** $\frac{100}{8}$

Check your answers in the Answer Key.

Table of Equivalent Fractions

This table lists equivalent fractions. All of the fractions in a row name the same number. For example, all the fractions in the last row are names for the number $\frac{11}{12}$.

Simplest Name	Equivalent Fraction Name								
0 (zero)	$\frac{0}{1}$	$\frac{0}{2}$	$\frac{0}{3}$	$\frac{0}{4}$	$\frac{0}{5}$	$\frac{0}{6}$	$\frac{0}{7}$	$\frac{0}{8}$	$\frac{0}{9}$
1 (one)	$\frac{1}{1}$	$\frac{2}{2}$	$\frac{3}{3}$	$\frac{4}{4}$	$\frac{5}{5}$	$\frac{6}{6}$	$\frac{7}{7}$	$\frac{8}{8}$	$\frac{9}{9}$
$\frac{1}{2}$	$\frac{2}{4}$	$\frac{3}{6}$	$\frac{4}{8}$	$\frac{5}{10}$	$\frac{6}{12}$	$\frac{7}{14}$	$\frac{8}{16}$	$\frac{9}{18}$	$\frac{10}{20}$
$\frac{1}{3}$	$\frac{2}{6}$	$\frac{3}{9}$	$\frac{4}{12}$	$\frac{5}{15}$	$\frac{6}{18}$	$\frac{7}{21}$	$\frac{8}{24}$	$\frac{9}{27}$	$\frac{10}{30}$
$\frac{2}{3}$	$\frac{4}{6}$	$\frac{6}{9}$	$\frac{8}{12}$	$\frac{10}{15}$	$\frac{12}{18}$	$\frac{14}{21}$	$\frac{16}{24}$	$\frac{18}{27}$	$\frac{20}{30}$
$\frac{1}{4}$	$\frac{2}{8}$	$\frac{3}{12}$	$\frac{4}{16}$	$\frac{5}{20}$	$\frac{6}{24}$	$\frac{7}{28}$	$\frac{8}{32}$	$\frac{9}{36}$	$\frac{10}{40}$
$\frac{3}{4}$	$\frac{6}{8}$	$\frac{9}{12}$	$\frac{12}{16}$	$\frac{15}{20}$	$\frac{18}{24}$	$\frac{21}{28}$	$\frac{24}{32}$	$\frac{27}{36}$	$\frac{30}{40}$
$\frac{1}{5}$	$\frac{2}{10}$	$\frac{3}{15}$	$\frac{4}{20}$	$\frac{5}{25}$	$\frac{6}{30}$	$\frac{7}{35}$	$\frac{8}{40}$	$\frac{9}{45}$	$\frac{10}{50}$
$\frac{2}{5}$	$\frac{4}{10}$	$\frac{6}{15}$	$\frac{8}{20}$	$\frac{10}{25}$	$\frac{12}{30}$	$\frac{14}{35}$	$\frac{16}{40}$	$\frac{18}{45}$	$\frac{20}{50}$
$\frac{3}{5}$	$\frac{6}{10}$	$\frac{9}{15}$	$\frac{12}{20}$	$\frac{15}{25}$	$\frac{18}{30}$	$\frac{21}{35}$	$\frac{24}{40}$	$\frac{27}{45}$	$\frac{30}{50}$
$\frac{4}{5}$	$\frac{8}{10}$	$\frac{12}{15}$	$\frac{16}{20}$	$\frac{20}{25}$	$\frac{24}{30}$	$\frac{28}{35}$	$\frac{32}{40}$	$\frac{36}{45}$	$\frac{40}{50}$
$\frac{1}{6}$	$\frac{2}{12}$	$\frac{3}{18}$	$\frac{4}{24}$	$\frac{5}{30}$	$\frac{6}{36}$	$\frac{7}{42}$	$\frac{8}{48}$	$\frac{9}{54}$	$\frac{10}{60}$
$\frac{5}{6}$	$\frac{10}{12}$	$\frac{15}{18}$	$\frac{20}{24}$	$\frac{25}{30}$	$\frac{30}{36}$	$\frac{35}{42}$	$\frac{40}{48}$	$\frac{45}{54}$	$\frac{50}{60}$
$\frac{1}{8}$	$\frac{2}{16}$	$\frac{3}{24}$	$\frac{4}{32}$	$\frac{5}{40}$	$\frac{6}{48}$	$\frac{7}{56}$	$\frac{8}{64}$	$\frac{9}{72}$	$\frac{10}{80}$
$\frac{3}{8}$	$\frac{6}{16}$	$\frac{9}{24}$	$\frac{12}{32}$	$\frac{15}{40}$	$\frac{18}{48}$	$\frac{21}{56}$	$\frac{24}{64}$	$\frac{27}{72}$	$\frac{30}{80}$
$\frac{5}{8}$	$\frac{10}{16}$	$\frac{15}{24}$	$\frac{20}{32}$	$\frac{25}{40}$	$\frac{30}{48}$	$\frac{35}{56}$	$\frac{40}{64}$	$\frac{45}{72}$	$\frac{50}{80}$
$\frac{7}{8}$	$\frac{14}{16}$	$\frac{21}{24}$	$\frac{28}{32}$	$\frac{35}{40}$	$\frac{42}{48}$	$\frac{49}{56}$	$\frac{56}{64}$	$\frac{63}{72}$	$\frac{70}{80}$
$\frac{1}{12}$	$\frac{2}{24}$	$\frac{3}{36}$	$\frac{4}{48}$	$\frac{5}{60}$	$\frac{6}{72}$	$\frac{7}{84}$	$\frac{8}{96}$	$\frac{9}{108}$	$\frac{10}{120}$
$\frac{5}{12}$	$\frac{10}{24}$	$\frac{15}{36}$	$\frac{20}{48}$	$\frac{25}{60}$	$\frac{30}{72}$	$\frac{35}{84}$	$\frac{40}{96}$	$\frac{45}{108}$	$\frac{50}{120}$
$\frac{7}{12}$	$\frac{14}{24}$	$\frac{21}{36}$	$\frac{28}{48}$	$\frac{35}{60}$	$\frac{42}{72}$	$\frac{49}{84}$	$\frac{56}{96}$	$\frac{63}{108}$	$\frac{70}{120}$
$\frac{11}{12}$	$\frac{22}{24}$	$\frac{33}{36}$	$\frac{44}{48}$	$\frac{55}{60}$	$\frac{66}{72}$	$\frac{77}{84}$	$\frac{88}{96}$	$\frac{99}{108}$	$\frac{110}{120}$

Note Every fraction in the first column is in *simplest form.* This means that there is no equivalent fraction with a smaller numerator and smaller denominator.

Every fraction is either in simplest form or is equivalent to a fraction in simplest form.

Lowest terms means the same as *simplest form.*

Check Your Understanding

1. True or false?

a. $\frac{1}{3} = \frac{5}{15}$ **b.** $\frac{4}{4} = \frac{8}{8}$ **c.** $\frac{6}{10} = \frac{15}{25}$ **d.** $\frac{5}{8} = \frac{40}{72}$

2. Use the table to list five fractions that are equivalent to $\frac{2}{3}$.

Check your answers in the Answer Key.

Renaming Fractions Greater Than One

When the numerator of a fraction is greater than the denominator, the fraction is greater than one. You can rename fractions greater than one as mixed numbers or as whole numbers.

Note When the numerator and denominator of a fraction are the same, the fraction is equal to 1. For example, the fraction $\frac{6}{6} = 1$.

Different mixed-number names for a fraction can be useful for different reasons. For example, the name with the greatest whole-number part, such as $4\frac{2}{3}$, is useful for comparing the mixed number to whole numbers or placing it on a number line. A name with a fraction part that is greater than one, such as $3\frac{5}{3}$, can be useful when you are subtracting mixed numbers.

One way to rename a fraction greater than one is to use **unit fractions** to make wholes.

Example

Rename $\frac{19}{4}$ as a mixed number.

Show nineteen-fourths of a circle.

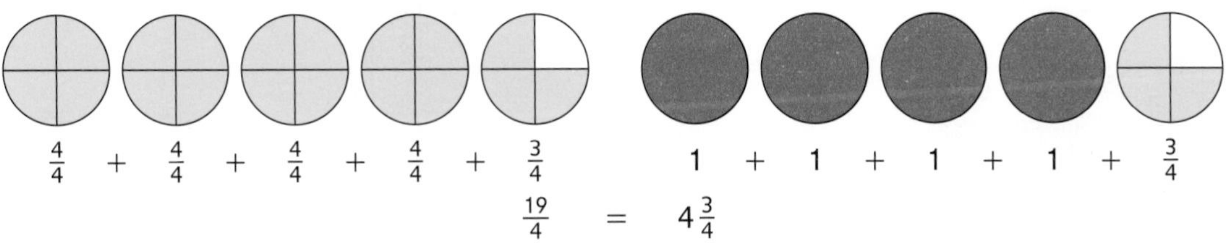

$\frac{4}{4} + \frac{4}{4} + \frac{4}{4} + \frac{4}{4} + \frac{3}{4}$ $1 + 1 + 1 + 1 + \frac{3}{4}$

$\frac{19}{4} = 4\frac{3}{4}$

- Trade a group of 4 fourths $\left(\frac{4}{4}\right)$ for 1 whole. Now you have 1 whole and 15 fourths. $1\frac{15}{4}$ is one mixed-number name for $\frac{19}{4}$.
- Trade another group of 4 fourths for 1 whole to get the name $2\frac{11}{4}$.
- Trade more groups to get $3\frac{7}{4}$ and $4\frac{3}{4}$.

$\frac{19}{4}$ can have different mixed-number names: $\frac{19}{4} = 1\frac{15}{4} = 2\frac{11}{4} = 3\frac{7}{4} = 4\frac{3}{4}$

Another way to rename a fraction as a mixed number is to divide the numerator by the denominator. The remainder in the division problem can be rewritten as a fraction. The remainder is the numerator, and the divisor is the denominator. For example, to rename $\frac{37}{5}$ as a mixed number divide 37 by 5: $37 \div 5 \longrightarrow 7$ R2, or $7\frac{2}{5}$.

Example

Rename $\frac{19}{4}$ as a mixed number.

Think about $\frac{19}{4}$ as dividing 19 by 4. $19 \div 4 \longrightarrow 4$ R3

- The quotient, 4, is the whole-number part of the mixed number. It tells the number of wholes in $\frac{19}{4}$.
- The remainder, 3, is the numerator of the fraction part of the mixed number. It tells how many fourths are left.

$\frac{19}{4} = 4\frac{3}{4}$

Renaming Mixed Numbers

You can rename mixed numbers as fractions greater than one.

Note You can think of making wholes and breaking apart wholes as fair trades. Putting 4 fourths together to make a whole is like trading $\frac{4}{4}$ for 1. This is a fair trade since $\frac{4}{4} = 1$. Breaking apart a whole into 5 fifths is like trading 1 for $\frac{5}{5}$. This is a fair trade since $1 = \frac{5}{5}$.

Example

Rename $3\frac{1}{2}$ as a fraction.

One way: Think about breaking apart the wholes.

If a circle is the whole, then $3\frac{1}{2}$ is three whole circles and $\frac{1}{2}$ of another circle.

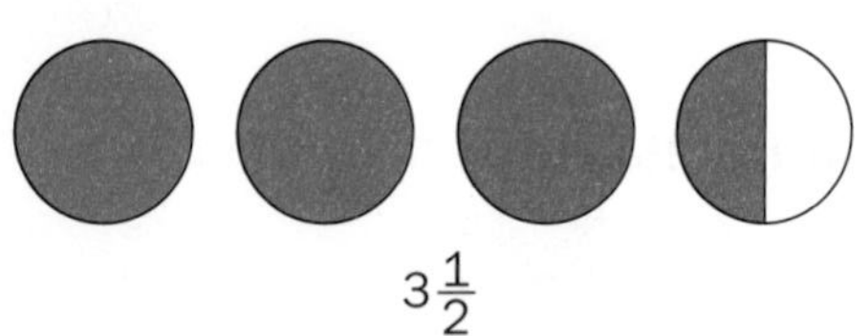

$3\frac{1}{2}$

If you break apart each of the whole circles into halves, then you can see that $3\frac{1}{2} = \frac{7}{2}$.

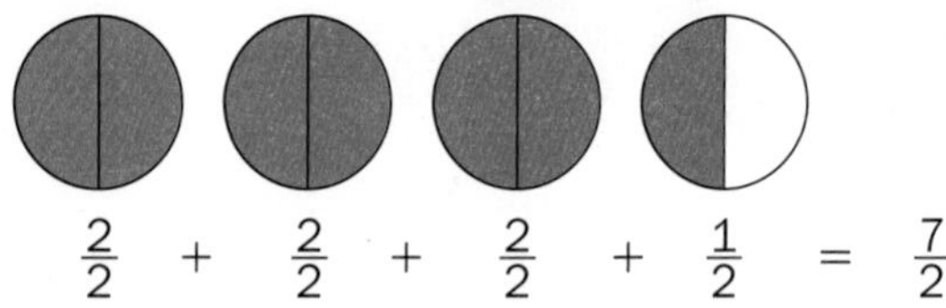

$\frac{2}{2} + \frac{2}{2} + \frac{2}{2} + \frac{1}{2} = \frac{7}{2}$

Another way: Think about a number line.

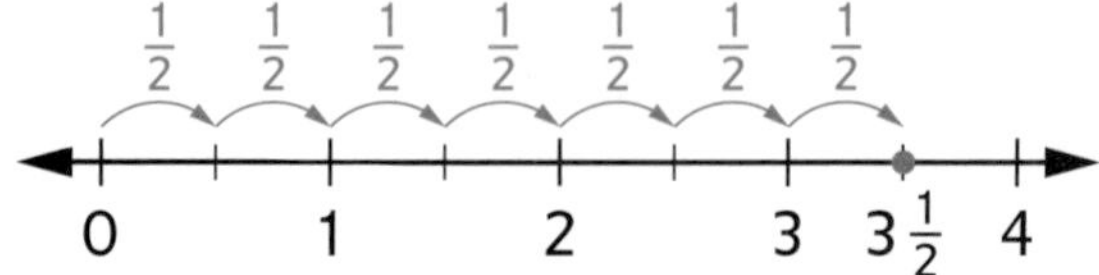

Count the halves. You count seven halves from 0 to $3\frac{1}{2}$, so $3\frac{1}{2} = \frac{7}{2}$.

Another way: Rename the whole number as a fraction with the same denominator as the fraction part, and add the numerators.

$3\frac{1}{2} = 3 + \frac{1}{2}$

Rename 3 as $\frac{6}{2}$.

So, $3\frac{1}{2} = \frac{6}{2} + \frac{1}{2} = \frac{7}{2}$.

You can also rename many mixed numbers as other mixed numbers that have the same denominator.

Example

Write as many equivalent names as you can for $3\frac{2}{5}$ using the denominator 5.

Think about breaking apart each whole into fifths, one at a time.

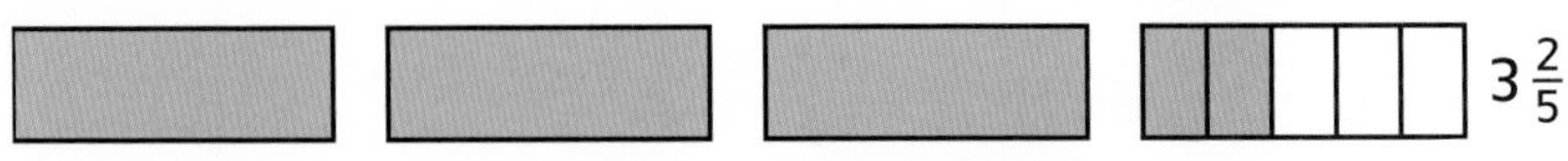

$3\frac{2}{5}$

$2\frac{7}{5}$

$1\frac{12}{5}$

$\frac{17}{5}$

$3\frac{2}{5} = 2\frac{7}{5} = 1\frac{12}{5} = \frac{17}{5}$

Check Your Understanding

Write each fraction as a mixed number.

1. $\frac{51}{4}$ **2.** $\frac{26}{3}$ **3.** $\frac{34}{5}$ **4.** $\frac{60}{16}$

Write each mixed number as a fraction.

5. $4\frac{3}{4}$ **6.** $3\frac{2}{3}$ **7.** $4\frac{5}{6}$ **8.** $1\frac{4}{3}$

9. Write as many equivalent names as you can for $6\frac{2}{3}$ using 3 as the denominator.

Check your answers in the Answer Key.

Comparing Fractions

You can compare fractions when they name parts of the same whole.
You can use fraction representations to compare fractions.

Example

Compare $\frac{5}{8}$ and $\frac{3}{4}$.

Find $\frac{5}{8}$ and $\frac{3}{4}$ on the Fraction Number Lines Poster.

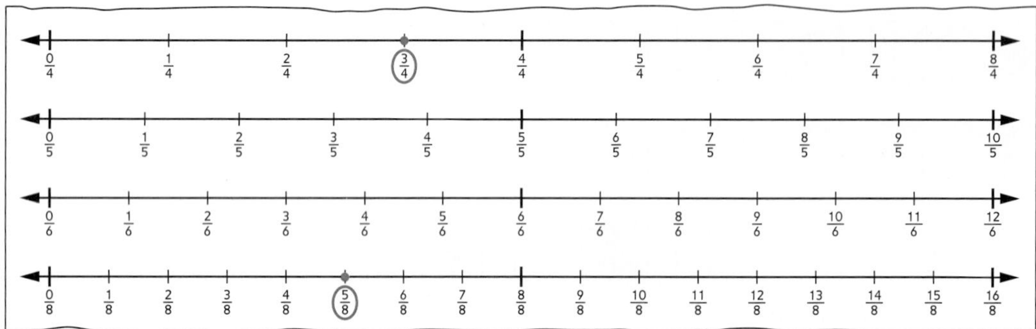

The whole is the same distance from 0 on both number lines.
$\frac{3}{4}$ is to the right of $\frac{5}{8}$, so $\frac{3}{4}$ is a greater distance from 0 than $\frac{5}{8}$.

Write $\frac{3}{4} > \frac{5}{8}$ or $\frac{5}{8} < \frac{3}{4}$.

Example

Which is greater, $\frac{2}{3}$ or $\frac{7}{8}$?

You can use the red fraction circle as the whole.
Show $\frac{2}{3}$ and $\frac{7}{8}$ with fraction circle pieces.
Two-thirds covers less space than seven-eighths.

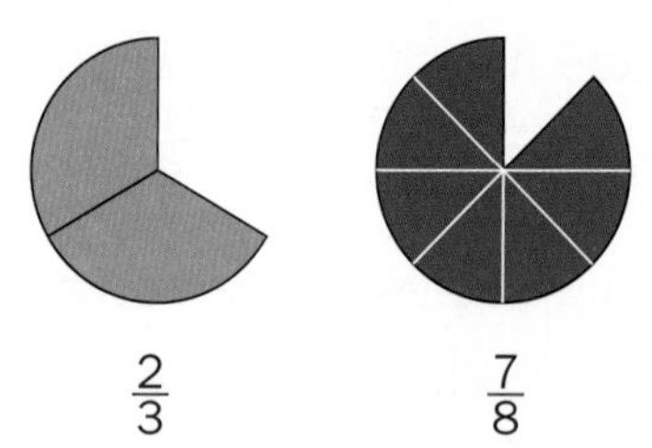

$\frac{2}{3}$ $\frac{7}{8}$

$\frac{2}{3}$ is less than $\frac{7}{8}$, so $\frac{7}{8}$ is greater than $\frac{2}{3}$.

Example

Kiara and Tania each had a small pizza for lunch. Kiara ate $\frac{5}{6}$ of her pizza. Tania ate $\frac{3}{4}$ of her pizza. Who ate more pizza?

You can model the pizza that each girl ate using fraction circle pieces.
You can use the red fraction circle to represent the whole pizza.
The fraction circle pieces show the amount each girl ate.
The empty parts of the circles show the amount that each girl didn't eat.
The piece of pizza that Tania *did not eat* is larger than the piece of pizza that Kiara *did not eat*. That means the amount that Kiara ate is closer to a whole pizza.

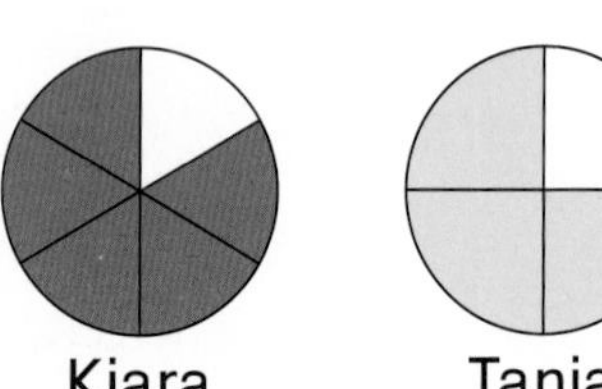

Kiara Tania

So Kiara ate more pizza than Tania.

When you compare fractions, you have to pay attention to both the numerator and the denominator.

Like Denominators

You can compare fractions that have the same denominator by comparing the numerators. To decide whether $\frac{7}{8}$ or $\frac{5}{8}$ is larger, think of them as 7 eighths and 5 eighths. Because the eighths are the same size, 7 eighths is more than 5 eighths. The fraction with the larger numerator is the larger number.

<	is less than
>	is greater than
=	is equal to

Note Fractions with **like denominators** have the same denominator.

$\frac{1}{4}$ and $\frac{3}{4}$ have like denominators.

Fractions with **like numerators** have the same numerator.

$\frac{2}{3}$ and $\frac{2}{5}$ have like numerators.

$\frac{5}{8}$ $\frac{7}{8}$

$\frac{5}{8} < \frac{7}{8}$ or $\frac{7}{8} > \frac{5}{8}$

Examples

$\frac{4}{5} > \frac{3}{5}$ because $4 > 3$ and the denominators are both fifths.

$\frac{2}{9} < \frac{7}{9}$ because $2 < 7$ and the denominators are both ninths.

Like Numerators

If the numerators of two fractions are the same, then the fraction with the smaller denominator is the larger number. Remember, a smaller denominator means the whole has fewer parts, so each part is bigger. For example, $\frac{3}{5} > \frac{3}{8}$ because fifths are bigger than eighths, so 3 fifths is more than 3 eighths.

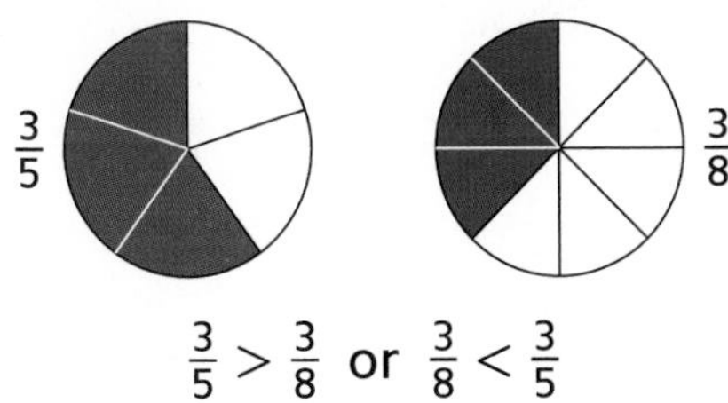

$\frac{3}{5} > \frac{3}{8}$ or $\frac{3}{8} < \frac{3}{5}$

Examples

$\frac{1}{2} > \frac{1}{3}$ because halves are bigger than thirds of the same size whole.

$\frac{3}{8} < \frac{3}{4}$ because eighths are smaller than fourths of the same size whole.

Unlike Numerators and Unlike Denominators

You can use several strategies to compare fractions when both the numerators and the denominators are different.

Comparing to Benchmarks. You can compare fractions by comparing each fraction to a **benchmark,** or familiar reference point. Numbers such as 0, $\frac{1}{2}$, 1, $1\frac{1}{2}$, and 2 are often used as benchmarks because they are easy to visualize and to compare to other fractions.

Examples

Compare $\frac{2}{5}$ and $\frac{5}{8}$.

Think: How does each fraction compare to $\frac{1}{2}$?

Notice that $\frac{5}{8}$ is more than $\frac{1}{2}$ and $\frac{2}{5}$ is less than $\frac{1}{2}$. So, $\frac{2}{5} < \frac{5}{8}$ and $\frac{5}{8} > \frac{2}{5}$.

Compare $\frac{7}{8}$ and $\frac{3}{4}$.

Think: Both fractions are less than 1. Which fraction is closer to 1?

$\frac{7}{8}$ is $\frac{1}{8}$ away from 1, and $\frac{3}{4}$ is $\frac{1}{4}$ away from 1. Since eighths are smaller than fourths, $\frac{7}{8}$ is closer to 1.

So, $\frac{7}{8} > \frac{3}{4}$ and $\frac{3}{4} < \frac{7}{8}$.

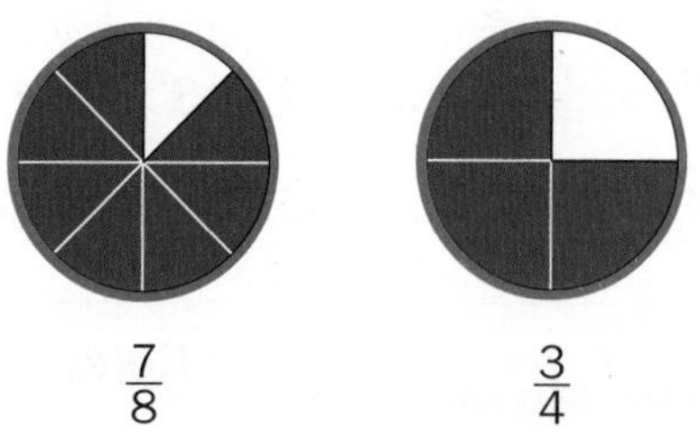

$\frac{7}{8}$ $\frac{3}{4}$

Using Equivalent Fractions. One way to compare fractions that always works is to find equivalent fractions that have the same denominator.

Example

Compare $\frac{5}{8}$ and $\frac{3}{5}$.

Look at the table on page 170. Both fifths and eighths can be written as *fortieths*:

$\frac{5}{8} = \frac{25}{40}$ and $\frac{3}{5} = \frac{24}{40}$. Since $\frac{25}{40} > \frac{24}{40}$, you know that $\frac{5}{8} > \frac{3}{5}$.

Using Decimal Equivalents. Using decimal equivalents is another way to compare fractions that always works. Think of the fractions as division problems: $\frac{a}{b} = a \div b$.

Example

Compare $\frac{2}{5}$ and $\frac{3}{8}$. (A calculator is an appropriate tool for this problem.)

Change both fractions to decimals: $\frac{2}{5} = 2 \div 5 = 0.4$ $\quad$ $\frac{3}{8} = 3 \div 8 = 0.375$

0.4 is the same as 0.400, and 0.400 > 0.375. Since 0.4 > 0.375, you know that $\frac{2}{5} > \frac{3}{8}$.

Common Denominators

To solve problems with unlike denominators, it sometimes helps to rename the fractions with the same denominator, or a **common denominator.** Many common denominators exist for any set of fractions. There are several ways to find possible common denominators.

Example

Rename $\frac{3}{4}$ and $\frac{1}{6}$ with a common denominator.

Strategy 1: List Equivalent Fractions

List equivalent fractions for $\frac{3}{4}$ and $\frac{1}{6}$. Look for denominators that appear in both lists.

$$\frac{3}{4} = \frac{6}{8} = \mathbf{\frac{9}{12}} = \frac{12}{16} = \frac{15}{20} = \mathbf{\frac{18}{24}} = \frac{21}{28} = \frac{24}{32} = \mathbf{\frac{27}{36}} = \ldots$$

$$\frac{1}{6} = \mathbf{\frac{2}{12}} = \frac{3}{18} = \mathbf{\frac{4}{24}} = \frac{5}{30} = \mathbf{\frac{6}{36}} = \frac{7}{42} = \ldots$$

You can rename $\frac{3}{4}$ and $\frac{1}{6}$ with a denominator of 12, so 12 is a possible common denominator.

$\frac{3}{4} = \frac{9}{12}$ and $\frac{1}{6} = \frac{2}{12}$

Strategy 2: List Multiples of the Denominators

List multiples of each denominator. Look for multiples common to both denominators.

Multiples of 4: 4, 8, **12**, 16, 20, **24**, . . .
Multiples of 6: 6, **12**, 18, **24**, 30, 36, . . .

12 and 24 are the first two common multiples for 4 and 6. Since 12 is the smallest, 12 is called the **least common multiple.**

Rename the fractions so that their denominator is a common multiple.

Think: What number times 4 will give 12 as a denominator? $4 * 3 = 12$

So multiply the numerator and denominator of $\frac{3}{4}$ by 3 to find an equivalent fraction. $\frac{3}{4} = \frac{(3 * 3)}{(4 * 3)} = \frac{9}{12}$

Think: What number times 6 will give 12 as a denominator? $6 * 2 = 12$

So multiply the numerator and denominator of $\frac{1}{6}$ by 2. $\frac{1}{6} = \frac{(1 * 2)}{(6 * 2)} = \frac{2}{12}$

Strategy 3: Multiply the Denominators

Multiply the two denominators and use the product as a common denominator. Use the equivalent fractions rule to rename each fraction with the common denominator. This strategy gives what *Everyday Mathematics* calls the **quick common denominator.**

$4 * 6 = 24$, so use 24 as a common denominator.

$\frac{3}{4} = \frac{3 * 6}{4 * 6} = \frac{18}{24}$ $\frac{1}{6} = \frac{1 * 4}{6 * 4} = \frac{4}{24}$

Solving Problems with Fractions

You can use fraction circles, fraction strips, number lines, or pictures to model and solve problems with fractions. Problems can often be modeled in different ways. Use models that make sense to you.

Example

Five small pizzas are shared equally among 4 people. How much pizza does each person get?

One way:

Show 5 pizzas using 5 red fraction circles.

Trade fraction circle pieces to divide each pizza into 4 equal parts.

Each person gets $\frac{1}{4}$ of each pizza. There are 5 pizzas, so each person gets $\frac{1}{4} + \frac{1}{4} + \frac{1}{4} + \frac{1}{4} + \frac{1}{4}$, which is the same as $5 * \frac{1}{4}$, or $\frac{5}{4}$ pizzas.

Since $\frac{4}{4}$ makes 1 whole pizza, $\frac{5}{4}$ is the same as 1 whole with $\frac{1}{4}$ left over, or $1\frac{1}{4}$ pizzas.

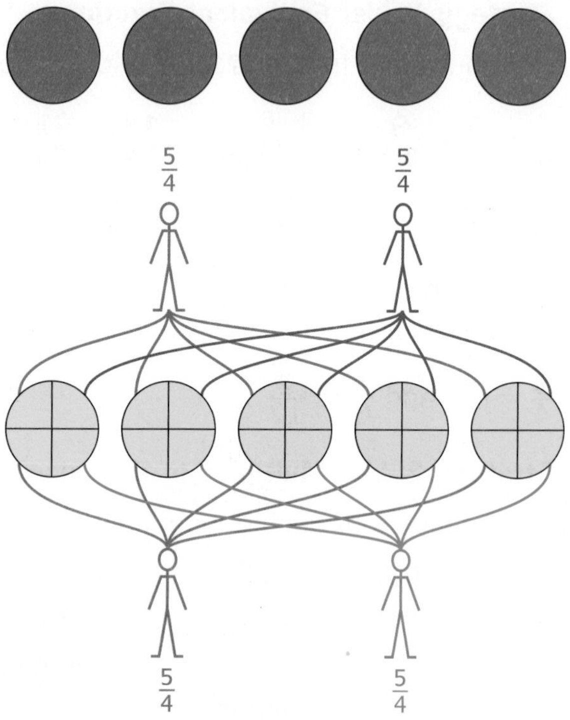

$\frac{5}{4}$ $\frac{5}{4}$

$\frac{5}{4}$ $\frac{5}{4}$

Another way:

Show 5 pizzas using 5 red fraction circles, as above.

Imagine giving each person 1 whole pizza.

Trade fraction circle pieces to divide the last pizza into 4 equal parts. Each person gets another $\frac{1}{4}$ pizza.

All together, each person gets 1 and $\frac{1}{4}$ pizza. This is the same as $1\frac{1}{4}$ pizzas.

Both models show that each person gets $1\frac{1}{4}$ pizzas.

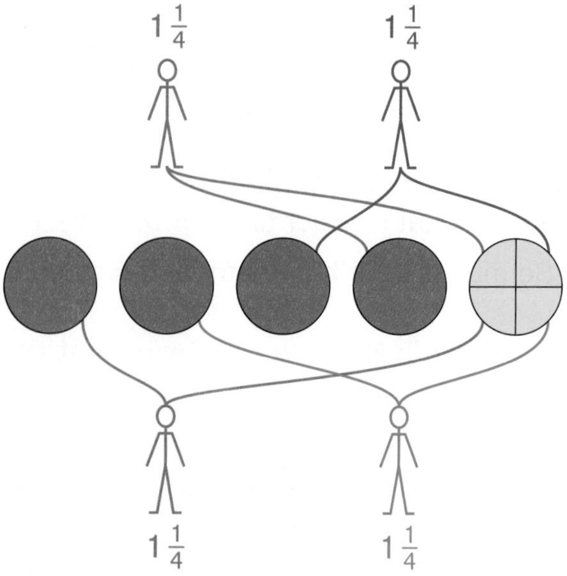

$1\frac{1}{4}$ $1\frac{1}{4}$

$1\frac{1}{4}$ $1\frac{1}{4}$

Example

Derrick has 3 gallons of paint. He uses $\frac{1}{2}$ gallon to paint one wall of his bedroom. How many gallons of paint does Derrick have left?

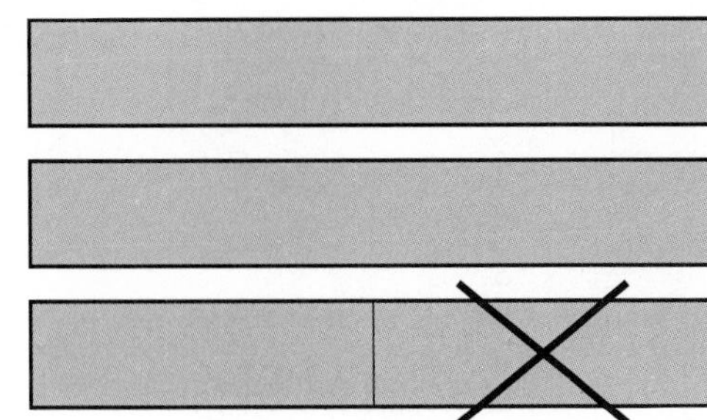

You can draw 3 rectangles to represent 3 gallons of paint.

Cross out $\frac{1}{2}$ of one rectangle to show that Derrick used $\frac{1}{2}$ gallon of paint.

Think: How much is left?

There are still 2 whole rectangles and $\frac{1}{2}$ rectangle remaining.

Derrick has $2\frac{1}{2}$ gallons of paint left.

Example

Melinda ran on a track that is $\frac{1}{8}$ mile long. She ran 5 laps, or $\frac{5}{8}$ mile, and walked 4 laps, or $\frac{4}{8}$ mile. How far did she go all together?

You can use a number line to represent the distance Melinda ran and walked. Make a number line that goes from 0 to 2. Label the whole numbers. Make tick marks for eighths between the wholes.

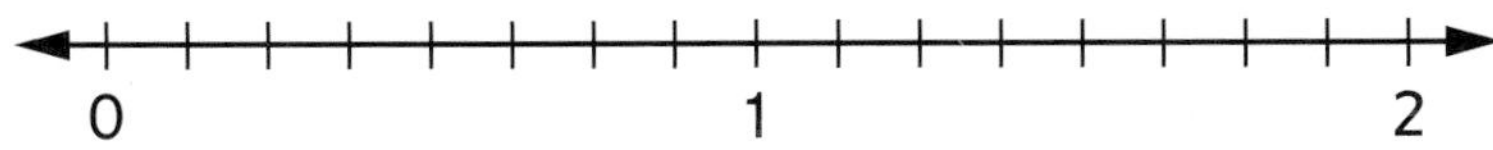

Travel 5 eighths from 0 to stop at $\frac{5}{8}$. This is how far Melinda ran.

Then travel 4 more eighths from $\frac{5}{8}$. This is how far Melinda walked.

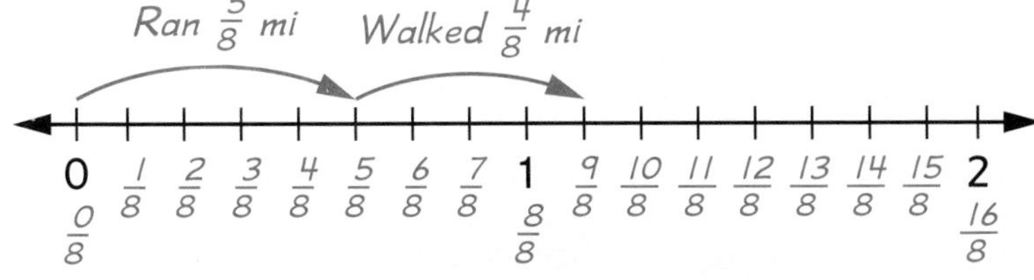

The endpoint is at $\frac{9}{8}$, so Melinda traveled a total of $\frac{9}{8}$ miles.

$\frac{8}{8}$ makes 1 whole mile, so $\frac{9}{8}$ miles is the same as 1 mile with $\frac{1}{8}$ mile left over. Melinda traveled a total of $1\frac{1}{8}$ miles.

Example

Ryan has 4 boxes of books. All of the boxes are the same size, and each box is only $\frac{2}{3}$ full. If Ryan combines the partially filled boxes, how many boxes of books will he have?

You can draw a picture to show 4 boxes, each $\frac{2}{3}$ full.

Count the number of thirds that are shaded.

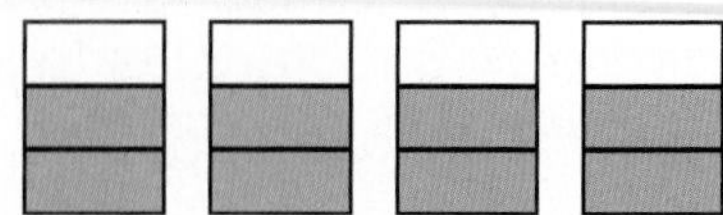

There are 4 groups of 2 thirds. $4 * \frac{2}{3}$ makes 8 thirds $\left(\frac{8}{3}\right)$ all together.

Think: How many wholes are in $\frac{8}{3}$?

Draw another picture to group 8 thirds into wholes.

$\frac{8}{3}$ is the same as $\frac{3}{3} + \frac{3}{3} + \frac{2}{3}$, or 2 wholes and $\frac{2}{3}$.

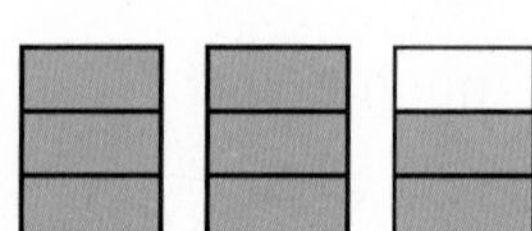

Ryan will have $2\frac{2}{3}$ boxes of books.

Example

Garrett has $4\frac{1}{2}$ feet of yarn for an art project. If he cuts the yarn into $\frac{1}{2}$-foot pieces, how many pieces will he have?

Use a number line to show $4\frac{1}{2}$ feet of yarn. Draw a number line that goes to at least 5 wholes. Make tick marks to show each $\frac{1}{2}$-foot interval.

Locate $4\frac{1}{2}$ on the number line.

Count the number of $\frac{1}{2}$-foot lengths starting from 0 and ending at $4\frac{1}{2}$. There are 9.

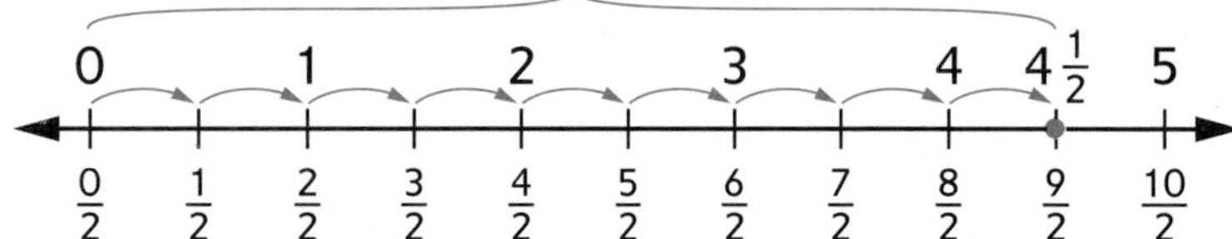

Garrett will have nine $\frac{1}{2}$-foot pieces of yarn. This makes sense because 9 halves, or $\frac{9}{2}$, is equal to $4\frac{1}{2}$.

Check Your Understanding

Solve the problems below. You can use fraction circles, fraction strips, number lines, or drawings to model the problems. Show your thinking.

1. Two melons are shared equally among 8 people. What fraction of a whole melon does each person get?
2. Mara has a stack of 7 books. Each book is $\frac{1}{2}$ inch thick. How tall is the stack?

Check your answers in the Answer Key.

Estimating with Fractions

You can use estimates to help make sense of a situation, approximate a calculation, and check that answers are reasonable.

Using Visual Representations

One way to estimate with fractions is to think about visual representations, such as fraction circles, fraction strips, or number lines.

Example

Estimate: Will the sum of $\frac{4}{5} + \frac{1}{10}$ be greater than or less than 1?

Use the red circle as the whole.

Picture each amount with fraction circles.

Picture $\frac{4}{5}$ as 4 dark green pieces. $\frac{4}{5}$ is almost 1 whole.

Picture $\frac{1}{10}$ as 1 purple piece. $\frac{1}{10}$ is a small sliver.

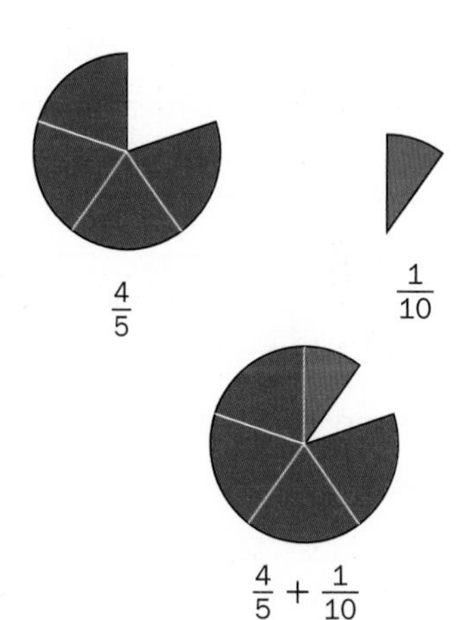

Adding the small $\frac{1}{10}$ sliver to $\frac{4}{5}$ would not complete the whole circle, so the sum is less than 1.

Example

Estimate: Will the result of $2\frac{1}{4} - \frac{5}{6}$ be greater than or less than 1?

Visualize the starting amount, $2\frac{1}{4}$.

Think about the amount being subtracted.

$\frac{5}{6}$ is a little less than 1 whole, so about 1 whole will be subtracted.

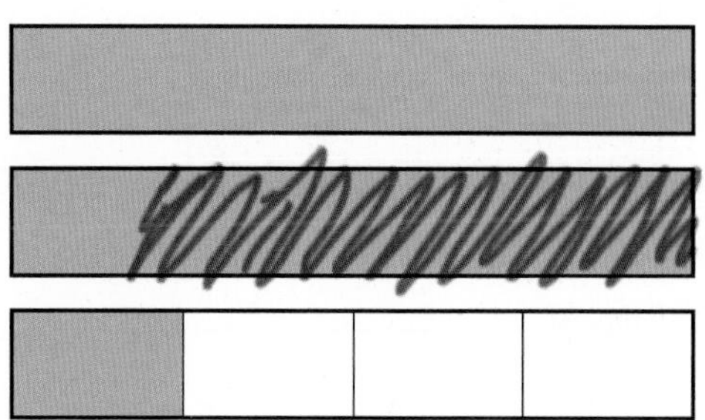

At least 1 whole and the $\frac{1}{4}$ piece remain, so $2\frac{1}{4} - \frac{5}{6}$ will be greater than 1.

Using Benchmarks to Estimate

A **benchmark** is a familiar reference point. Numbers such as 0, $\frac{1}{2}$, 1, $1\frac{1}{2}$, and 2 are often used as benchmarks because they are easy to visualize and to compare with other fractions. You can use benchmarks as close-but-easier numbers to make estimates for fraction sums and differences. Use or picture fraction tools in your head to choose benchmarks for one or more numbers in your problem.

Example

Make an estimate for $\frac{11}{12} + 1\frac{5}{8}$.

You can picture $\frac{11}{12}$ on a number line.

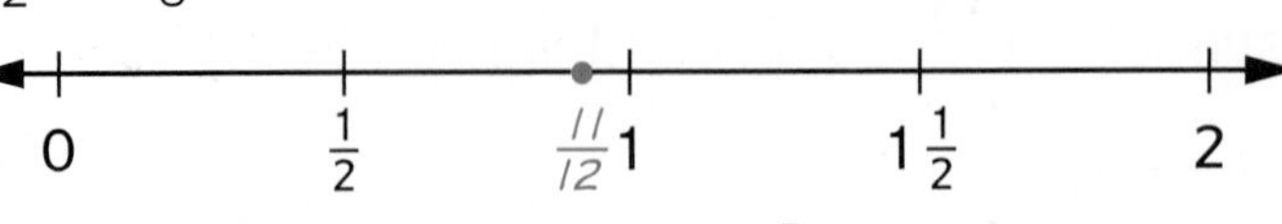

$\frac{11}{12}$ is close to 1. You can add $1 + 1\frac{5}{8}$ mentally. The sum is $2\frac{5}{8}$.

So, $\frac{11}{12} + 1\frac{5}{8}$ is a little less than $2\frac{5}{8}$, or close to $2\frac{1}{2}$.

Example

Make an estimate for $1\frac{1}{3} - \frac{3}{5}$.

You can picture fraction circles in your head.

$1\frac{1}{3}$ is close to $1\frac{1}{2}$.

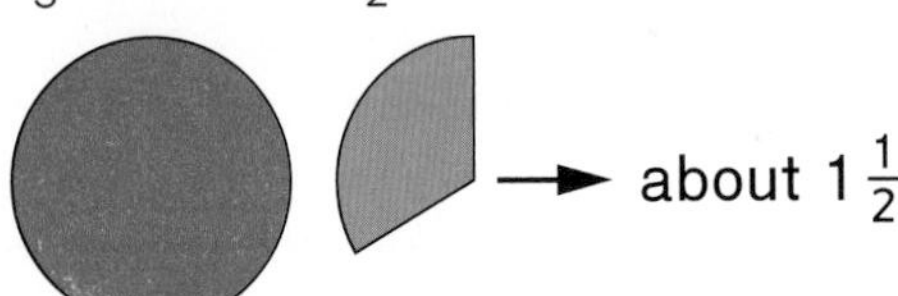

$\frac{3}{5}$ is close to $\frac{1}{2}$.

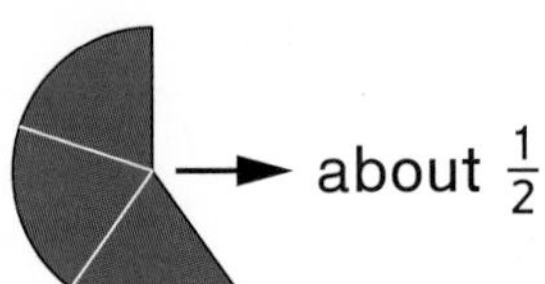

Subtract the close-but-easier numbers: $1\frac{1}{2} - \frac{1}{2} = 1$.

So, $1\frac{1}{3} - \frac{3}{5}$ is about 1.

When estimating sums and differences of mixed numbers, it is sometimes helpful to work with wholes and fractions separately.

Example

Estimate the sum of $2\frac{6}{10} + 3\frac{7}{8}$.

Add the wholes first.	$2 + 3 = 5$
Estimate the sum of the fractions using benchmarks.	$\frac{6}{10}$ is close to $\frac{1}{2}$. $\frac{7}{8}$ is close to 1. $\frac{1}{2} + 1 = 1\frac{1}{2}$
Combine the sum of the wholes with your estimate for the sum of the fractions.	$5 + 1\frac{1}{2} = 6\frac{1}{2}$

$2\frac{6}{10} + 3\frac{7}{8}$ is about $6\frac{1}{2}$.

Using Rounding to Estimate

You can estimate sums and differences of fractions by rounding one or more of the numbers to a whole number. This method can be helpful when you work with mixed numbers.

To round a fraction or mixed number to the nearest whole number, think: *What two whole numbers is this number between?* For example, $3\frac{1}{5}$ is between 3 and 4. Next look at the fraction part of the number you are rounding to decide which whole number is closer.

- If the fraction is greater than $\frac{1}{2}$, the number being rounded is closer to the higher whole number, so round up.
- If the fraction is less than $\frac{1}{2}$, the number being rounded is closer to the lower whole number, so round down.
- If the fraction is exactly $\frac{1}{2}$, use the problem situation to decide whether to round up or down. If there is nothing in the problem to help decide, many people use a rule to always round up.

Example

Estimate the sum of $14\frac{4}{5} + 15\frac{1}{3}$.

Round $14\frac{4}{5}$ to the nearest whole number.

$14\frac{4}{5}$ is between 14 and 15, and $\frac{4}{5}$ is greater than $\frac{1}{2}$.
So, $14\frac{4}{5}$ rounded to the nearest whole number is 15.

Next round $15\frac{1}{3}$ to the nearest whole number.

$15\frac{1}{3}$ is between 15 and 16, and $\frac{1}{3}$ is less than $\frac{1}{2}$.
So, $15\frac{1}{3}$ rounded to the nearest whole number is 15.

Add the rounded numbers: $15 + 15 = 30$. The sum of $14\frac{4}{5} + 15\frac{1}{3}$ is about 30.

When working with large mixed numbers, it often makes sense to ignore the fraction parts and round the whole-number parts to close-but-easier whole numbers.

Example

Make an estimate for $48\frac{4}{5} - 23\frac{9}{10}$.

Ignore the fractions and estimate with the whole numbers.

One way: Round 48 to 50 and 23 to 20. Estimate: $50 - 20 = 30$

Another way: Round 48 to 50 and 23 to 25. Estimate: $50 - 25 = 25$

Both 30 and 25 are reasonable estimates for $48\frac{4}{5} - 23\frac{9}{10}$.

You can combine rounding with other strategies.

Example

Estimate the sum of $2\frac{8}{9} + 1\frac{3}{5}$.

One way: Round each addend to the nearest whole number.

$2\frac{8}{9}$ is between 2 and 3, and $\frac{8}{9}$ is greater than $\frac{1}{2}$.

So, $2\frac{8}{9}$ rounded to the nearest whole number is 3.

$1\frac{3}{5}$ is between 1 and 2, and $\frac{3}{5}$ is greater than $\frac{1}{2}$.

So, $1\frac{3}{5}$ rounded to the nearest whole number is 2.

Add the rounded numbers: $3 + 2 = 5$.

The sum of $2\frac{8}{9} + 1\frac{3}{5}$ is about 5.

Another way: You can round one addend to the nearest whole number and use a benchmark for the other addend.

You know from above that $2\frac{8}{9}$ rounded to the nearest whole number is 3.

Next identify the benchmark that is closest to $1\frac{3}{5}$.

$1\frac{3}{5}$ is close to $1\frac{1}{2}$.

Add the estimated numbers: $3 + 1\frac{1}{2} = 4\frac{1}{2}$.

The sum of $2\frac{8}{9} + 1\frac{3}{5}$ is about $4\frac{1}{2}$.

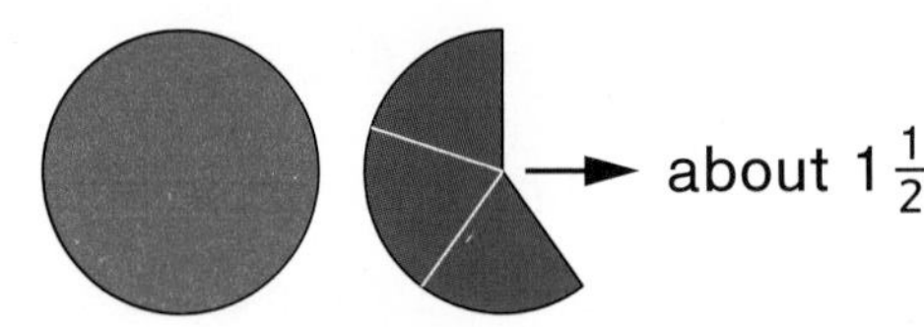

Another way: You can round one addend to the nearest whole number and if the mental math is easy, you do not need to change the other addend.

You know from above that $2\frac{8}{9}$ rounded to the nearest whole number is 3.

Adding $3 + 1\frac{3}{5}$ is easy to do mentally. $3 + 1\frac{3}{5} = 4\frac{3}{5}$

The sum of $2\frac{8}{9} + 1\frac{3}{5}$ is about $4\frac{3}{5}$.

NOTE: The sum has to be a little less than $4\frac{3}{5}$ because one addend was rounded up.

The three estimates for $2\frac{8}{9} + 1\frac{3}{5}$ are 5, $4\frac{1}{2}$, and $4\frac{3}{5}$.

All of the strategies produced estimates that are reasonably close to each other.

Check Your Understanding

Make estimates for the following problems. Use the method of your choice.

1. $2\frac{1}{4} + \frac{5}{6}$ **2.** $3\frac{7}{8} - 1\frac{2}{3}$ **3.** $14\frac{2}{5} - 3\frac{3}{4}$ **4.** $23\frac{3}{8} + 25\frac{1}{6}$

5. Explain your estimation strategy for Problem 1.

Check your answers in the Answer Key.

Using Estimation to Check Answers

When using estimation to check whether an answer is reasonable, it can help to compare the fractions in the original problem to the answer.

Example

Amani added $\frac{3}{8} + \frac{3}{4}$ and got $\frac{6}{12}$ as the sum. Does her answer make sense?

Compare the addends, $\frac{3}{8}$ and $\frac{3}{4}$, with the sum, $\frac{6}{12}$.

One way: Represent the problem by showing each of the addends with fraction circle pieces. Compare the total amount to Amani's answer.

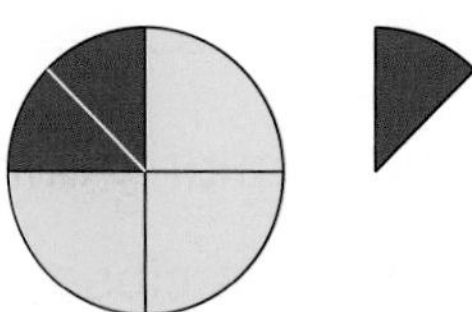

$\frac{3}{8} + \frac{3}{4}$ covers more than the whole.

When $\frac{3}{8}$ and $\frac{3}{4}$ are combined, the sum covers more than one whole circle. Amani's answer of $\frac{6}{12}$ is less than 1 whole. Her answer does not make sense.

Another way: Reason about the numbers.

The numerator (6) in Amani's answer is half of the denominator (12), so Amani's answer is equivalent to $\frac{1}{2}$. Adding $\frac{3}{8}$ to $\frac{3}{4}$ should make a sum greater than $\frac{3}{4}$. Since $\frac{1}{2}$ is less than the addend $\frac{3}{4}$, Amani's answer doesn't make sense.

Both strategies show that $\frac{6}{12}$ does not make sense as the sum of $\frac{3}{8} + \frac{3}{4}$.

You can also check whether an answer is reasonable by making an estimate and then comparing the answer you are checking to your estimate.

Example

Richa solved $1\frac{7}{8} - \frac{2}{5}$ and got $1\frac{5}{3}$. Is $1\frac{5}{3}$ a reasonable answer?

Make an estimate.	$1\frac{7}{8}$ is close to 2, and $\frac{2}{5}$ is close to $\frac{1}{2}$. $2 - \frac{1}{2} = 1\frac{1}{2}$ So the difference of $1\frac{7}{8} - \frac{2}{5}$ will be close to $1\frac{1}{2}$.
Compare your estimate to the answer you are checking.	The estimate is less than 2, but Richa's answer, $1\frac{5}{3}$, is greater than 2. When compared to the estimate, the answer $1\frac{5}{3}$ does not make sense.

$1\frac{5}{3}$ is not a reasonable answer for $1\frac{7}{8} - \frac{2}{5}$. Richa should try the problem again.

Check Your Understanding

1. Kira found that $2\frac{1}{8} - \frac{4}{5} = \frac{3}{4}$. Does her answer make sense? How do you know?

2. Brandon found that $3\frac{1}{2} + \frac{2}{3} = 3\frac{3}{5}$. Does his answer make sense? How do you know?

Check your answers in the Answer Key.

Adding and Subtracting Fractions with Like Denominators

When adding or subtracting fractions, it often helps to represent the problem with a visual model such as fraction circles, a number line, or a picture. Ask: *Are all of the pieces the same size?* If all of the pieces are the same size, or the fractions have **like denominators,** then you can just add or subtract the number of pieces, which is indicated by the numerators.

Example

Find the sum of $\frac{3}{5}$ and $\frac{4}{5}$.

Estimate: $\frac{3}{5}$ is close to $\frac{1}{2}$, and $\frac{4}{5}$ is close to 1. $\frac{1}{2} + 1 = 1\frac{1}{2}$ The sum should be close to $1\frac{1}{2}$.

Represent the problem with fraction circles.

Think: Are all of the pieces the same size?

Yes, all of the pieces are fifths.

Add to find the number of pieces: 3 fifths + 4 fifths = 7 fifths.

Name the result: 7 fifths = $\frac{7}{5}$.

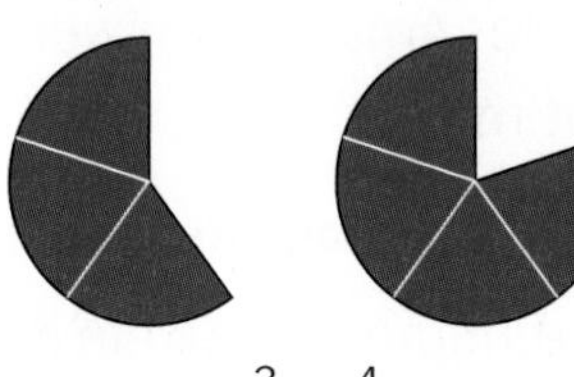

$\frac{3}{5} + \frac{4}{5}$

Since $\frac{5}{5}$ makes 1 whole, $\frac{7}{5}$ can also be written as $1\frac{2}{5}$.

So, $\frac{3}{5} + \frac{4}{5} = \frac{7}{5}$, or $1\frac{2}{5}$. This makes sense because $1\frac{2}{5}$ is close to the estimate of $1\frac{1}{2}$.

Example

Kendrick has $\frac{5}{8}$ pound of clay. He uses $\frac{2}{8}$ pound for a school project. How many pounds of clay does Kendrick have left?

Estimate: $\frac{5}{8}$ is close to $\frac{1}{2}$. $\frac{2}{8}$ is between 0 and $\frac{1}{2}$. $\frac{5}{8} - \frac{2}{8}$ will result in a number less than $\frac{1}{2}$.

You can represent the problem on a number line.

Kendrick starts with $\frac{5}{8}$ pound of clay. Subtract $\frac{2}{8}$ pound to show the amount of clay he used.

Kendrick has $\frac{3}{8}$ pound of clay left. This makes sense because $\frac{3}{8}$ is less than $\frac{1}{2}$, which matches the estimate.

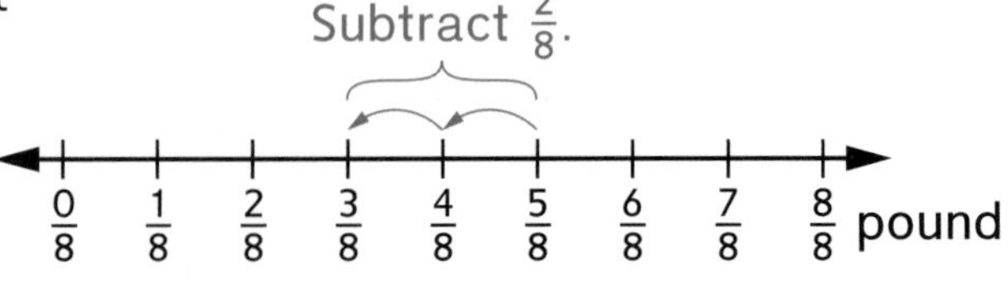

Check Your Understanding

Estimate and solve.

1. $\frac{1}{4} + \frac{2}{4} = ?$ **2.** $\frac{4}{6} - \frac{2}{6} = ?$ **3.** $\frac{2}{3} + \frac{1}{3} = ?$ **4.** $\frac{9}{10} - \frac{3}{10} = ?$

Check your answers in the Answer Key.

Adding Mixed Numbers with Like Denominators

> **Note** When adding mixed numbers, you will get the same answer whether you add the whole number or fraction parts first.

One way to add mixed numbers with common denominators is to add the fractions and the whole numbers separately. You can rename the sum as an equivalent mixed number to make your answer easier to check.

Example

$1\frac{3}{8} + 2\frac{7}{8} = \textbf{?}$

Estimate: $2\frac{7}{8}$ is close to 3. $1\frac{3}{8} + 3 = 4\frac{3}{8}$, so the sum should be close to $4\frac{3}{8}$.

Add the whole numbers.

$$\begin{array}{r} \mathbf{1}\frac{3}{8} \\ +\ \mathbf{2}\frac{7}{8} \\ \hline \mathbf{3} \end{array}$$

Add the fractions.

$$\begin{array}{r} 1\frac{\mathbf{3}}{\mathbf{8}} \\ +\ 2\frac{\mathbf{7}}{\mathbf{8}} \\ \hline 3\frac{\mathbf{10}}{\mathbf{8}} \end{array}$$

Rename the sum so it is easier to check.

Remember that $\frac{8}{8} = 1$ whole, so $\frac{10}{8}$ is the same as 1 whole with $\frac{2}{8}$ left over.

$3\frac{10}{8} = 3 + 1 + \frac{2}{8} = 4\frac{2}{8}$

$1\frac{3}{8} + 2\frac{7}{8} = 3\frac{10}{8}$, or $\mathbf{4\frac{2}{8}}$

This makes sense because $4\frac{2}{8}$ is close to the estimate of $4\frac{3}{8}$.

You can also add mixed numbers by counting up.

Example

$1\frac{2}{3} + 2\frac{1}{3} = \textbf{?}$

Estimate: $1\frac{2}{3}$ is a little less than 2. $2\frac{1}{3}$ is a little more than 2. The sum should be about 4.

Represent $1\frac{2}{3}$ on a number line.

Count up $2\frac{1}{3}$. You land at 4.

$1\frac{2}{3} + 2\frac{1}{3} = \mathbf{4}$

This matches the estimate.

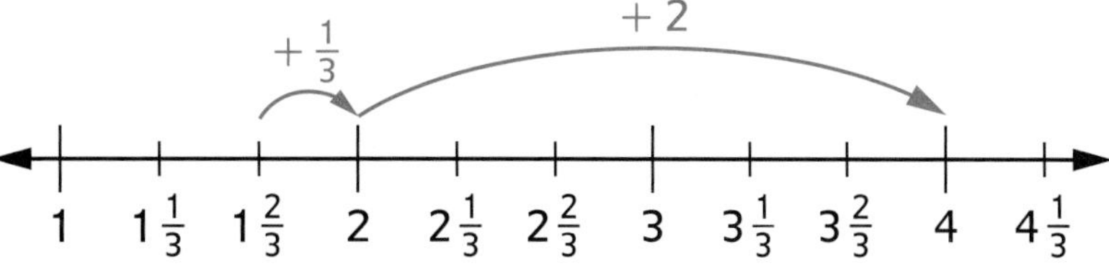

Subtracting Mixed Numbers with Like Denominators

When subtracting mixed numbers, you can sometimes subtract the fractions and whole numbers separately.

Example

$4\frac{3}{4} - 2\frac{1}{4} = \textbf{?}$

Estimate: $4\frac{3}{4}$ rounds to 5. $2\frac{1}{4}$ rounds to 2. $5 - 2 = 3$, so the difference is about 3.

Subtract the fractions.

$$\begin{array}{r} 4\mathbf{\frac{3}{4}} \\ -\ 2\mathbf{\frac{1}{4}} \\ \hline \mathbf{\frac{2}{4}} \end{array}$$

Subtract the whole numbers.

$$\begin{array}{r} \mathbf{4}\frac{3}{4} \\ -\ \mathbf{2}\frac{1}{4} \\ \hline \mathbf{2}\frac{2}{4} \end{array}$$

$4\frac{3}{4} - 2\frac{1}{4} = \mathbf{2\frac{2}{4}}$ Since $2\frac{2}{4}$ is close to the estimate of 3, the answer makes sense.

When the fraction part of the mixed number you are taking away is greater than the fraction part of the starting number, rename the starting number as an equivalent mixed number with a larger fraction part before you subtract.

Example

$5\frac{1}{6} - 3\frac{5}{6} = \textbf{?}$

Estimate: $5\frac{1}{6}$ is close to 5. $3\frac{5}{6}$ is close to 4. An estimate is $5 - 4 = 1$.

$\frac{1}{6}$ is less than $\frac{5}{6}$, so rename $5\frac{1}{6}$ as a mixed number with a larger fraction part.

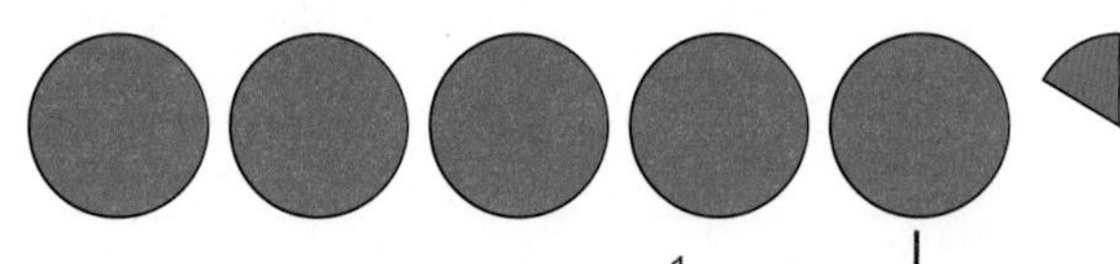

Show $5\frac{1}{6}$.

Break up 1 whole into 6 sixths. You now have 4 wholes and 7 sixths, or $4\frac{7}{6}$.

$5\frac{1}{6} \rightarrow 4\frac{7}{6}$

Rewrite the problem, then subtract $3\frac{5}{6}$.

$$\begin{array}{rcr} 5\frac{1}{6} & \rightarrow & 4\frac{7}{6} \\ -\ 3\frac{5}{6} & & -\ 3\frac{5}{6} \\ \hline & & 1\frac{2}{6} \end{array}$$

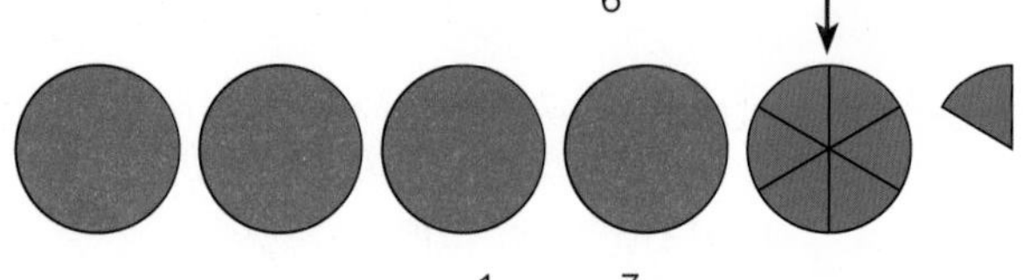

Take away $3\frac{5}{6}$. $1\frac{2}{6}$ is left.

$5\frac{1}{6} - 3\frac{5}{6} = \mathbf{1\frac{2}{6}}$ This makes sense because $1\frac{2}{6}$ is close to the estimate of 1.

Adding and Subtracting Fractions with Unlike Denominators

To add or subtract fractions with unlike denominators, you can think about a visual representation.

Example

$\frac{1}{3} + \frac{1}{6} = \textbf{?}$

Estimate: $\frac{1}{3}$ is a little less than $\frac{1}{2}$, and $\frac{1}{6}$ is very small. The sum should be close to $\frac{1}{2}$.

Imagine representing the problem with fraction circles.

Think: Can I name the amount that is covered?

One orange piece and one light blue piece cover exactly one-half of the circle.

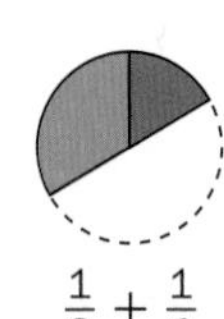
$\frac{1}{3} + \frac{1}{6}$

$\frac{1}{3} + \frac{1}{6} = \mathbf{\frac{1}{2}}$

This makes sense because the estimated sum was close to $\frac{1}{2}$.

When you are adding or subtracting fractions with unlike denominators, it often helps to make trades so that you are working with all same-size pieces.

Example

$\frac{9}{10} - \frac{1}{2} = \textbf{?}$

Estimate: $\frac{9}{10}$ is close to 1. $1 - \frac{1}{2} = \frac{1}{2}$, so the difference should be about $\frac{1}{2}$.

Show $\frac{9}{10}$ with fraction circles.

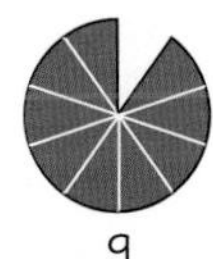
$\frac{9}{10}$

You need to subtract $\frac{1}{2}$. *Think:* How many tenths make $\frac{1}{2}$?

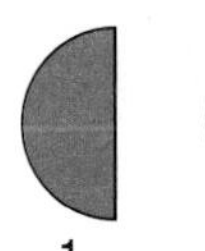
$\frac{1}{2} = \frac{5}{10}$

5 purple pieces (5 tenths) cover the same area as 1 pink piece (1 half), so subtract 5 tenths from 9 tenths.

$\frac{9}{10} - \frac{5}{10} = \frac{4}{10}$

Name the result: There are 4 tenths, or $\frac{4}{10}$, remaining.

$\frac{9}{10} - \frac{1}{2} = \mathbf{\frac{4}{10}}$

This makes sense because $\frac{4}{10}$ is close to $\frac{1}{2}$, which was the estimated difference.

You can add or subtract fractions with unlike denominators by renaming the fractions with common denominators. Before you find a common denomintor, think about whether you solve the problem mentally. If you cannot, use common denominators.

Example

$\frac{2}{3} + \frac{3}{4} = \mathbf{?}$

Estimate: $\frac{2}{3}$ and $\frac{3}{4}$ are both greater than $\frac{1}{2}$. So the sum will be greater than $\frac{1}{2} + \frac{1}{2} = 1$.

Think: Can I solve this problem mentally? It may be difficult to add thirds and fourths.

12 is a multiple of both 3 and 4, so use 12 as a common denominator. Use the equivalent fractions rule to rename $\frac{2}{3}$ and $\frac{3}{4}$ with a common denominator of 12.

$\frac{2}{3} = \frac{2 * 4}{3 * 4} = \frac{8}{12}$ $\frac{3}{4} = \frac{3 * 3}{4 * 3} = \frac{9}{12}$

Rewrite the problem using the renamed fractions. $\frac{8}{12} + \frac{9}{12} = ?$

Now you can add the numerators and keep the denominators the same. 8 twelfths + 9 twelfths = 17 twelfths

Return to the original problem. $\frac{2}{3} + \frac{3}{4} = \mathbf{\frac{17}{12}}$, or $\mathbf{1\frac{5}{12}}$

The answer is reasonable because $1\frac{5}{12}$ is greater 1.

Example

$\frac{7}{8} - \frac{1}{4} = \mathbf{?}$

Estimate: $\frac{7}{8}$ is just less than 1. $\frac{1}{4}$ is less than $\frac{1}{2}$. The answer should be around $1 - \frac{1}{2} = \frac{1}{2}$.

Think: Can I solve this problem mentally? If not, rename the fractions using common denominators.

Look at the denominators. 8 is a multiple of 4, so use 8 as a common denominator.

Rename $\frac{1}{4}$ as an equivalent fraction with a denominator of 8. $\frac{1}{4} = \frac{(1 * 2)}{(4 * 2)} = \frac{2}{8}$

Rewrite the problem using the renamed fraction. $\frac{7}{8} - \frac{2}{8} = ?$

Now you can subtract the numerators and keep the denominators the same. 7 eighths − 2 eighths = 5 eighths

Return to the original problem. $\frac{7}{8} - \frac{1}{4} = \mathbf{\frac{5}{8}}$

This makes sense because $\frac{5}{8}$ is close to $\frac{1}{2}$, which was the estimated difference.

Adding Mixed Numbers with Unlike Denominators

You can add mixed numbers with unlike denominators by combining what you know about adding mixed numbers and what you know about finding common denominators.

- Consider whether you can solve a problem mentally before you find a common denominator. If you can, solve it mentally. If you cannot, use common denominators.
- Sometimes it is helpful to rename the sum in order to check your answer.

Example

$4\frac{2}{5} + 1\frac{9}{10} = \mathbf{?}$

Estimate: 4 + 1 is 5, and $\frac{2}{5} + \frac{9}{10}$ is more than 1, so the sum is greater than 6.

Think: Can I solve this problem mentally? If not, rename the fractions using common denominators.

You can list equivalent fractions to find a common denominator.

$\frac{2}{5} = \mathbf{\frac{4}{10}} = \frac{6}{15} = \mathbf{\frac{8}{20}}$ $\mathbf{\frac{9}{10}} = \mathbf{\frac{18}{20}} = \frac{27}{30} = \frac{36}{40}$

Notice that both lists have fractions with denominators 10 and 20. Use 10 as a common denominator because only one fraction will need to be renamed.

Rewrite the problem with the equivalent fraction. $4\frac{4}{10} + 1\frac{9}{10} = ?$

Add the fractions.

$$\begin{array}{r} 4\mathbf{\frac{4}{10}} \\ +\ 1\mathbf{\frac{9}{10}} \\ \hline \mathbf{\frac{13}{10}} \end{array}$$

Add the whole numbers.

$$\begin{array}{r} \mathbf{4}\frac{4}{10} \\ +\ \mathbf{1}\frac{9}{10} \\ \hline \mathbf{5}\frac{13}{10} \end{array}$$

Rename the sum.

$$5\frac{13}{10} = 5 + \frac{10}{10} + \frac{3}{10}$$
$$= 5 + 1 + \frac{3}{10} = 6\frac{3}{10}$$

Return to the original problem. $4\frac{2}{5} + 1\frac{9}{10} = \mathbf{6\frac{3}{10}}$

$6\frac{3}{10}$ is a reasonable answer because it is greater than 6, which matches the estimate.

Check Your Understanding

Estimate and add.

1. $2\frac{1}{2} + 1\frac{1}{4}$ **2.** $4\frac{1}{3} + 6\frac{3}{4}$ **3.** $5\frac{5}{6} + 3\frac{2}{3}$

4. Explain your estimation strategy for Problem 3.

Check your answers in the Answer Key.

Subtracting Mixed Numbers with Unlike Denominators

You can subtract mixed numbers with unlike denominators by combining what you know about subtracting mixed numbers with what you know about finding common denominators.

- Consider whether you can solve a problem mentally before you find a common denominator. If you can, solve it mentally. If you cannot, use common denominators.
- Sometimes it is helpful to rename the starting number as an equivalent mixed number with a larger fraction part before you subtract.

Example

$5\frac{1}{3} - 2\frac{7}{8} =$ **?**

Estimate: $5\frac{1}{3}$ rounds to 5, and $2\frac{7}{8}$ rounds to 3. The difference should be about $5 - 3 = 2$.

Think: Can I solve this problem mentally? Most people cannot easily combine thirds and eighths mentally, so they use common denominators to solve the problem.

You can multiply the denominators to find the quick common denominator.

$3 * 8 = 24$, so use 24 as the common denominator.

Rename each fraction with a denominator of 24.

$\frac{1}{3} = \frac{(1 * 8)}{(3 * 8)} = \frac{8}{24}$ $\frac{7}{8} = \frac{(7 * 3)}{(8 * 3)} = \frac{21}{24}$

Rewrite the problem with the renamed fractions. $5\frac{8}{24} - 2\frac{21}{24} = ?$

Notice that the fraction part of the starting amount $\left(\frac{8}{24}\right)$ is less than the fraction part of the number being subtracted $\left(\frac{21}{24}\right)$. Rename the starting amount as an equivalent mixed number with a larger fraction part.

Trade 1 whole for $\frac{24}{24}$.

$5\frac{8}{24} = 4 + 1 + \frac{8}{24}$

$= 4 + \frac{24}{24} + \frac{8}{24} = 4\frac{32}{24}$

Rewrite the problem. $4\frac{32}{24} - 2\frac{21}{24} = ?$

Subtract the fractions.

$$\begin{array}{r} 4\mathbf{\frac{32}{24}} \\ -\ 2\mathbf{\frac{21}{24}} \\ \hline \mathbf{\frac{11}{24}} \end{array}$$

Subtract the whole numbers.

$$\begin{array}{r} \mathbf{4}\frac{32}{24} \\ -\ \mathbf{2}\frac{21}{24} \\ \hline \mathbf{2}\frac{11}{24} \end{array}$$

Return to the original problem. $5\frac{1}{3} - 2\frac{7}{8} = \mathbf{2\frac{11}{24}}$

$2\frac{11}{24}$ is a reasonable answer because it is close to the estimate of 2.

Example

$5 - 2\frac{3}{5} = $ **?**

Estimate: $2\frac{3}{5}$ is close to $2\frac{1}{2}$, and $5 - 2\frac{1}{2} = 2\frac{1}{2}$. The difference should be around $2\frac{1}{2}$.

One way:
Solve the problem mentally by imagining visual representations.

Start by picturing 5 whole circles. Imagine crossing off 2 whole circles and then crossing off 3 out of the 5 equal parts in another circle. That leaves 2 whole circles and $\frac{2}{5}$ of a circle, or $2\frac{2}{5}$.

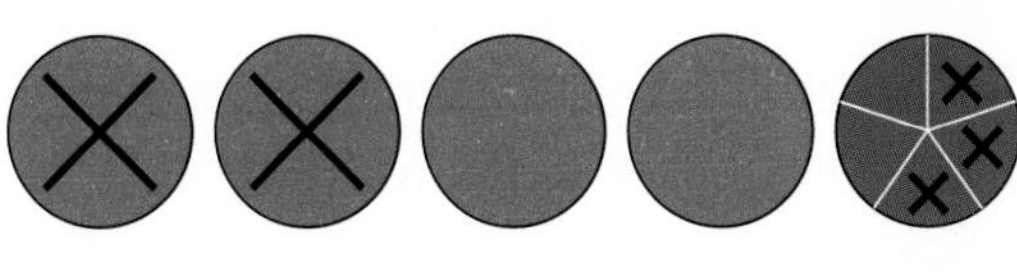

Another way:
Rename the whole number as a mixed number. Use fifths to match the denominator of the number being subtracted.

$5 = 4 + 1$
$= 4 + \frac{5}{5} = 4\frac{5}{5}$

Rewrite the problem.

$4\frac{5}{5} - 2\frac{3}{5} = ?$

Subtract the fractions.

$$\begin{array}{r} 4\mathbf{\frac{5}{5}} \\ -\ 2\mathbf{\frac{3}{5}} \\ \hline \mathbf{\frac{2}{5}} \end{array}$$

Subtract the whole numbers.

$$\begin{array}{r} \mathbf{4}\frac{5}{5} \\ -\ \mathbf{2}\frac{3}{5} \\ \hline \mathbf{2}\frac{2}{5} \end{array}$$

$5 - 2\frac{3}{5} = \mathbf{2\frac{2}{5}}$ This makes sense because $2\frac{2}{5}$ is close to the estimate of $2\frac{1}{2}$.

You can subtract (or add) mixed numbers by renaming the mixed numbers as fractions greater than one.

Example

$4\frac{1}{6} - 2\frac{2}{3} = $ **?**

Estimate: $4\frac{1}{6}$ rounds to 4. $2\frac{2}{3}$ rounds to 3. $4 - 3 = 1$, so the difference should be about 1.

Rename the mixed numbers as fractions. $4\frac{1}{6} = \frac{25}{6}$ $\quad 2\frac{2}{3} = \frac{8}{3}$

Rename the fractions with a common denominator. Use 6 as a common denominator, since 6 is a multiple of 3. $\frac{8}{3} = \frac{(8 * 2)}{(3 * 2)} = \frac{16}{6}$

Rewrite the problem. $\frac{25}{6} - \frac{16}{6} = ?$

Subtract. 25 sixths − 16 sixths = 9 sixths

Rename the result as a mixed number. $\frac{9}{6} = \frac{6}{6} + \frac{3}{6} = 1\frac{3}{6}$

So, $4\frac{1}{6} - 2\frac{2}{3} = \mathbf{\frac{9}{6}}$, or $\mathbf{1\frac{3}{6}}$. This makes sense with an estimate of about 1.

Fractions as Multiples of Unit Fractions

Fractions with a numerator of 1, such as $\frac{1}{2}$, $\frac{1}{4}$, and $\frac{1}{10}$ are called **unit fractions.** Unit fractions name 1 part of the whole. You can build other fractions from multiples of unit fractions.

Example

Rename $\frac{3}{4}$ as a multiple of a unit fraction.

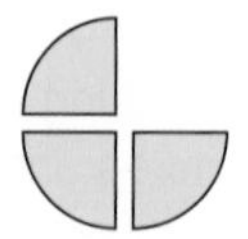

Represent $\frac{3}{4}$ with fraction circle pieces. If the red circle is one whole, then the yellow piece represents one-fourth.

It takes 3 one-fourth pieces to make three-fourths. So, $\frac{3}{4}$ is the same as $3[\frac{1}{4}\text{s}]$.

You can rename $\frac{3}{4}$ as $\frac{1}{4} + \frac{1}{4} + \frac{1}{4}$, or $3 * \frac{1}{4}$.

Example

Rename $\frac{3}{8}$ as a multiple of a unit fraction.

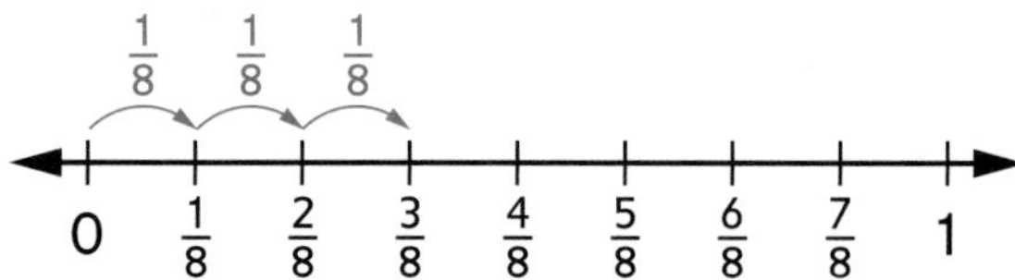

Locate $\frac{3}{8}$ on a number line. If the distance between 0 and 1 is divided into 8 equal parts, $\frac{3}{8}$ is 3 one-eighth jumps from 0 towards 1.

The 3 one-eighth jumps can be represented with addition or with multiplication:

Addition: $\frac{1}{8} + \frac{1}{8} + \frac{1}{8} = \frac{3}{8}$ Multiplication: $3 * \frac{1}{8} = \frac{3}{8}$

The number line and number sentences all show that three-eighths is the same as 3 copies of $\frac{1}{8}$.

Example

Use what you know about unit fractions and multiples to solve each multiplication problem.

$9 * \frac{1}{2} = ?$ 9 times $\frac{1}{2}$ is the same as 9 halves, or $\frac{9}{2}$. $9 * \frac{1}{2} = \frac{9}{2}$

$4 * \frac{1}{5} = ?$ 4 times $\frac{1}{5}$ is the same as 4 fifths, or $\frac{4}{5}$. $4 * \frac{1}{5} = \frac{4}{5}$

$7 * \frac{1}{3} = ?$ 7 times $\frac{1}{3}$ is the same as 7 thirds, or $\frac{7}{3}$. $7 * \frac{1}{3} = \frac{7}{3}$

Check Your Understanding

1. Write an addition number sentence and a multiplication number sentence to describe the picture.

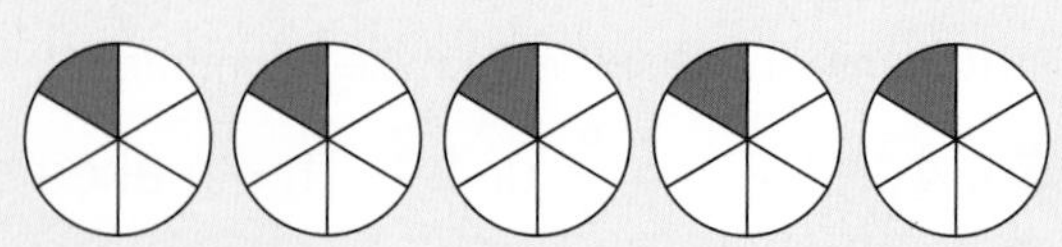

2. Draw one picture to represent these two number sentences:

$\frac{1}{3} + \frac{1}{3} + \frac{1}{3} + \frac{1}{3} = \frac{4}{3}$ $4 * \frac{1}{3} = \frac{4}{3}$

Check your answers in the Answer Key.

Finding a Fraction of a Number

Many problems with fractions involve finding a fraction of a number.

Note "$\frac{1}{3}$ of 24" has the same meaning as "$\frac{1}{3} * 24$." When you find a fraction of a number, you can replace the word *of* with a multiplication symbol. For example, $\frac{1}{6}$ of 18 means $\frac{1}{6} * 18$, and $\frac{3}{4}$ of 40 means $\frac{3}{4} * 40$.

Example

There are 24 students in Ms. Dunning's class. $\frac{1}{3}$ of the students are participating in a school performance. How many students in Ms. Dunning's class are participating in the performance?

Estimate: $\frac{1}{3}$ is less than $\frac{1}{2}$, so $\frac{1}{3}$ of 24 will be less than $\frac{1}{2}$ of 24. Half of 24 is 12, so the answer will be less than 12.

To find the number of students participating in the performance, find $\frac{1}{3}$ of 24.

You can draw a picture to model the problem. Each X represents one student. Divide 24 Xs into 3 equal groups.

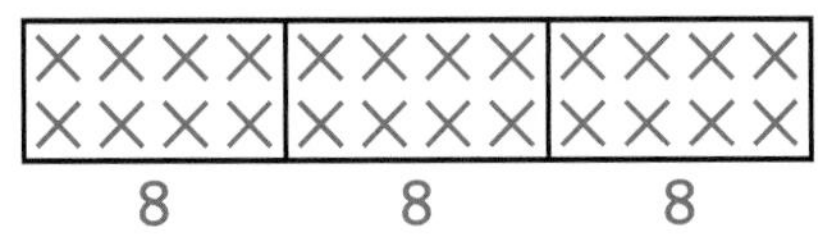

Each group has $\frac{1}{3}$ of the Xs.
There are 8 Xs in each group.
So, $\frac{1}{3}$ of 24 is 8.

Eight of Ms. Dunning's students are participating in the school performance. This answer makes sense because 8 is less than 12, which matches the estimate.

Example

A shirt that costs \$13 is on sale for $\frac{1}{2}$ the regular price. What is the sale price?

Estimate: \$13 is close to \$12, and half of \$12 is \$6. The sale price should be close to \$6.

To find the sale price, find $\frac{1}{2}$ of 13.

You can use counters to model the problem.

Divide 13 counters into 2 equal groups.

There are 6 counters in each group, with 1 counter left over. Imagine splitting the 1 leftover counter in half. Each group now has $6\frac{1}{2}$ counters.

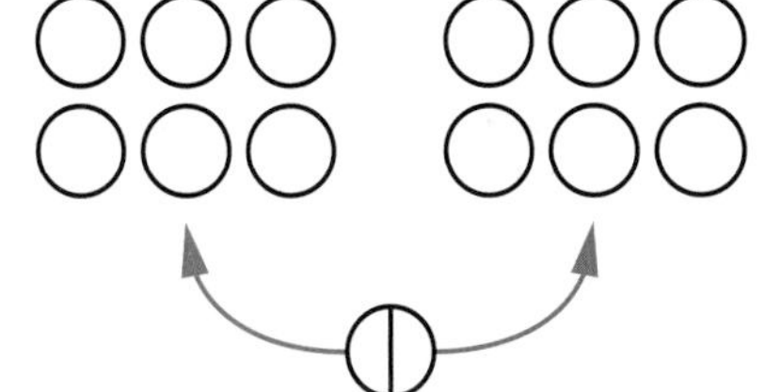

So, $\frac{1}{2}$ of 13 is $6\frac{1}{2}$. We can write $6\frac{1}{2}$ dollars as \$6.50.

The sale price is \$6.50. This makes sense because it is close to the estimate of \$6.

Sometimes it can be helpful to find a unit fraction of a number in order to find another fraction of that number. Remember that a **unit fraction** is a fraction with a numerator of 1, such as $\frac{1}{2}$, $\frac{1}{4}$, and $\frac{1}{10}$.

Example

Alex borrowed 12 books from the library. $\frac{3}{4}$ of the books he borrowed were books about planets. How many books about planets did Alex borrow?

Estimate: $\frac{3}{4}$ of 12 is a part of 12, so it is less than 12. $\frac{3}{4}$ is more than $\frac{1}{2}$, so $\frac{3}{4}$ of 12 is more than $\frac{1}{2}$ of 12, or more than 6. The answer is more than 6 and less than 12.

To find the number of books about planets, find $\frac{3}{4}$ of 12.

You can use counters to model the problem.

Divide 12 counters into 4 equal groups.

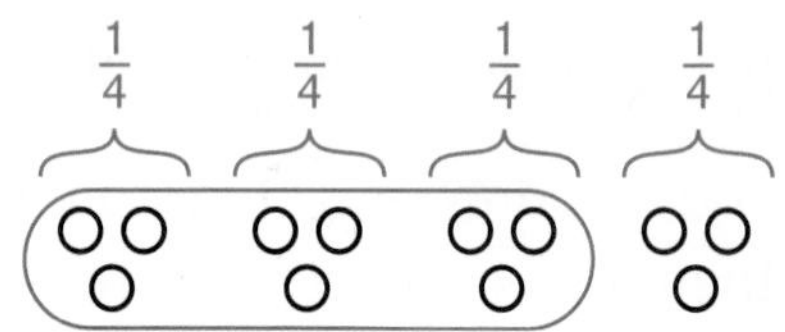

Think: What is $\frac{1}{4}$ of 12?

Each group represents $\frac{1}{4}$ of 12. There are 3 counters in each group, so $\frac{1}{4}$ of 12 is 3.

Use $\frac{1}{4}$ of 12 to find $\frac{3}{4}$ of 12.

Imagine taking 3 of the $\frac{1}{4}$ groups to find $\frac{3}{4}$ of 12. Three groups of 3 counters is 9 counters, so $\frac{3}{4}$ of 12 is 9.

Alex borrowed 9 books about planets. This answer is reasonable because it matches the estimate of more than 6 and less than 12.

Check Your Understanding

Estimate and solve.

1. What is $\frac{1}{4}$ of 32? **2.** What is $\frac{3}{4}$ of 32? **3.** What is $\frac{1}{3}$ of 25? **4.** What is $\frac{2}{3}$ of 25?

5. Rita and Hunter earned $20 raking lawns. Since Rita did most of the work, they decided that Rita should get $\frac{4}{5}$ of the money. How much does each person get?

Check your answers in the Answer Key.

Predicting the Size of Products

You can often predict the size of a product before you multiply. To predict the size of a product, examine the size of each factor.

Multiplying by Numbers Greater Than One

Example

Predict the size of the product for $8 * 1\frac{1}{2}$.

Look at the size of each factor. 8 is a whole number greater than 1.

$1\frac{1}{2}$ is a mixed number and is greater than 1.

Think about what multiplication means. You can think of $8 * 1\frac{1}{2}$ as making 8 groups with $1\frac{1}{2}$ in each group. Just thinking of 2 groups of $1\frac{1}{2}$ gives 3, so 8 groups of $1\frac{1}{2}$ will certainly be more than $1\frac{1}{2}$.

You can also think of $8 * 1\frac{1}{2}$ as making 1 full group of 8 and then adding on half of 8. Making more than 1 group of 8 means that the product will be greater than 8.

The product of $8 * 1\frac{1}{2}$ will be greater than each of the factors, 8 and $1\frac{1}{2}$.

The problem above is an example of a general rule. When you multiply two numbers greater than one, the product is greater than both numbers.

Multiplying by Numbers Equal to One

Example

Will the product of $\frac{1}{2} * \frac{3}{3}$ be greater than, equal to, or less than $\frac{1}{2}$?

Look at the size of each factor. $\frac{1}{2}$ is a fraction less than 1. $\frac{3}{3}$ is a fraction equal to 1.

Think about what multiplication means. Multiplying by a fraction is like finding a fraction of a number. Finding $\frac{1}{2}$ of $\frac{3}{3}$ is equivalent to finding $\frac{1}{2}$ of 1 whole, which still results in $\frac{1}{2}$.

Multiplying any number by 1 gives that number. Since $\frac{3}{3} = 1$ and $\frac{1}{2} * 1 = \frac{1}{2}$, you know $\frac{1}{2} * \frac{3}{3} = \frac{1}{2}$.

The product of $\frac{1}{2} * \frac{3}{3}$ will be equal to $\frac{1}{2}$.

The problem above is an example of another rule. When you multiply a given number by a number equal to one, the product is equal to the given number.

Multiplying by Numbers Less Than One

Example

How will the product of $12 * \frac{2}{3}$ compare to 12?

Look at the size of each factor. 12 is a whole number greater than 1. $\frac{2}{3}$ is a fraction less than 1.

Think about what multiplication means. $12 * \frac{2}{3}$ has the same product as $\frac{2}{3} * 12$. You can think of $\frac{2}{3} * 12$ as finding $\frac{2}{3}$ of 12. Taking a fractional part of 12 will result in a number less than 12.

The product of $12 * \frac{2}{3}$ will be less than 12.

Example

Complete the number sentences with >, <, or =. $\frac{1}{3} * \frac{3}{4} \square \frac{3}{4}$ $\frac{1}{3} * \frac{3}{4} \square \frac{1}{3}$

Look at the size of each factor. $\frac{1}{3}$ is a fraction less than 1. $\frac{3}{4}$ is a fraction less than 1.

Think about what multiplication means. Think of $\frac{1}{3} * \frac{3}{4}$ as finding $\frac{1}{3}$ of $\frac{3}{4}$. Taking part of $\frac{3}{4}$ will result in an amount less than $\frac{3}{4}$.

$\frac{1}{3} * \frac{3}{4} = \frac{3}{4} * \frac{1}{3}$, so think of finding $\frac{3}{4}$ of $\frac{1}{3}$. Taking part of $\frac{1}{3}$ will result in an amount less than $\frac{1}{3}$.

So, $\frac{1}{3} * \frac{3}{4} < \frac{3}{4}$, and $\frac{1}{3} * \frac{3}{4} < \frac{1}{3}$.

The problems above are examples of another rule. When you multiply a given number (greater than zero) by a number less than one, the product is less than the given number.

Check Your Understanding

Complete each number sentence with >, <, or =.

1. $\frac{7}{8} * 16 \square 16$ **2.** $\frac{6}{1} * \frac{3}{4} \square \frac{3}{4}$ **3.** $3\frac{1}{2} * 2\frac{5}{6} \square 3\frac{1}{2}$ **4.** $\frac{2}{3} * \frac{2}{2} \square \frac{2}{3}$

Check your answers in the Answer Key.

Multiplying Fractions by Whole Numbers

There are several ways to think about multiplying a whole number and a fraction.

Using Visual Representations

You can think about multiplying a whole number and a fraction as making copies of a fraction. You can represent copies of fractions with fraction circles, pictures, and number lines.

Example

Use visual representations to solve $4 * \frac{2}{3}$.

One way: Use fraction circle pieces to make 4 groups of $\frac{2}{3}$.

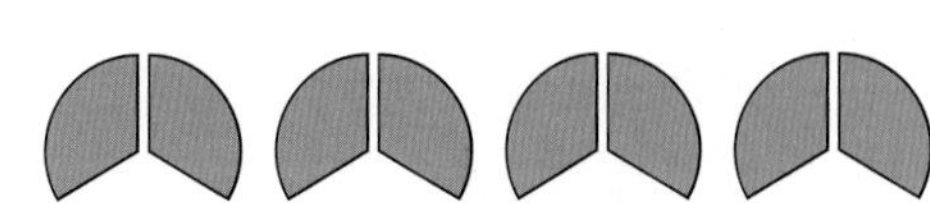

There are eight $\frac{1}{3}$ pieces in all, so $4 * \frac{2}{3} = \frac{8}{3}$.

Another way: Draw rectangles to represent wholes. Split each rectangle into three equal parts to show thirds. Shade 2 thirds of each rectangle.

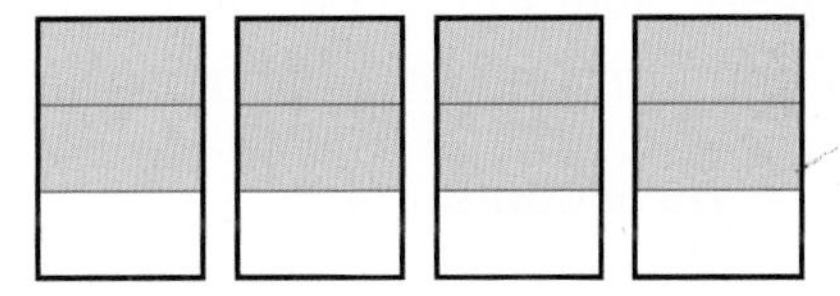

There are 8 thirds shaded, so $4 * \frac{2}{3} = \frac{8}{3}$.

Another way: Sketch a number line that shows thirds. Starting at 0, make 4 hops that are each $\frac{2}{3}$-unit long.

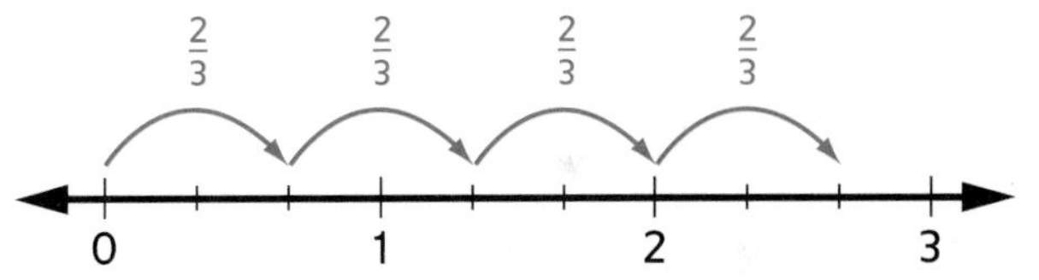

In total, you have traveled eight $\frac{1}{3}$-units from 0. You end at $2\frac{2}{3}$.

All three representations show that $4 * \frac{2}{3} = \frac{8}{3}$, or $2\frac{2}{3}$.

Using Repeated Addition

You can also think about multiplying a fraction and a whole number as repeated addition.

Example

Use addition to solve $3 * \frac{5}{6}$.

$3 * \frac{5}{6}$ is like making 3 copies of $\frac{5}{6}$. So, $3 * \frac{5}{6} = \frac{5}{6} + \frac{5}{6} + \frac{5}{6}$.

Adding $\frac{5}{6}$ at a time, you get 5 sixths, 10 sixths, 15 sixths.

$\frac{5}{6} + \frac{5}{6} + \frac{5}{6} = \frac{15}{6}$, so $3 * \frac{5}{6} = \frac{15}{6}$.

Thinking about Fraction-Of Problems

You can think about fraction multiplication as fraction-of problems.

Example

Use what you know about fraction-of problems to solve $15 * \frac{4}{5}$.

Estimate: Since the fraction is less than 1, the fraction of 15 should be less than 15.

You can rewrite the problem with the fraction as the first factor if it seems easier to you. $15 * \frac{4}{5}$ is the same as $\frac{4}{5} * 15$.

Remember: Changing the order of factors does not affect the product.

Think of the problem as a fraction-of problem. What is $\frac{4}{5}$ of 15?

To find $\frac{4}{5}$ of 15, first find $\frac{1}{5}$ of 15.

When 15 Xs are split into 5 equal groups, there are 3 Xs in each group. So, $\frac{1}{5}$ of $15 = 3$.

XXX|XXX|XXX|XXX|XXX

$\frac{4}{5}$ is the same as 4 copies of $\frac{1}{5}$.

Take 4 [$\frac{1}{5}$s] of 15, or 4 copies of 3.

XXX|XXX|XXX|XXX|XXX

You can add 3 four times, or multiply 3 by 4. $3 + 3 + 3 + 3 = 12$ or $4 * 3 = 12$

$\frac{4}{5}$ of 15 is 12, so $15 * \frac{4}{5} = 12$. This makes sense because 12 is less than 15.

Thinking About Fractions as Division

Thinking about fractions as division can help you multiply whole numbers and fractions.

Example

Use your understanding of fractions as division to solve $10 * \frac{5}{8}$.

Estimate: $\frac{5}{8}$ is about $\frac{1}{2}$, so the product should be about $\frac{1}{2}$ of 10, or 5.

Think of $\frac{5}{8}$ as $5 \div 8$. Rewrite the problem. $10 * 5 \div 8$

To evaluate the expression, multiply then divide. $10 * 5 = 50$

$50 \div 8 \rightarrow 6$ R2, or $6\frac{2}{8}$

So, $10 * \frac{5}{8} = 6\frac{2}{8}$. This makes sense because $6\frac{2}{8}$ is close to the estimate of 5.

Check Your Understanding

Estimate and solve using any method.

1. $8 * \frac{3}{4}$ **2.** $3 * \frac{2}{5}$ **3.** $\frac{7}{8} * 16$ **4.** $24 * \frac{4}{6}$

Check your answers in the Answer Key.

Finding Fractions of Fractions

Finding a fraction of a fraction is similar to finding a fraction of a whole number. You are still finding part of a given amount, except now the given amount is a fraction.

You can use visual representations to help you find a fraction of a fraction.

Example

Use fraction circles to find $\frac{1}{3}$ of $\frac{6}{8}$.

Estimate: $\frac{1}{3}$ of $\frac{6}{8}$ should be less than $\frac{6}{8}$.

Show $\frac{6}{8}$ using 6 one-eighth fraction circle pieces.

Split the 6 one-eighth pieces into 3 equal groups.

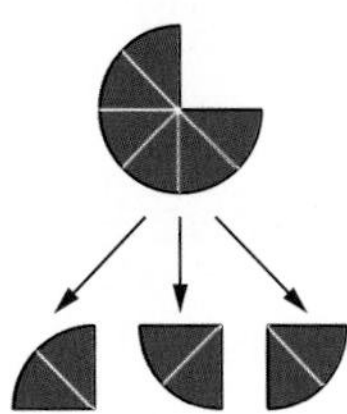

Each group represents $\frac{1}{3}$ of $\frac{6}{8}$. There are 2 one-eighth pieces in each group, so $\frac{1}{3}$ of $\frac{6}{8}$ is $\frac{2}{8}$.

$\frac{1}{3}$ of $\frac{6}{8} = \frac{2}{8}$ This makes sense because $\frac{2}{8}$ is less than $\frac{6}{8}$.

Folding paper is another way to find a fraction of a fraction.

Example

What is $\frac{3}{4}$ of $\frac{2}{3}$?

Estimate: Taking a fraction of $\frac{2}{3}$ will result in an amount less than $\frac{2}{3}$.

Imagine starting with $\frac{2}{3}$ of a sheet of paper and taking $\frac{3}{4}$ of that $\frac{2}{3}$.

Show $\frac{2}{3}$ by folding a piece of paper into thirds and shading 2 of them.

Fold the paper in the opposite direction to make fourths. Show $\frac{3}{4}$ of $\frac{2}{3}$ by shading 3 of the 4 rows in a darker color.

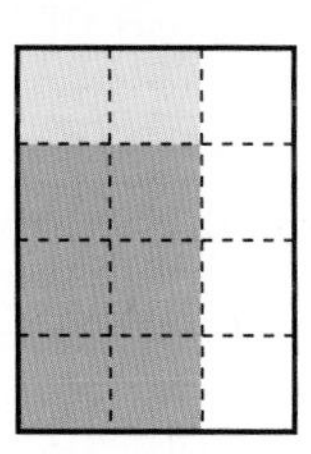

Name the area that is shaded in the darker color. There are 6 parts shaded out of 12 equal parts in the whole piece of paper, so the shaded area is $\frac{6}{12}$ of the whole.

$\frac{3}{4}$ of $\frac{2}{3} = \frac{6}{12}$ This answer makes sense because $\frac{6}{12}$ equals $\frac{1}{2}$, which is less than $\frac{2}{3}$.

Check Your Understanding

Use fraction circles, fold paper, or make drawings to solve.

1. Find $\frac{1}{2}$ of $\frac{4}{5}$. **2.** Find $\frac{1}{4}$ of $\frac{2}{3}$. **3.** Find $\frac{3}{4}$ of $\frac{1}{6}$.

Check your answers in the Answer Key.

Multiplying Fractions

Using an Area Model

One way to multiply fractions is to use an **area model.** In an area model, the side lengths of a rectangle represent the **factors.** The area of a rectangle represents the **product** of the factors because the area of a rectangle can be found by multiplying the length times the width.

Example

Use an area model to solve $\frac{2}{3} * \frac{3}{5} = ?$

Estimate: $\frac{2}{3}$ of $\frac{3}{5}$ is a fraction of $\frac{3}{5}$. The product should be less than $\frac{3}{5}$.

Sketch a rectangle that is $\frac{2}{3}$ unit long and $\frac{3}{5}$ unit wide.

- Start with a unit square. The number lines show that the square is 1 unit long and 1 unit wide.
- Partition the distance from 0 to 1 on one number line into thirds. Label $\frac{1}{3}$ and $\frac{2}{3}$.
- Partition the distance from 0 to 1 on the other number line into fifths. Label $\frac{1}{5}$, $\frac{2}{5}$, $\frac{3}{5}$, and $\frac{4}{5}$.
- Shade a rectangle with one side that goes from 0 to $\frac{2}{3}$ on one number line and with another side that goes from 0 to $\frac{3}{5}$ on the other number line.

Find the area of the shaded rectangle.

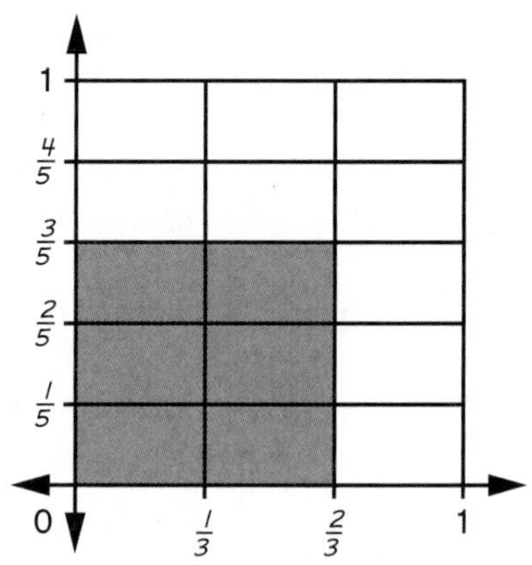

Area model for $\frac{2}{3} * \frac{3}{5}$

There are 6 shaded parts out of 15 parts in the unit square, so the area of the shaded rectangle is $\frac{6}{15}$ square unit.

The area of this rectangle is $\frac{6}{15}$, which equals the product of its length and width.

$\frac{2}{3} * \frac{3}{5} = \frac{6}{15}$

The answer, $\frac{6}{15}$, matches the estimate of less than $\frac{3}{5}$, because $\frac{6}{15}$ is the same as $\frac{2}{5}$.

The problem above is an example of the following:

- Multiplying the numerators of the fractions gives the number of shaded parts in an area model, and that is the numerator of the product.

  $2 * 3 = 6$

  There are 6 shaded parts in the area model.

- Multiplying the denominators of the fractions gives the number of parts in the whole square, and that is the denominator of the product.

  $3 * 5 = 15$

  There are 15 parts in the whole square.

- The product of two fractions can be found by multiplying the numerators and multiplying the denominators.

  $\frac{2}{3} * \frac{3}{5} = \frac{(2 * 3)}{(3 * 5)} = \frac{6}{15}$

Fraction Multiplication Method

The example on the previous page illustrates the following method:
To multiply fractions, multiply the numerators and multiply the denominators.

Or, for two fractions $\frac{a}{b}$ and $\frac{c}{d}$, where b and d are not 0: $\frac{a}{b} * \frac{c}{d} = \frac{a * c}{b * d}$.

Examples

$\frac{3}{4} * \frac{2}{3} =$ **?**

Estimate: Taking a fraction of $\frac{2}{3}$ means the product will be less than $\frac{2}{3}$.

$\frac{3}{4} * \frac{2}{3} = \frac{(3 * 2)}{(4 * 3)} = \mathbf{\frac{6}{12}}$, or $\mathbf{\frac{1}{2}}$

Check: $\frac{1}{2}$ is less than $\frac{2}{3}$.

$\frac{2}{5} * \frac{6}{7} =$ **?**

Estimate: Taking a fraction of $\frac{6}{7}$ means the product will be less than $\frac{6}{7}$.

$\frac{2}{5} * \frac{6}{7} = \frac{(2 * 6)}{(5 * 7)} = \mathbf{\frac{12}{35}}$

Check: $\frac{12}{35}$ is less than half, so it is less than $\frac{6}{7}$.

$\frac{8}{5} * \frac{1}{2} =$ **?**

Estimate: Since $\frac{8}{5}$ is greater than 1, the product should be more than $1 * \frac{1}{2}$, or more than $\frac{1}{2}$.

$\frac{8}{5} * \frac{1}{2} = \frac{(8 * 1)}{(5 * 2)} = \mathbf{\frac{8}{10}}$

Check: $\frac{8}{10}$ is more than $\frac{1}{2}$.

This method can also be used to multiply a whole number and a fraction. First rename the whole number as a fraction with 1 in the denominator. For example, to solve $5 * \frac{2}{3}$, rename 5 as $\frac{5}{1}$, then multiply $\frac{5}{1} * \frac{2}{3}$.

Example

Solve $4 * \frac{7}{9}$.

Estimate: The answer will be more than $4 * \frac{1}{2} = 2$ but less than $4 * 1 = 4$.

One way: Solve the problem mentally.

Think about $4 * \frac{7}{9}$ as $\frac{7}{9} + \frac{7}{9} + \frac{7}{9} + \frac{7}{9}$ then mentally calculate $\frac{(4 * 7)}{9} = \frac{28}{9}$, or $3\frac{1}{9}$.

Another way: Rename 4 as a fraction. $4 = \frac{4}{1}$

Rewrite the problem. $\frac{4}{1} * \frac{7}{9}$

Multiply the numerators and multiply the denominators. $\frac{4}{1} * \frac{7}{9} = \frac{(4 * 7)}{(1 * 9)} = \frac{28}{9}$, or $3\frac{1}{9}$

Using either method, $4 * \frac{7}{9} = \mathbf{\frac{28}{9}}$, or $\mathbf{3\frac{1}{9}}$.

The answer is reasonable. It matches the estimate that the product is between 2 and 4.

Check Your Understanding

Estimate and multiply.

1. $\frac{1}{2} * \frac{2}{3}$ **2.** $\frac{2}{5} * \frac{3}{4}$ **3.** $\frac{5}{6} * \frac{3}{8}$ **4.** $\frac{1}{4} * \frac{1}{3}$

5. Explain your estimation strategy for Problem 2.

Check your answers in the Answer Key.

Mixed Number Multiplication

Multiplying Mixed Numbers Using Partial Products

You can use partial products to multiply mixed numbers by whole numbers, fractions, or mixed numbers. Each part of the mixed number must be multiplied by each part of the other factor. You can use an **area model** to help find partial products.

Example

Solve $4 * 2\frac{3}{4}$.

Estimate: Since $2\frac{3}{4}$ is greater than 2, the product of $4 * 2\frac{3}{4}$ should be greater than $4 * 2$, or 8.

Use an area model to find partial products.

- Draw a rectangle with a length of $2\frac{3}{4}$ and a width of 4.
- Think of $2\frac{3}{4}$ as $2 + \frac{3}{4}$ and partition the rectangle into two sections.
- Find the area of the section on the left by multiplying $2 * 4$.
- Find the area of the section on the right by multiplying $\frac{3}{4} * 4$.

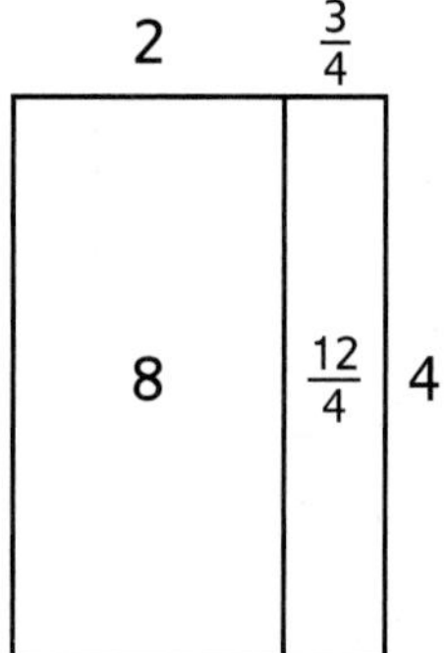

Partial products: $2 * 4 = 8$ $\quad \frac{3}{4} * 4 = \frac{12}{4}$, or 3

Add the partial products: $8 + 3 = 11$.

$4 * 2\frac{3}{4} = \mathbf{11}$

This makes sense because the product is more than 8, which matches the estimate.

Example

Solve $5\frac{1}{2} * \frac{2}{3}$.

Estimate: $\frac{2}{3}$ is a fraction less than 1. So, $\frac{2}{3}$ of $5\frac{1}{2}$ should be less than $5\frac{1}{2}$.

Use an area model to find partial products.

- Draw a rectangle with a length of $5\frac{1}{2}$ and a width of $\frac{2}{3}$.
- Think of $5\frac{1}{2}$ as $5 + \frac{1}{2}$ and partition the rectangle into two sections.
- Find the area of the section on the left by multiplying $5 * \frac{2}{3}$.
- Find the area of the section on the right by multiplying $\frac{1}{2} * \frac{2}{3}$.

5 $\quad \frac{1}{2}$

$\frac{10}{3}$ $\quad \frac{2}{6}$ $\quad \frac{2}{3}$

Partial products: $5 * \frac{2}{3} = \frac{10}{3}$

$\frac{1}{2} * \frac{2}{3} = \frac{2}{6}$

Rename the partial products with a common denominator, then add.

$\frac{10}{3} = 3\frac{1}{3}$, or $3\frac{2}{6}$

$3\frac{2}{6} + \frac{2}{6} = 3\frac{4}{6}$

$5\frac{1}{2} * \frac{2}{3} = \mathbf{3\frac{4}{6}}$

This makes sense because $3\frac{4}{6}$ is less than $5\frac{1}{2}$, which matches the estimate.

Example

Solve $2\frac{3}{5} * 7\frac{1}{2}$.

Estimate: $2\frac{3}{5}$ is close to 3 and $7\frac{1}{2}$ is close to 7. $3 * 7 = 21$, so the product should be close to 21.

Use an area model to find partial products.

- Draw a rectangle with a width of $2\frac{3}{5}$ and a length of $7\frac{1}{2}$, as shown below.
- Think of $2\frac{3}{5}$ as $2 + \frac{3}{5}$ and $7\frac{1}{2}$ as $7 + \frac{1}{2}$. Partition the rectangle into four sections.
- Find the area of the top left section by multiplying $7 * 2$.
- Find the area of the top right section by multiplying $\frac{1}{2} * 2$.
- Find the area of the bottom left section by multiplying $7 * \frac{3}{5}$.
- Find the area of the bottom right section by multiplying $\frac{1}{2} * \frac{3}{5}$.

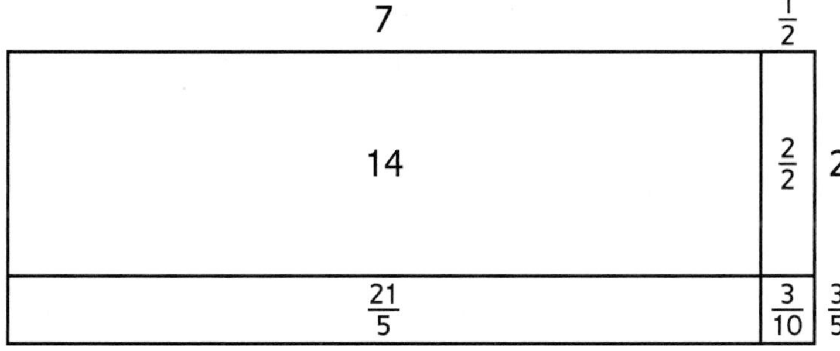

Partial products: $7 * 2 = 14$

$\frac{1}{2} * 2 = \frac{2}{2}$, or 1

$7 * \frac{3}{5} = \frac{21}{5}$

$\frac{1}{2} * \frac{3}{5} = \frac{3}{10}$

Rename the partial products with a common denominator, then add.

$\frac{21}{5} = \frac{42}{10}$, or $4\frac{2}{10}$

$14 + 1 + 4\frac{2}{10} + \frac{3}{10} = 19\frac{5}{10}$

$7\frac{1}{2} * 2\frac{3}{5} = \mathbf{19\frac{5}{10}}$

This makes sense because $19\frac{5}{10}$ is reasonably close to the estimate of 21.

Multiplying Mixed Numbers by Renaming Them as Fractions

You can multiply with mixed numbers by renaming any whole or mixed numbers as fractions, and then use a fraction multiplication method.

Example

Find $5 * 2\frac{1}{4}$.

Estimate: $2\frac{1}{4}$ is a little more than 2. The answer should be a little more than $5 * 2 = 10$.

Rename any whole or mixed numbers as fractions. $5 = \frac{5}{1}$ $2\frac{1}{4} = \frac{9}{4}$

Rewrite the problem, then multiply. $\frac{5}{1} * \frac{9}{4} = \frac{(5 * 9)}{(1 * 4)} = \frac{45}{4}$, or $11\frac{1}{4}$

$5 * 2\frac{1}{4} = \mathbf{11\frac{1}{4}}$

This makes sense because $11\frac{1}{4}$ is a little more than 10, which matches the estimate.

Example

Solve $3\frac{1}{4} * 1\frac{5}{6}$.

Estimate: $3\frac{1}{4}$ is close to 3, and $1\frac{5}{6}$ is close to 2. $3 * 2 = 6$, so the product should be close to 6.

Rename the mixed numbers as fractions. $3\frac{1}{4} = \frac{13}{4}$ $1\frac{5}{6} = \frac{11}{6}$

Rewrite the problem, then multiply. $\frac{13}{4} * \frac{11}{6} = \frac{(13 * 11)}{(4 * 6)} = \frac{143}{24}$

Rename the product as a mixed or whole number. $143 \div 24 \rightarrow 5$ R23, or $5\frac{23}{24}$

$3\frac{1}{4} * 1\frac{5}{6} = \mathbf{5\frac{23}{24}}$

This makes sense because $5\frac{23}{24}$ is close to the estimate of 6.

Check Your Understanding

Estimate and multiply using the method of your choice.

1. $6 * 1\frac{2}{3}$ **2.** $1\frac{1}{2} * \frac{1}{4}$ **3.** $3\frac{2}{5} * 2\frac{1}{2}$ **4.** $4\frac{5}{6} * \frac{1}{2}$

5. Explain your estimate for Problem 4.

Check your answers in the Answer Key.

Division with Fractions

When you solve division problems with fractions, it is helpful to consider the different meanings of division.

Equal-Sharing Division

In equal-sharing division problems, you can think of the dividend as an initial amount that is being split into a number of equal shares. For example, splitting $\frac{1}{3}$ of a watermelon among four people is an equal-sharing division problem. This situation can be modeled by the expression $\frac{1}{3} \div 4$.

Problems in which a fraction is divided by a whole number can usually be interpreted as equal-sharing division. One way to solve equal-sharing division problems is to use visual models, such as pictures, fraction circles, or number lines.

Example

A $\frac{1}{2}$ loaf of bread is shared equally among 3 people. What part of a whole loaf does each person get?

Estimate: $\frac{1}{2}$ loaf is being split 3 ways, so the amount that each person gets should be small compared to $\frac{1}{2}$.

Solve this equal-sharing division problem by finding $\frac{1}{2} \div 3$.

You can draw a picture to model the problem.

Use a rectangle to represent a loaf of bread.

Show $\frac{1}{2}$ of the loaf of bread.

Split each half of the loaf into 3 equal parts.

Each person's share is one of the equal parts.

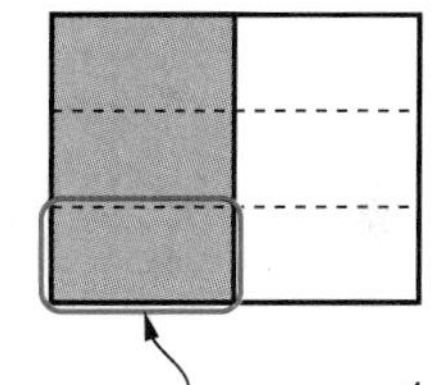

Each person gets $\frac{1}{6}$ loaf.

Think: What part of a whole loaf is one person's share?

One share would fit into the whole 6 times, so one share is $\frac{1}{6}$ loaf.

Each person gets $\frac{1}{6}$ loaf of bread. This makes sense because $\frac{1}{6}$ loaf is small compared to $\frac{1}{2}$ loaf.

You can check division problems using multiplication. For example, you can check that $6 \div 2 = 3$ by multiplying $3 * 2 = 6$. You can check division problems with fractions in a similar way.

Example

Use multiplication to check that $\frac{1}{2} \div 3 = \frac{1}{6}$.

Think: Does $\frac{1}{6} * 3 = \frac{1}{2}$? Do 3 [$\frac{1}{6}$s] make $\frac{1}{2}$ a loaf?

Yes, $\frac{1}{6} * 3 = \frac{3}{6}$, which is the same as $\frac{1}{2}$.

Another way to divide a fraction by a whole number is to think of finding a fraction of a number. In the example on the previous page, $\frac{1}{2} \div 3$ is like dividing $\frac{1}{2}$ into 3 equal parts, or finding $\frac{1}{3}$ of $\frac{1}{2}$. Remember, $\frac{1}{3}$ of $\frac{1}{2}$ is the same as $\frac{1}{3} * \frac{1}{2}$, so you can use what you know about fraction multiplication to solve fraction division problems.

Example

Eli has $\frac{1}{4}$ cup of raisins. If he splits the raisins equally with a friend, how many cups of raisins will each person get?

Estimate: Since $\frac{1}{4}$ cup of raisins is being divided, the answer will be smaller than $\frac{1}{4}$ cup.

One Way:

Solve this equal-sharing division problem by finding $\frac{1}{4} \div 2$.

Use fraction circles to model the problem.

Start with 1 yellow piece to represent $\frac{1}{4}$ cup of raisins.

Exchange 1 yellow piece for 2 dark blue pieces. This makes sense because $\frac{1}{4} = \frac{2}{8}$.

Split the initial amount into 2 equal parts.

$\frac{2}{8} \div 2 = \frac{1}{8}$

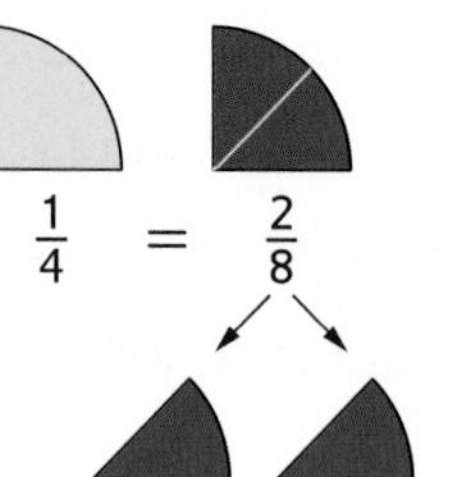

Each person gets $\frac{1}{8}$ cup.

Another Way:

You can also solve this equal-sharing problem by using "multiplication of" and the fraction multiplication method:

Each friend will get $\frac{1}{2}$ of the $\frac{1}{4}$ cup of raisins.

What is $\frac{1}{2}$ of $\frac{1}{4}$?

$$\begin{aligned} \frac{1}{2} \text{ of } \frac{1}{4} &= \frac{1}{2} * \frac{1}{4} \\ &= \frac{(1 * 1)}{(2 * 4)} \\ &= \frac{1}{8} \end{aligned}$$

Using either method, each person will get $\frac{1}{8}$ cup of raisins.

You can check your answer in two ways.

Check using your estimate: Each person will get $\frac{1}{8}$ cup of raisins. This makes sense because the estimate was an amount smaller than $\frac{1}{4}$.

Check using multiplication: Do 2 [$\frac{1}{8}$s] make $\frac{1}{4}$ cup? Does $\frac{1}{8} * 2 = \frac{1}{4}$?
Yes, $\frac{1}{8} * 2 = \frac{2}{8}$, which is the same as $\frac{1}{4}$.

Equal-Grouping Division

One way to interpret a division problem written $a \div b = ?$ is to think, "How many groups of b are there in a?" For example, if you think about equal-grouping division, the problem $6 \div 3 = ?$ asks, "How many groups of 3 are in 6?" The figure at the right shows that there are two groups of 3 in 6, so $6 \div 3 = 2$.

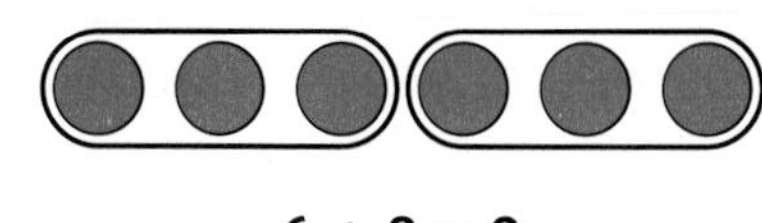

$6 \div 3 = 2$

The problem $6 \div \frac{1}{3}$ asks, "How many $\frac{1}{3}$s are in 6?" The figure at the right shows that there are 18 thirds in 6, so $6 \div \frac{1}{3} = 18$. Problems in which a whole number is divided by a fraction can often be interpreted as equal-grouping division and often result in a quotient (in this case 18) larger than the dividend (6).

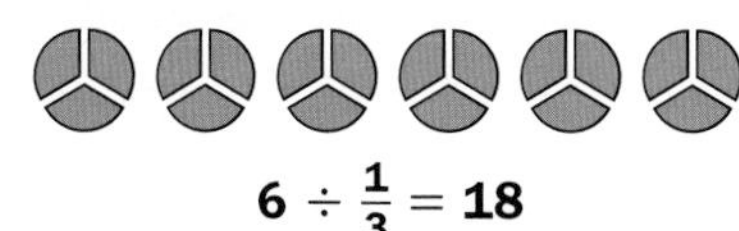

$6 \div \frac{1}{3} = 18$

Example

Frank is walking 5 miles for a walk-a-thon. The course is a $\frac{1}{2}$-mile loop around a park. How many times will Frank have to go around the $\frac{1}{2}$-mile loop to reach 5 miles?

This problem can be solved by finding how many $\frac{1}{2}$s are in 5, which is the same as finding $5 \div \frac{1}{2}$.

Sketch a number line to show 5 miles. Make tick marks to show half-miles.

Count the number of $\frac{1}{2}$-mile jumps from 0 to 5.

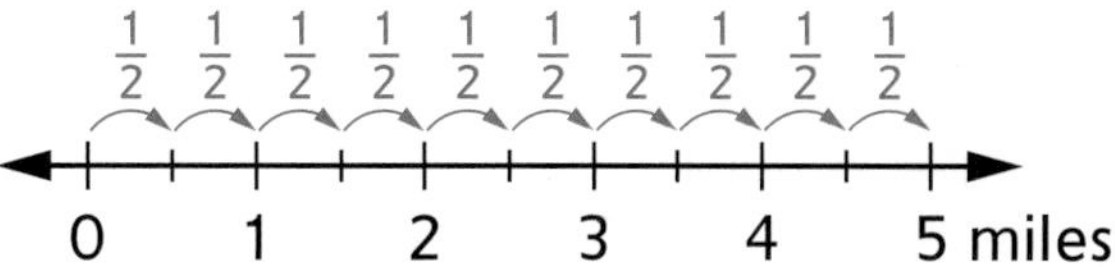

There are 10 [$\frac{1}{2}$s] in 5, so Frank has to go around the $\frac{1}{2}$-mile loop 10 times to reach 5 miles. This makes sense because $10 * \frac{1}{2} = \frac{10}{2}$, or 5.

You can also think about equal-grouping division problems as multiplication problems with a missing factor. In the example above, you can think about $5 \div \frac{1}{2}$ as $\square * \frac{1}{2} = 5$. Since $10 * \frac{1}{2} = 5$, you know that $5 \div \frac{1}{2} = 10$.

Example

Find $2 \div \frac{1}{6}$.

Think: How many $\frac{1}{6}$s are in 2?

This is the same as $\square * \frac{1}{6} = 2$.

Represent the problem with fraction circles.

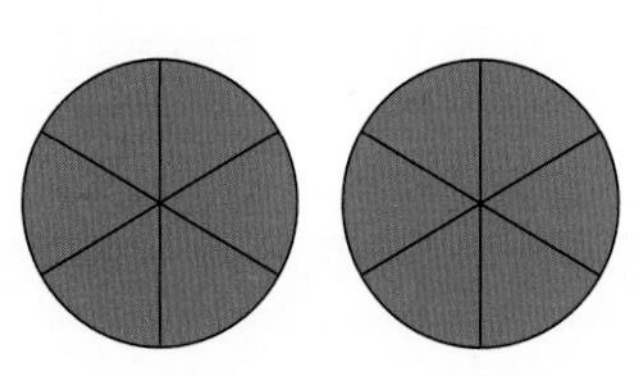

There are 6 [$\frac{1}{6}$s] in 1 whole and 12 [$\frac{1}{6}$s] in 2 wholes, so $12 * \frac{1}{6} = 2$ and $2 \div \frac{1}{6} = 12$.

$2 \div \frac{1}{6} = \mathbf{12}$

Dividing with Common Denominators

One way to solve a fraction division problem is to rename both the dividend and divisor as fractions with a common denominator.

Example

Use common denominators to solve $3 \div \frac{1}{6}$.

Rename 3 as a number of sixths.
Use fraction circles to help.

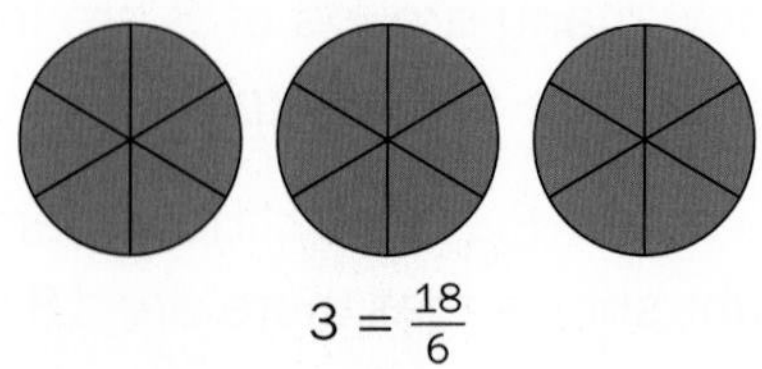

$3 = \frac{18}{6}$

Rewrite the problem with a common denominator: $\frac{18}{6} \div \frac{1}{6}$.
Think: How many $\frac{1}{6}$s are in $\frac{18}{6}$? There are 18 [$\frac{1}{6}$s] in $\frac{18}{6}$, so $\frac{18}{6} \div \frac{1}{6} = 18$.
Because 3 is another name for $\frac{18}{6}$, you know that $3 \div \frac{1}{6} = 18$.
This makes sense because there are 6 sixths in 1 whole, and 3 groups of 6 sixths is 18 sixths, so there are 18 sixths in 3 wholes.

If you look carefully at the example above, you will notice that the same result could be found by dividing the numerators of the rewritten problem: $18 \div 1 = 18$ gives the same result as $\frac{18}{6} \div \frac{1}{6} = 18$. This suggests a shortcut for fraction division: Rename the dividend and divisor as fractions with a common denominator. Then divide the numerators.

Examples

Use the common denominator shortcut to solve each problem.

$8 \div \frac{1}{3} = \mathbf{?}$
Rename 8 as $\frac{24}{3}$.
The problem becomes $\frac{24}{3} \div \frac{1}{3}$.
Divide the numerators:
$24 \div 1 = 24$.
Return to the original problem:
$8 \div \frac{1}{3} = \mathbf{24.}$
This makes sense because $24 * \frac{1}{3} = \frac{24}{3}$, which is the same as 8.

$\frac{1}{6} \div 5 = \mathbf{?}$
Rename 5 as $\frac{30}{6}$.
The problem becomes $\frac{1}{6} \div \frac{30}{6}$.
Divide the numerators:
$1 \div 30 = \frac{1}{30}$.
Return to the original problem:
$\frac{1}{6} \div 5 = \mathbf{\frac{1}{30}}$.
This makes sense because $\frac{1}{30} * 5 = \frac{5}{30}$, which is the same as $\frac{1}{6}$.

Check Your Understanding

Solve.

1. $4 \div \frac{1}{5} = ?$ **2.** $\frac{1}{3} \div 5 = ?$ **3.** $\frac{1}{8} \div 4 = ?$ **4.** $10 \div \frac{1}{4} = ?$

Check your answers in the Answer Key.

Measurement and Data

Natural Measures and Standard Units

Systems of weights and measures have been used in many parts of the world since ancient times. People measured lengths and weights long before they had rulers and scales.

> **Did You know?**
>
> Millet was raised as a grain crop in ancient China. The Chinese used millet seeds to define a unit of weight called the *zhu*. One zhu was the weight of 100 millet seeds, which is about $\frac{1}{50}$ ounce.

Ancient Measures of Weight

Shells and grains, such as wheat or rice, were often used as units of weight. For example, a small item might be said to weigh 300 grains of rice. Large weights were often compared to the load that could be carried by a man or a pack animal.

Ancient Measures of Length

People used natural measures based on the human body to measure length and distance. Some of these units are shown below.

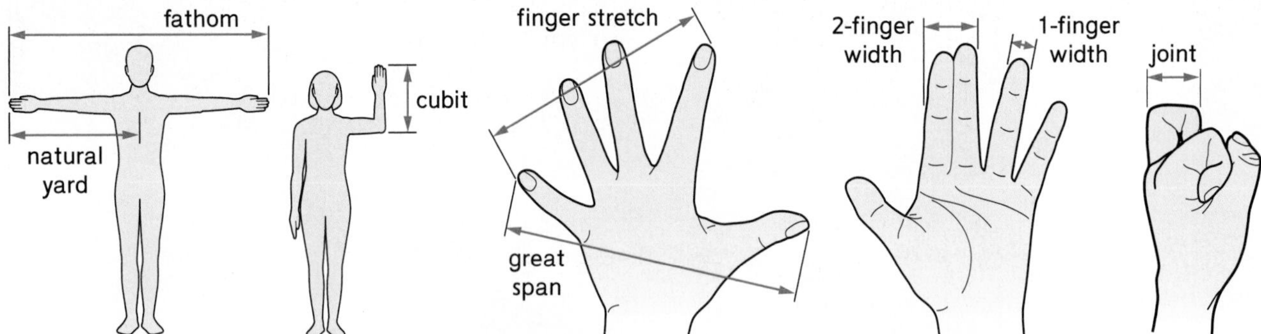

Standard Units of Length and Weight

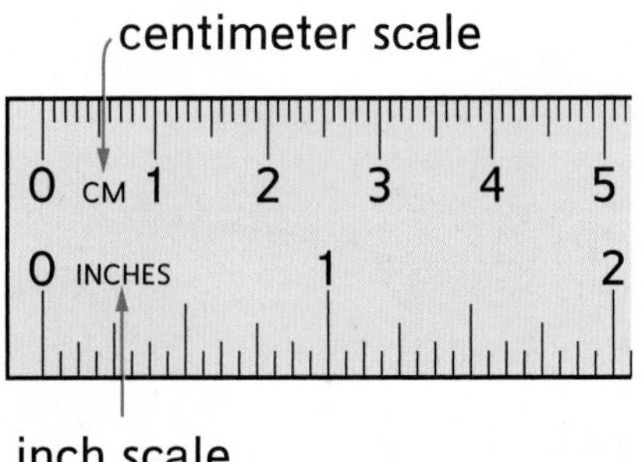

Using shells and grains to measure weight is not exact. Even if the shells and grains are of the same type, they vary in size and weight.

Using body lengths to measure length is not exact either. Body measures depend upon the person who is doing the measuring. The problem is that different persons have hands and arms of different lengths.

One way to solve this problem is to make **standard units** of length and weight. Most rulers are marked off using inches and centimeters as standard units. Bath scales are labeled using pounds and kilograms as standard units. Standard units never change and are the same for everyone. If two people measure the same object using standard units, their measurements will be the same or almost the same.

The Metric System

About 200 years ago, a system of weights and measures called the **metric system** was developed. The metric system uses standard units for length, mass and weight, capacity and liquid volume, and temperature. In the metric system:

- The **meter** is the standard unit for length. The symbol for a meter is **m.** A meter is about the width of a front door. Other commonly used metric units of length include the **millimeter, centimeter,** and **kilometer.**
- The **gram** is the standard unit for mass and weight. The symbol for a gram is **g.** A large paper clip weighs about 1 gram. Another commonly used metric unit of mass is the **kilogram.**
- The **liter** is the standard unit for capacity and liquid volume. The symbol for a liter is **L.** Three regular-size canned drinks have a volume of about 1 liter. Another commonly used metric unit of volume is the **milliliter.**
- The **Celsius degree** is the standard unit for temperature. The symbol for degrees Celsius is **°C.** Water freezes at 0°C and boils at 100°C. Normal room temperature is about 21°C.

Scientists almost always use the metric system. It is easy to use because it is a base-10 system. Larger and smaller units are defined by multiplying or dividing the standard units by powers of 10 (10, 100, 1,000, and so on).

All metric units of length are based on the meter. Each unit is defined by multiplying or dividing the meter by a power of 10.

Units of Length Based on the Meter		Prefix	Meaning
10 decimeters (dm) = 1 meter	1 dm = $\frac{1}{10}$ m	deci-	$\frac{1}{10}$
100 centimeters (cm) = 1 meter	1 cm = $\frac{1}{100}$ m	centi-	$\frac{1}{100}$
1,000 millimeters (mm) = 1 meter	1 mm = $\frac{1}{1{,}000}$ m	milli-	$\frac{1}{1{,}000}$
$\frac{1}{1{,}000}$ kilometer (km) = 1 meter	1 km = 1,000 m	kilo-	1,000

Because the metric system is based on powers of 10, it is relatively easy to convert between metric units.

Example

How many millimeters are in 4.5 meters?

1 m = 1,000 mm, so 4.5 m is 4.5 groups of 1,000 mm.

4.5 ∗ 1,000 = 4,500

So, 4.5 m = 4,500 mm.

This involves converting a larger unit of length to a smaller unit. That means it takes a greater number of smaller units to represent the same length. So, 4,500 mm makes sense.

The U.S. Customary System

The metric system is used in most countries around the world. In the United States, however, the **U.S. customary system** is used for everyday purposes. In the U.S. customary system:

- The **yard** is the standard unit for length. The symbol for a yard is **yd.** A yard is about the width of a front door. Other commonly used U.S. customary units of length include the **inch, foot,** and **mile.**
- The **pound** is the standard unit for weight. The symbol for a pound is **lb.** Four sticks of butter weigh about 1 pound. Other commonly used U.S. customary units of weight include the **ounce** and **ton.**
- The **gallon** is the standard unit for capacity and liquid volume. The symbol for a gallon is **gal.** A gallon of milk is commonly used as an example of a gallon. Other commonly used U.S. customary units of capacity include the **cup, pint,** and **quart.**
- The **Fahrenheit degree** is the standard unit for temperature. The symbol for degrees Fahrenheit is **°F.** Water freezes at 32°F and boils at 212°F. Normal room temperature is about 70°F.

Note The U.S. customary system is not based on powers of 10, so it is more difficult to use than the metric system. For example, to change a measurement from inches to yards, you must know that 12 inches equal 1 foot, and that 3 feet equal 1 yard.

About 1 gallon

About 1 pound

Water boils at 212°F

Check Your Understanding

1. Which of these units are in the U.S. customary system?
   foot millimeter pound inch gram meter centimeter yard
2. What does the prefix *milli-* mean?
3. 3 meters = ? centimeters

Check your answers in the Answer Key.

Converting Units: Two-Column Tables

When you rename a measurement using a different unit, you are making a *unit conversion.* To convert a measurement from one unit to another, you must identify the relationship between the units you are starting with and the units you are converting to. For example, to convert 4 miles to feet, you need to know how many feet are in 1 mile. You can often find information about unit relationships in reference tables like the one below.

Note You can identify the relationship between metric units by paying close attention to prefixes. You can use these prefixes to help you convert units of length, volume, mass, and weight.

- *kilo-* means 1,000
- *deci-* means $\frac{1}{10}$
- *centi-* means $\frac{1}{100}$
- *milli-* means $\frac{1}{1,000}$

Units of Length	
Metric Units	U.S. Customary Units
1 kilometer (km) = 1,000 meters (m)	1 mile (mi) = 1,760 yards (yd)
1 meter = 10 decimeters (dm)	1 mile = 5,280 feet (ft)
1 meter = 100 centimeters (cm)	1 yard = 3 feet
1 meter = 1,000 millimeters (mm)	1 foot = 12 inches (in.)

One way to make conversions is to create a two-column table.

Example

How many millimeters are in 4 meters?

meters (m)	millimeters (mm)
1	1,000
2	2,000
3	3,000
4	4,000

Identify the relationship between meters and millimeters.
The prefix *milli-* means $\frac{1}{1,000}$, so 1 mm = $\frac{1}{1,000}$ m.
1 millimeter is 1 of 1,000 equal parts of a meter, so 1,000 mm make 1 m.
Use the unit relationship to make a two-column table.
Begin with 1 unit of the starting unit: 1 m = 1,000 mm.
Skip count or multiply to fill in the missing values:
If 1 meter is 1,000 mm, then 2 meters equals 2,000 mm, and so on. Continue the table until you reach the starting measurement, 4 m.
Read across the row to find the converted measurement: 4 m = 4,000 mm.

There are 4,000 millimeters in 4 meters.

Example

36 inches is equal to how many feet?

feet (ft)	inches (in.)
1	12
2	24
3	36

Identify the relationship between inches and feet: 1 foot = 12 inches.
Use the relationship to make a two-column table.
The starting measurement is in inches. Use the table to count up by groups of 12 inches until you reach 36 inches.
Read across the row to find the converted measurement: 36 inches = 3 feet.

36 inches is equal to 3 feet.

Converting Units: Using a Rule

Another way to make conversions is by writing a rule based on the relationship between the two units.

Example

How many ounces are in 8 pounds?

Identify the relationship between ounces and pounds.
1 pound = 16 ounces
Write a rule to convert from pounds to ounces.
Think: What operation relates pounds to ounces?
If 1 pound is 16 ounces, then 8 pounds is 8 groups of 16 ounces.
Multiplication makes sense for multiple groups.
Rule: number of pounds ∗ 16 ounces in a pound = number of ounces
Use the rule to convert the units: 8 pounds ∗ 16 ounces in a pound = 128 ounces

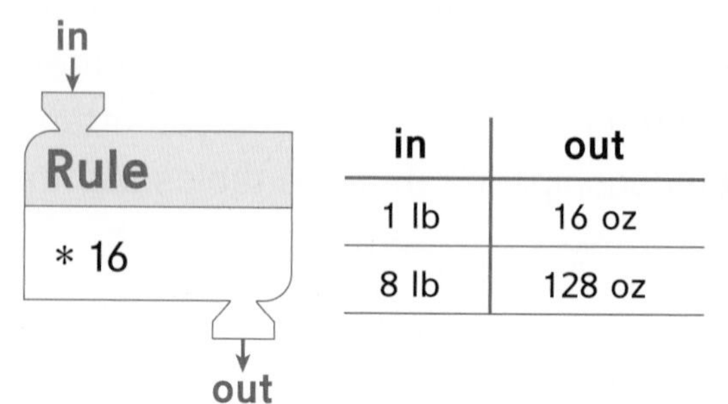

in	out
1 lb	16 oz
8 lb	128 oz

There are 128 ounces in 8 pounds.

Example

6,200 grams is equal to how many kilograms?

Identify the relationship between grams and kilograms.
Kilo- means 1,000, so 1 kilogram = 1,000 grams.
Write a rule to convert from grams to kilograms.
Think: What operation relates grams to kilograms?
If 1,000 grams make 1 kilogram, then you need to find out how many groups of 1,000 are in 6,200 grams.
Division makes sense for splitting an amount into equal groups.
Rule: number of grams ÷ 1,000 grams in a kilogram = number of kilograms
Use the rule to convert the units: 6,200 grams ÷ 1,000 grams in a kilogram = 6.2 kilograms

in
Rule
÷ 1,000
out

in	out
1,000 g	1 kg
6,200 g	6.2 kg

6,200 grams is equal to 6.2 kilograms.

Check Your Understanding

Use a two-column table or write a rule to convert the units.

1. How many meters are equal to 5 kilometers?

2. 16 fluid ounces is equal to how many cups? (Use the table on page 328.)

Check your answers in the Answer Key.

Multi-Step Unit Conversions

Sometimes unit conversions require more than one step. This happens when reference tables do not directly state the relationship between the units you need. For example, it is easy to find that 1 pint = 2 cups in a reference table, but the number of cups in 1 quart may not be listed. In these cases, you can convert to the desired units by using the unit relationships you know in steps.

Example

How many cups are in 2 quarts?

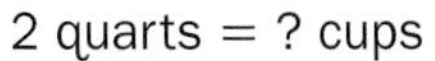

Identify unit relationships that you know.
1 quart = 2 pints
1 pint = 2 cups
Think: You don't know the relationship between quarts and cups, so you first need to convert quarts to pints, which is an intermediate unit.
If 1 quart = 2 pints, then 2 quarts = 4 pints.
Then convert pints to cups.
If 1 pint = 2 cups, then 4 pints = 8 cups.

There are 8 cups in 2 quarts.

Example

60,000 minutes equals how many days?

Identify unit relationships that you know.
1 hour = 60 minutes
1 day = 24 hours
Use what you know to convert to an intermediate unit. In this case, convert minutes to hours.
60 minutes make 1 hour. How many groups of 60 are in 60,000?
60,000 minutes ÷ 60 minutes in an hour = 1,000 hours
Use another unit relationship to convert the intermediate unit to the units you want. In this case, convert hours to days.
24 hours make 1 day. How many groups of 24 are in 1,000?
1,000 hours ÷ 24 hours in a day → 41 days R16 hours

60,000 minutes is equal to 41 days and 16 hours, or $41\frac{16}{24}$ days.

Check Your Understanding

1. How many inches are in 3 yards?
2. 64 fluid ounces are equal to how many pints? (Use the table on page 328.)

Check your answers in the Answer Key.

McGraw-Hill Education/Mark Steinmetz

Length

Length is the measure of the distance between two points. Length is a **1-dimensional** measurement.

Metric Units

In the metric system, units of length include meters (m), decimeters (dm), centimeters (cm), millimeters (mm), and kilometers (km). Length in the metric system is usually measured with a meterstick, the centimeter side of a ruler, or a metric tape measure.

Comparing Metric Units of Length				Symbols for Units of Length	
1 cm = 10 mm	1 m = 1,000 mm	1 m = 100 cm	1 km = 1,000 m	mm = millimeter	cm = centimeter
1 mm = $\frac{1}{10}$ cm	1 mm = $\frac{1}{1,000}$ m	1 cm = $\frac{1}{100}$ m	1 m = $\frac{1}{1,000}$ km	m = meter	km = kilometer

You can estimate lengths by using the lengths of common objects and distances that you know. These are called personal references. Some examples of personal references for metric units of length are given below.

Personal References for Metric Units of Length	
About 1 millimeter	About 1 centimeter
• thickness of a pushpin point • thickness of a dime	• width of a fingertip • thickness of a crayon
About 1 meter	About 1 kilometer
• one big step (for an adult) • width of a front door	• 1,000 big steps (for an adult) • length of 10 football fields (including the end zones)

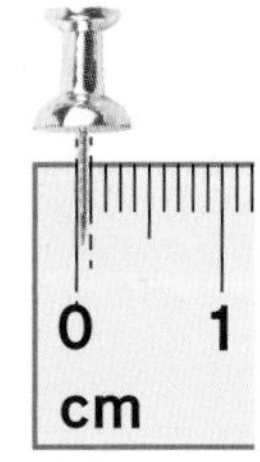

The point of the pushpin is about 1 millimeter thick.

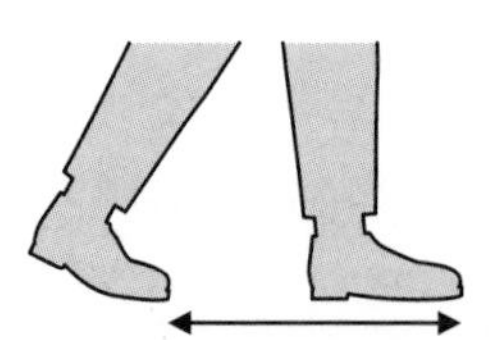
One big step for an adult is about 1 meter long. 1,000 big steps for an adult is about 1 kilometer long.

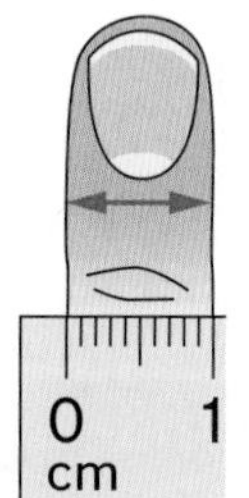

The witdh of your finger is about 1 centimeter

U.S. Customary Units

In the U.S. customary system, units of length include inches (in.), feet (ft), yards (yd) and miles (mi). Length in the U.S. customary system is usually measured with a yardstick, the inch side of a ruler, or a tape measure.

Comparing U.S. Customary Units of Length				Symbols for Units of Length	
1 ft = 12 in.	1 yd = 36 in.	1 yd = 3 ft	1 mi = 5,280 ft	in. = inch	ft = foot
1 in. = $\frac{1}{12}$ ft	1 in. = $\frac{1}{36}$ yd	1 ft = $\frac{1}{3}$ yd	1 ft = $\frac{1}{5,280}$ mi	yd = yard	mi = mile

Some examples of personal references for U.S. customary units of length are given below.

Personal References for U.S. Customary Units of Length	
About 1 inch	About 1 foot
• width (diameter) of a quarter • width of a man's thumb	• distance from elbow to wrist (for an adult) • length of a piece of paper
About 1 yard	About 1 mile
• one big step (for an adult) • width of a front door	• length of 15 football fields (including the end zones) • 2,000 big steps (for an adult)

Note The personal references for 1 meter can also be used for 1 yard. 1 yard equals 36 inches, while 1 meter is about 39.37 inches. One meter is often called a "fat yard," which means 1 yard plus 1 hand width.

The distance across a quarter is about 1 inch.

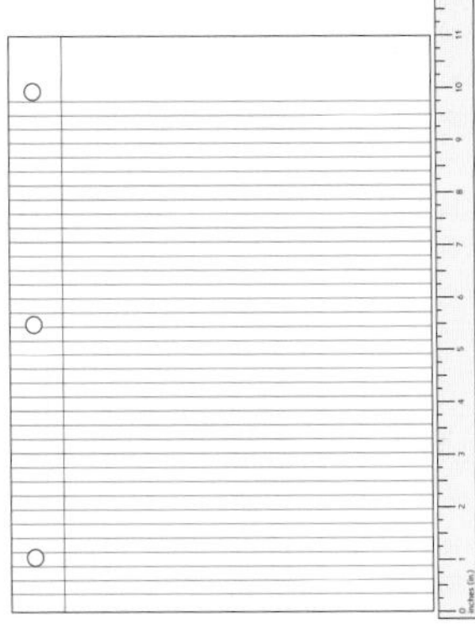

The length of a piece of paper is about 1 foot.

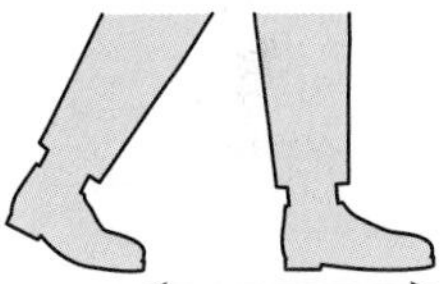

One big step for an adult is about 1 yard. 2,000 big steps for an adult is about 1 mile.

Perimeter

A concrete wall creates a boundary around the pond. The length of this boundary is the perimeter of the pond.

Sometimes we want to know the distance around a shape. **Perimeter** is the total length of the boundary of a closed shape. To measure perimeter, we use units of length such as inches, meters, or miles.

To find the perimeter of any polygon, add the length of each of its sides. Remember to name the unit of length used to measure the polygon's sides.

Example

Find the perimeter of polygon *ABCDE*.

3 m + 8 m + 5 m + 4 m + 10 m = 30 m

The perimeter is 30 meters.

B, C, D, E, A; 8 m, 5 m, 4 m, 10 m, 3 m

A rule, or **formula,** can be used to calculate the perimeter of polygons with known numbers of equal-length sides. To find the perimeter of rectangles or squares, you can use the formulas shown below.

Perimeter Formulas	
Rectangles	Squares
$p = 2l + 2w$ or $p = 2 * (l + w)$ *p* is the perimeter, *l* is the length, and *w* is the width of the rectangle.	$p = 4 * s$ *p* is the perimeter and *s* is the length of one of the sides of the square.

Examples

Find the perimeter of each polygon.

4 cm, 3 cm

Use the formula $p = 2 * (l + w)$:
length (l) = 4 cm
width (w) = 3 cm
perimeter (p) = 2 ∗ (4 cm + 3 cm)
= 2 ∗ 7 cm = 14 cm
The perimeter is 14 centimeters.

9 ft

Use the formula $p = 4 * s$:
length of a side (s) = 9 ft
perimeter (p) = 4 ∗ 9 ft = 36 ft

The perimeter is 36 feet.

Check Your Understanding

1. Find the perimeter of an equilateral triangle with a side length of 5 cm.
2. Find the perimeter of a rectangle with dimensions of 4 ft 2 in. and 9 ft 4 in.

Check your answers in the Answer Key.

Area

Sometimes we want to know the amount of surface inside a shape. The amount of surface inside the boundary of a closed shape is called the shape's **area.** Area is a **2-dimensional** measurement, which means it has length and width, but not thickness.

The images below represent two different backyards. The area of Carla's backyard is larger than Greg's. Greg's backyard could be placed inside of Carla's backyard with space left over.

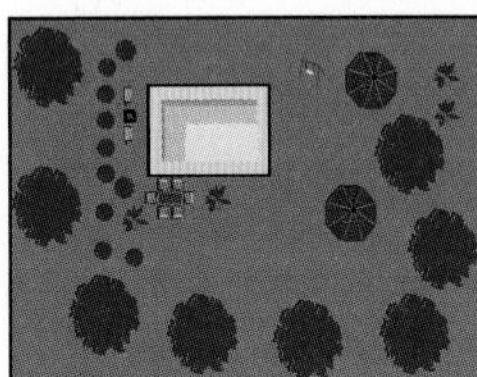

Carla's backyard

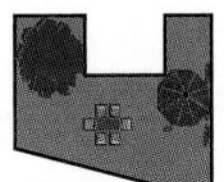

Greg's backyard

Area is measured in square units. A **square unit** is a square of consistent size and shape that is used to cover a surface. One square unit is called a **unit square.** You can find the area of a shape by counting the number of unit squares that cover the inside of the shape. The squares must cover the entire inside of the shape and must not overlap, have any gaps, or cover any surface outside of the shape.

Example

What is the area of the hexagon?

The hexagon at the right is covered by squares that measure one unit on each side. Each square is a **unit square.**

Five unit squares cover the hexagon.
The area of the hexagon is 5 square units.
This is written as 5 sq units or 5 units^2.

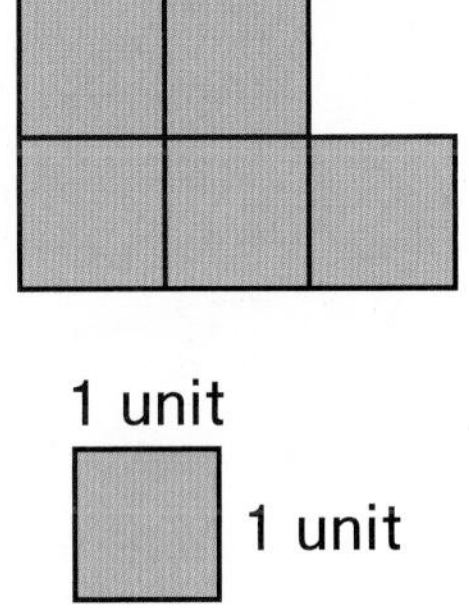

1 unit

1 unit

1 square unit

Sometimes a shape cannot be completely covered with whole square units. In that case, you can find the area of the shape by counting the number of whole squares and parts of squares that completely cover the inside of the shape.

Example

What is the area of the rectangle?

8 whole unit squares and 2 half-squares cover the inside of the rectangle.
2 half-squares make 1 unit square.
8 square units + 1 square unit = 9 square units

The area of the rectangle is 9 square units, or 9 units^2.

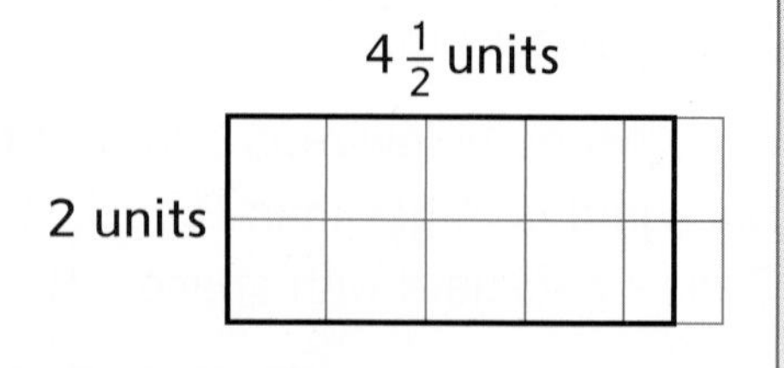

Units of Area

Area is reported in square units, such as square inches, square meters, or square miles. You can compare units of area by thinking about the related unit of length. For example, since 1 inch is longer than 1 centimeter, 1 square inch covers more area than 1 square centimeter. Similarly, since 1 yard is longer than 1 foot, 1 square yard covers more area than 1 square foot.

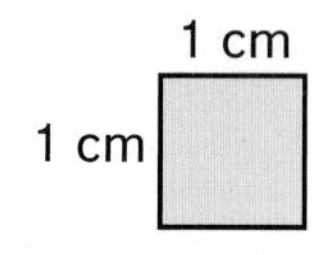

1 square centimeter (actual size)

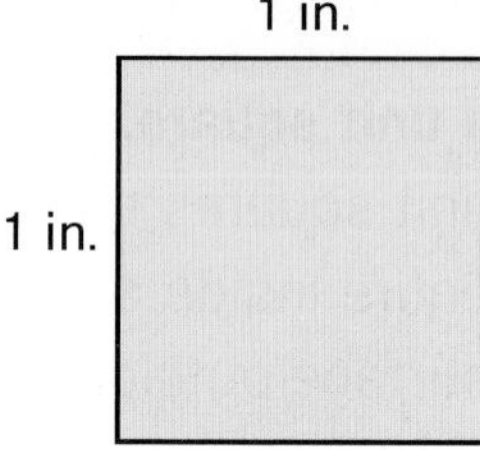

1 square inch (actual size)

We measure different objects using different units. Small regions, such as a book cover, might be measured in square inches (in.^{2}) or square centimeters (cm^2). The floor of a bedroom might be measured in square feet (ft^2), square yards (yd^2), or square meters (m^2). Large regions, such as farms or national parks, might be measured in square miles (mi^2) or square kilometers (km^2).

To compare units of area more precisely, you can tile a larger unit with a smaller unit.

Example

How does a square yard compare to a square foot?

A square yard is a square with a side length of 1 yard.
A square foot is a square with a side length of 1 foot.

1 yard = 3 feet, so a square yard is a 3-ft by 3-ft square.

When a square yard is tiled with square feet, it takes 9 square feet to cover 1 square yard.

1 square yard = 9 square feet, so a square yard is 9 times as large as 1 square foot.

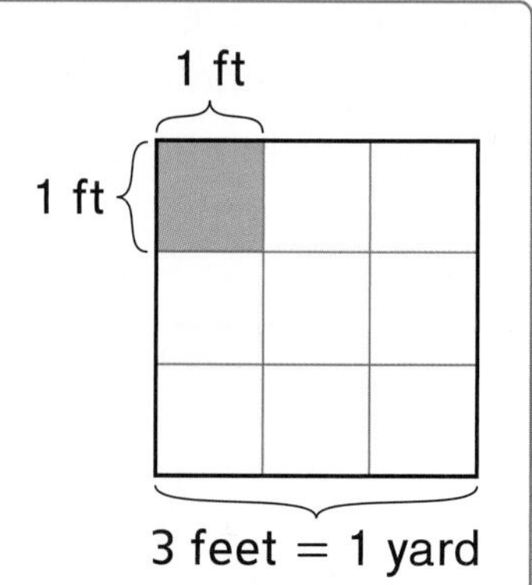

You may report area using any square units, but you should choose a square unit that makes sense for the region being measured.

Example

The area of a field-hockey field is reported below in three different ways.

The area of the field is 6,000 square yards.
area = 6,000 yd^2

The area of the field is 54,000 square feet.
area = 54,000 ft^2

The area of the field is 7,776,000 square inches.
area = 7,776,000 $in.^2$

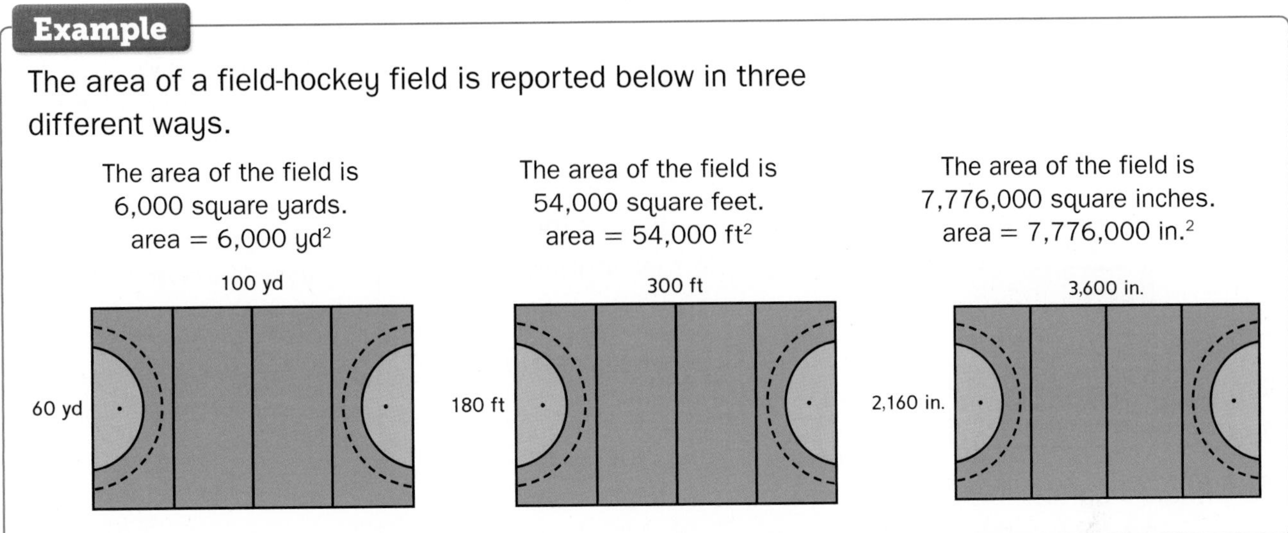

Although each of the measurements above is correct, reporting the area in square inches does not give a good idea of the size of the field. It is hard to imagine 7,776,000 of anything.

Note The **acre** is another unit of area. An acre was originally defined as the amount of land that could be plowed in one day. Now an acre is a standard unit and is defined as an area equal to 4,840 yd^2.

Check Your Understanding

1. Imagine that you are measuring the surface of your desk. Which square unit would make more sense: square inches or square yards? Why?
2. A teacher is measuring the area of her classroom rug. Should she measure the area in square centimeters or square meters? Why?

Check your answers in the Answer Key.

Finding Area with Rows and Columns

When you cover, or **tile,** a rectangular shape with unit squares, the squares can be arranged in rows and columns. You can use rows and columns to find the area of a rectangle by thinking about making copies of one complete row or one complete column.

Note When smaller units, such as unit squares, are grouped together, they can be called **composite units.** For example, if you group 5 unit squares into 1 row, the row is a composite unit. You can then find the area by adding the number of times you repeat the composite unit as you cover the inside of the shape.

Example

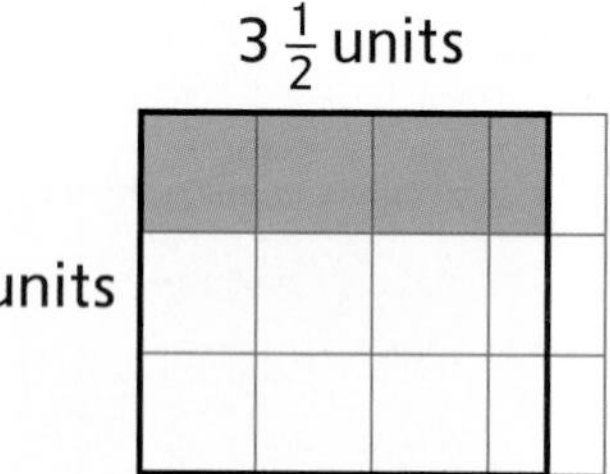

Inside the boundary of the rectangle, one row contains $3\frac{1}{2}$ square units.

There are 3 rows of unit squares inside the rectangle.

Imagine making 3 copies of $3\frac{1}{2}$ square units. How many square units is that in all?

$3\frac{1}{2} + 3\frac{1}{2} + 3\frac{1}{2} = 9\frac{3}{2}$, or $10\frac{1}{2}$

There are $10\frac{1}{2}$ square units inside the boundary of the rectangle, so the area is $10\frac{1}{2}$ units^2.

Sometimes it is helpful to draw lines for rows and columns if they are not already shown.

Example

Find the area of the rectangle.

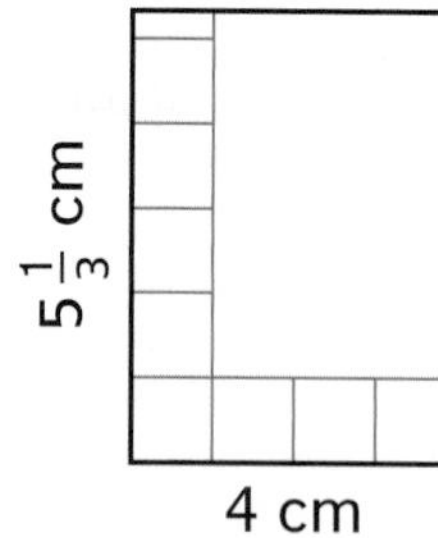

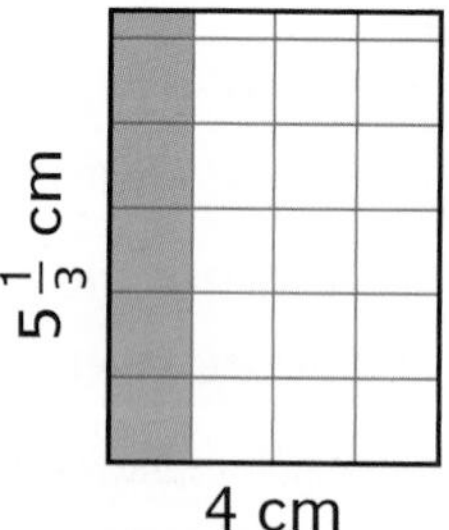

The rectangle is $5\frac{1}{3}$ cm tall and 4 cm across. This means that there are 4 columns, and each column contains $5\frac{1}{3}$ centimeter squares.

Show 4 copies of $5\frac{1}{3}$ square centimeters by extending the lines for each row and column. How many square centimeters cover the inside of the rectangle?

$5\frac{1}{3} + 5\frac{1}{3} + 5\frac{1}{3} + 5\frac{1}{3} = 20\frac{4}{3}$, or $21\frac{1}{3}$

The area of the rectangle is $21\frac{1}{3}$ square centimeters.

Area of a Rectangle

You can find the area of a rectangle by taking the number of squares in each row and multiplying the number of times you repeat each row as you cover the inside of a shape. Imagine sweeping or rolling across the surface of a shape with a row.

Example

Find the area of the rectangle.

There are 5 squares in each row, and 3 rows in the rectangle. There are 15 squares.

Area = 15 square units

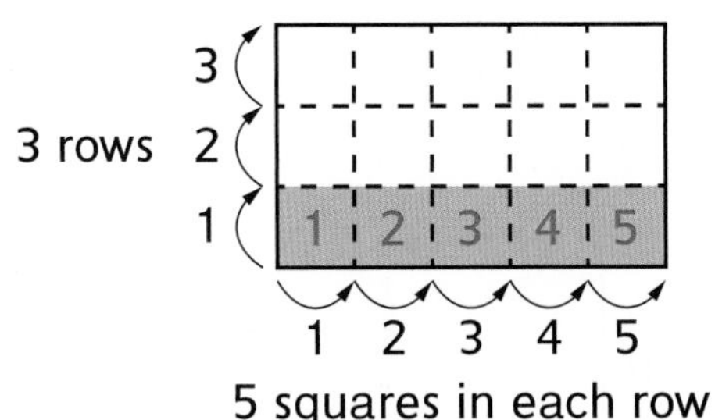

The number of squares in one row can be called the **length** of the rectangle, and the number of rows in the rectangle can be called the **width.** To find the area of a rectangle, use the following formula:

area	=	(the number of squares in one row)	∗	(the number of rows)
area	=	length	∗	width
A	=	l	∗	w

Example

Use the formula $A = l * w$:

Each row has $3\frac{1}{3}$ square units, so the length (l) of the rectangle is $3\frac{1}{3}$ units.

$3\frac{1}{3}$ units

2 units

There are 2 rows, so the width (w) of the rectangle is 2 units.

area (A) = $3\frac{1}{3}$ units ∗ 2 units

$3\frac{1}{3} * 2$ is the same as $3\frac{1}{3} + 3\frac{1}{3}$.

area (A) = $3\frac{1}{3}$ sq units + $3\frac{1}{3}$ sq units = $6\frac{2}{3}$ square units

The area of the rectangle is $6\frac{2}{3}$ square units.

Note The distance along one side of a rectangle is its length and the distance along the adjacent side is its width. Either dimension can be considered the length and then the adjacent side the width.

Check Your Understanding

Find the area of the following figures. Include the unit in your answers.

1.

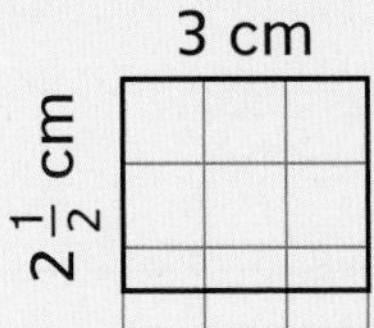

2.

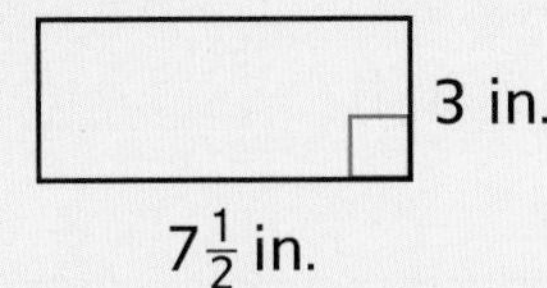

3.

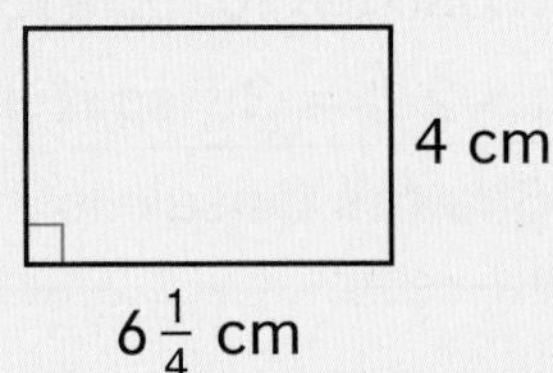

Check your answers in the Answer Key.

Tiling with Squares of Fractional Side Lengths

In everyday life, it is sometimes useful to tile rectangles with squares that are smaller than 1 square unit. For example, square tiles used in kitchens, bathrooms, and artwork often come in sizes that do not measure 1 full unit along each side.

One square foot is a square with sides that are each 1 foot long. You could tile a surface with square feet. You could use a smaller tile that is a square with $\frac{1}{2}$-foot side lengths.

Example

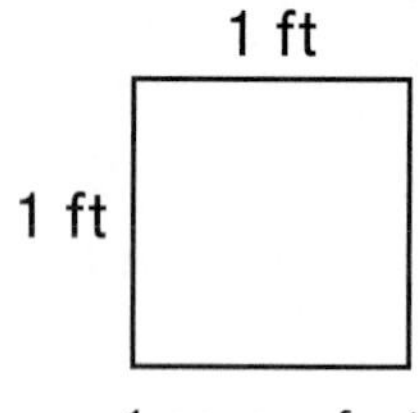

1 square foot
(1 square unit)

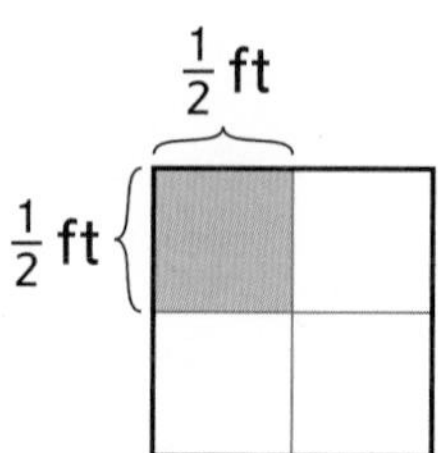

The shaded square is a $\frac{1}{2}$-ft by $\frac{1}{2}$-ft square. Since 4 of these smaller squares make 1 square foot, the area of each smaller square is $\frac{1}{4}$ square foot.

You can find the area of a rectangle by tiling with squares of fractional side lengths and thinking about how many smaller squares make 1 square unit.

You can check the area of the rectangle by multiplying its side lengths (using the area formula, $A = l * w$).

Example

Shawn is making a table-mosaic with $\frac{1}{2}$-ft by $\frac{1}{2}$-ft square tiles.

7 tiles fit along the length of the rectangle.
4 rows of tiles fit along the width of the rectangle.

What is the area of Shawn's mosaic?

4 rows ∗ 7 tiles in each row = 28 tiles

4 tiles have an area of 1 square foot

28 tiles ÷ 4 tiles in 1 square foot = 7 square feet

Check: The picture shows that the table is $3\frac{1}{2}$ ft by 2 ft. Use the formula $A = l * w$.

area = $3\frac{1}{2}$ ft ∗ 2 ft = $3\frac{1}{2}$ sq ft + $3\frac{1}{2}$ sq ft = 7 sq ft

The area of Shawn's mosaic is 7 square feet.

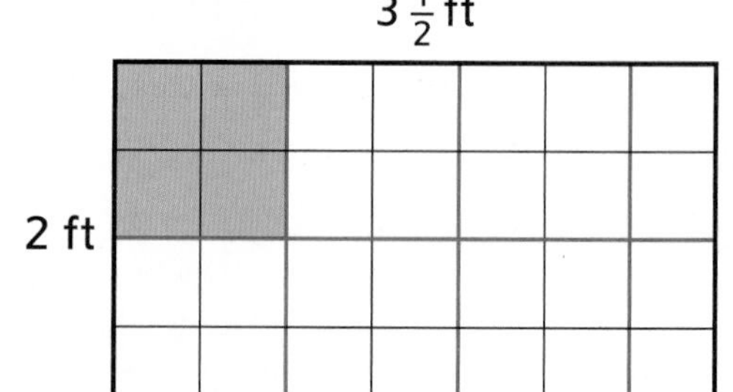

The highlighted area is 1 square foot. The rectangle is covered by 28 $\frac{1}{2}$-ft by $\frac{1}{2}$-ft tiles, which is equal to 7 square feet.

Area of Rectilinear Figures

The shapes below are called **rectilinear figures** because their sides are all line segments and the inside and outside corners of each shape are all right angles. Rectangles are rectilinear figures, but the figures on this page are not rectangles because they have more than four sides.

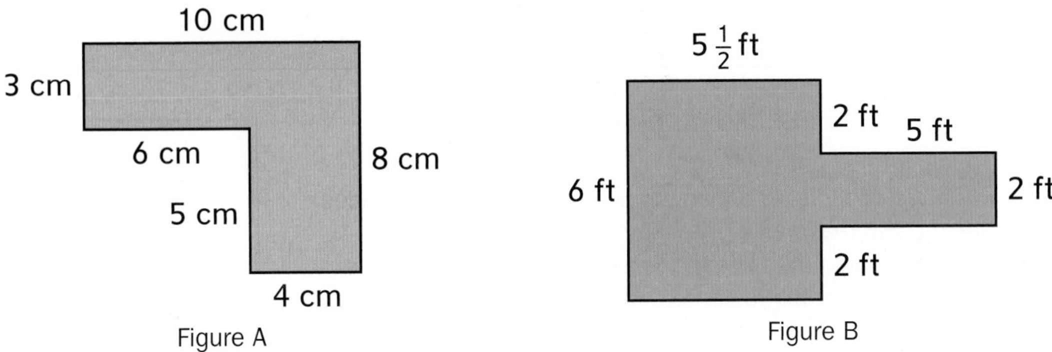

Figure A

Figure B

You can find the area of a rectilinear figure by decomposing it into non-overlapping rectangles, finding the area of each rectangle, and then adding up the smaller areas.

Example

Find the area of this rectilinear shape.

Each square is 1 square inch.

7 in.
3 in.
5 in.
3 in.
2 in.
4 in.

3 in.
3 in.
5 in.
4 in.

Step 1 Decompose the shape into non-overlapping rectangles.

Step 2 Find the area of each rectangle.
Use the formula $A = l * w$.
area of blue rectangle = 3 inches * 3 inches = 9 square inches
area of red rectangle = 4 inches * 5 inches = 20 square inches

Step 3 Add the areas of the two rectangles to find the area of the rectilinear shape.
area of shape = area of blue rectangle + area of red rectangle
= 9 square inches + 20 square inches
= 29 square inches

The area of the shape is 29 square inches.

Check Your Understanding

1. Copy Figure A above on grid paper or make a sketch. Divide it into two non-overlapping rectangles to find its area.
2. Copy Figure B above on grid paper or make a sketch. Divide it into two non-overlapping rectangles to find its area.

Check your answers in the Answer Key.

The Rectangle Method for Finding Area

Sometimes you will need to find the area of a polygon that is not a rectangle. Unit squares will not fit neatly inside the figure, and you will not be able to simply use the formula for the area of a rectangle. One approach is called the **rectangle method.** Rectangles are used to surround the figure or parts of the figure. Then you use what you know about finding the area of rectangles to calculate the area of the figure.

Example

What is the area of the triangle at the right?

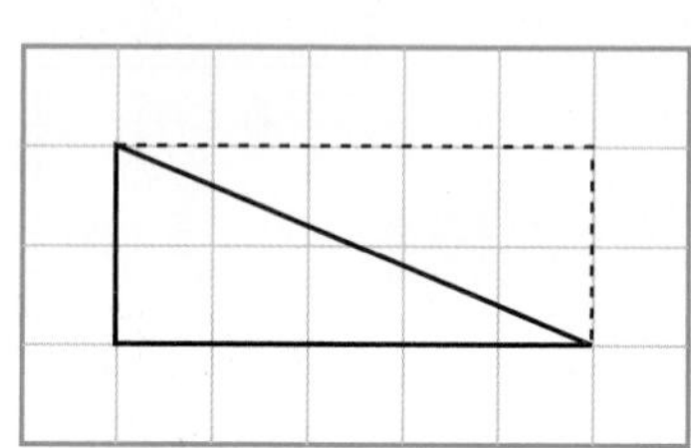

Draw a rectangle around the triangle.
The area of the rectangle is 10 square units. The diagonal line divides the rectangle into two triangles of the same size and shape.
Two copies of the triangle fit inside the rectangle, so the area of the triangle is half the area of the rectangle. Half of 10 square units is 5 square units.

The area of the triangle is 5 square units.

Example

What is the area of the trapezoid at the right?

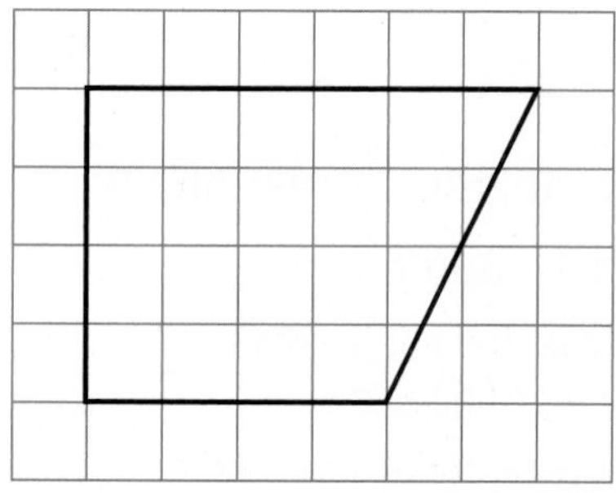

Step 1 Draw a rectangle around the trapezoid.
The area of the large rectangle is 24 square units ($A = 6 * 4 = 24$).

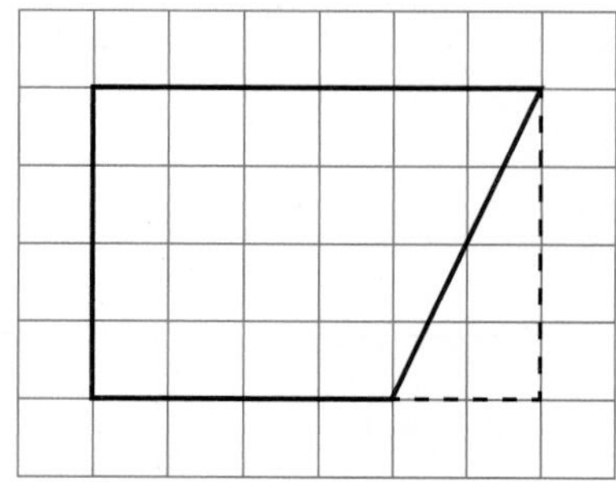

Step 2 Draw a rectangle around the triangular part of the trapezoid.
The area of the smaller rectangle is 8 square units ($A = 4 * 2 = 8$). The shaded triangle has half the area of the smaller rectangle. Half of 8 square units is 4 square units.

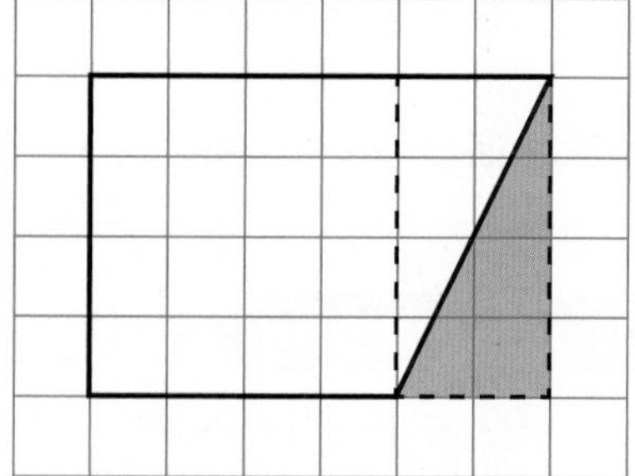

Step 3 Subtract the area of the extra triangle from the area of the larger rectangle:
$24 - 4 = 20$ square units.

The area of the trapezoid is 20 square units.

Example

What is the area of the triangle at the right?

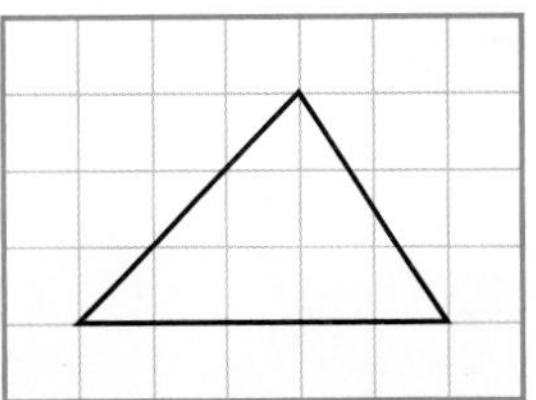

Step 1 Divide the triangle into two parts.

Step 2 Draw a rectangle around one of the shaded parts.

The area of the rectangle is 9 square units. The shaded area is $4\frac{1}{2}$ square units.

Step 3 Draw a rectangle around the other shaded part.

The area of the rectangle is 6 square units. The shaded area is 3 square units.

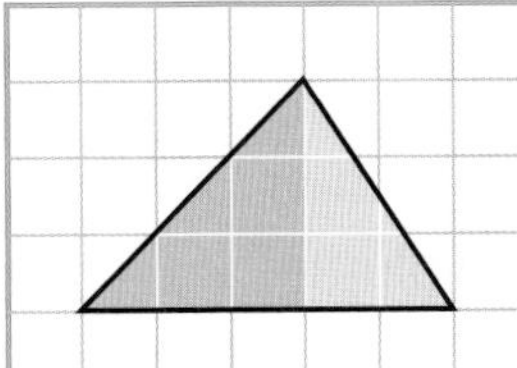

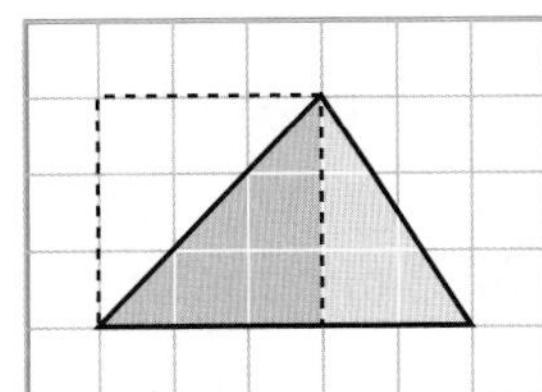

Step 4 Add the areas of the two shaded parts: $4\frac{1}{2} + 3 = 7\frac{1}{2}$ square units.

The area of the triangle is $7\frac{1}{2}$ square units.

Check Your Understanding

Use the rectangle method to find the area of each figure below.

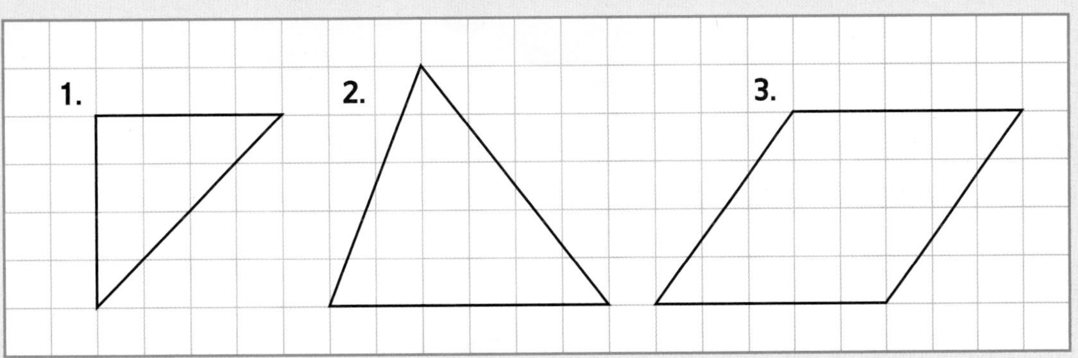

Check your answers in the Answer Key.

Volume

All **3-dimensional** objects take up space. When you measure how much space an object takes up, you are measuring **volume.**

The blue packing box has a greater volume than the wrapped box because it takes up more 3-dimensional space. A basketball has a greater volume than a baseball because it takes up more 3-dimensional space.

Some 3-dimensional objects are solid, like a book. Other 3-dimensional objects are hollow, or empty inside, like a box. Whether objects are filled-in or hollow, the object still takes up space. All 3-dimensional objects have volume.

All of the items below have volume.

One way to think about how much space something takes up is to see how much it holds. If objects are hollow, you can measure and compare their volumes by filling or packing both objects with the same material.

Example

Which container has a greater volume, the graduated cylinder or the glass bowl?

The amount of rice in both containers is the same.

The glass bowl has a greater volume because it can hold more rice than the graduated cylinder.

Measuring Volume with Unit Cubes

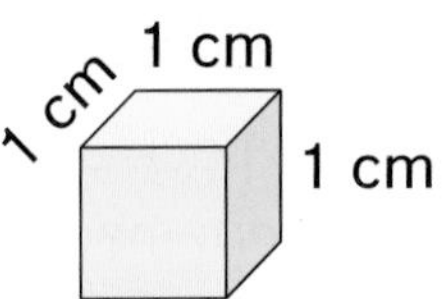

The volume of solid objects is measured in **cubic units,** such as cubic centimeters (cm^3), cubic inches (in.^{3}), and cubic feet (ft^3). A cubic centimeter (cm^3) is a cubic unit because every edge is 1 cm long. A single cube is called a **unit cube.**

You can find the volume of an object by counting the number of same-size cubes needed to fill the object. The cubes should be packed with no gaps and no overlaps. This method is particularly useful for finding the volume of **rectangular prisms.** Cracker boxes are examples of rectangular prisms.

Example

Count unit cubes to find the volume of each object.

Each cube is 1 cubic centimeter.

It takes 4 cubes to fill Object A with no gaps and no overlaps.

The volume of Object A is 4 cubic centimeters. This is written as 4 cm^3.

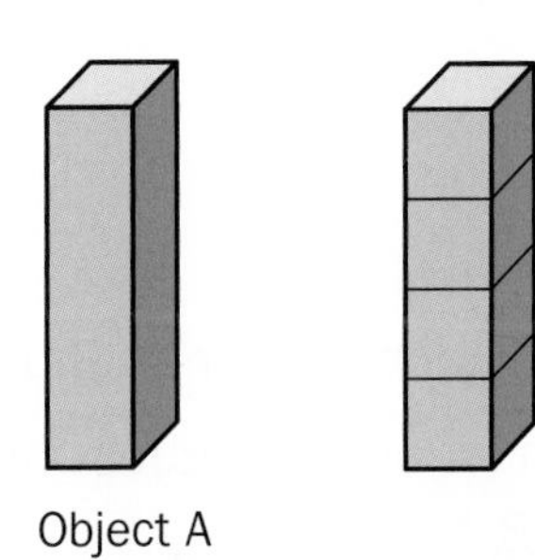

Object A

Each cube is 1 cubic centimeter.

It takes 30 cubes to fill Object B with no gaps and no overlaps.

The volume of Object B is 30 cm^3.

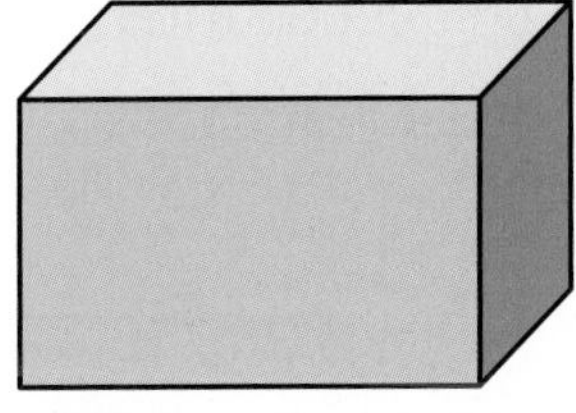

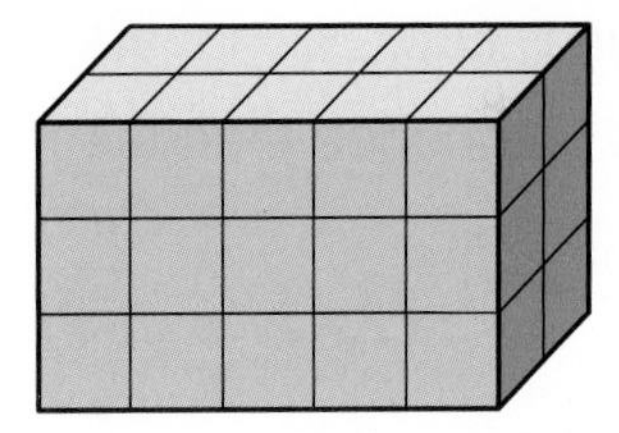

Object B

Check Your Understanding

Find the volume of each figure.

1.

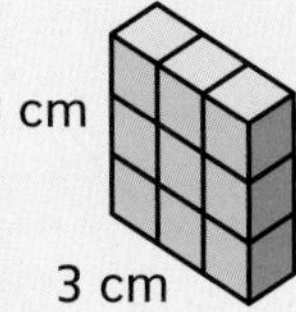

2.

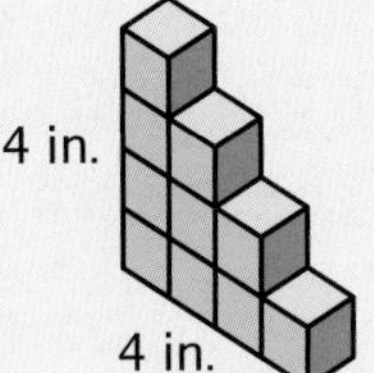

3.

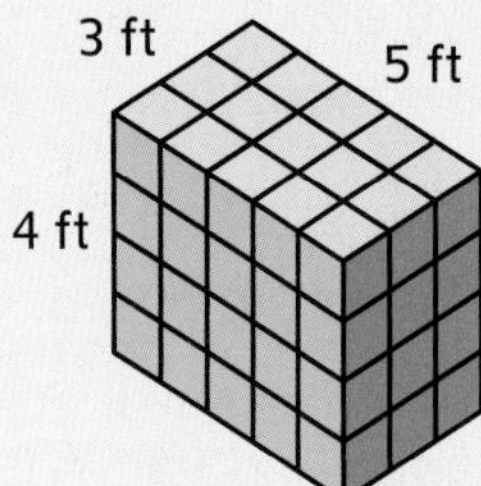

Check your answers in the Answer Key.

Measuring Volume Using Layers

If you pack a rectangular prism with unit cubes, the cubes will be arranged in layers made up of rows and columns. Since each layer has the same number of cubes, you can find the volume without having to count the cubes one-by-one. Instead, you can multiply the number of cubes in each layer by the number of layers. You can also use repeated addition.

Example

Find the volume of the prism.

The bottom layer has 2 rows, with 4 unit cubes in each row.

2 rows ∗ 4 unit cubes in each row = 8 unit cubes

There are 8 unit cubes in the bottom layer.

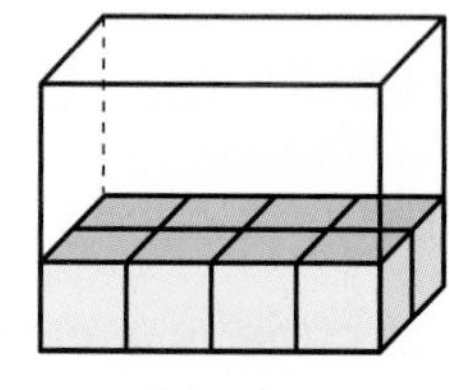

1 layer

There are 3 layers. Each layer has 8 unit cubes.

Picture repeating a layer of 8 cubes three times to find the total number of cubes: 8 + 8 + 8 = 24.

Or: 8 unit cubes in 1 layer ∗ 3 layers = 24 unit cubes total

Three layers of 8 unit cubes makes a total of 24 unit cubes.

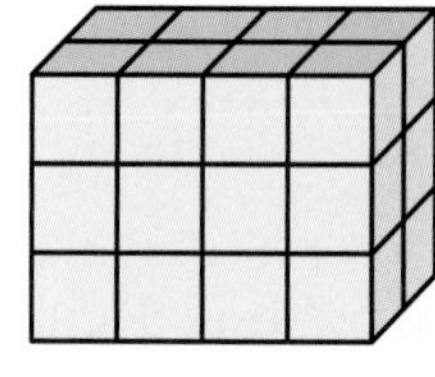

3 layers

Volume = 24 cubic units

If a prism is shown before it is completely packed with unit cubes, it can be helpful to imagine or build the prism with full rows, columns, and layers of cubes.

Example

Find the volume of the prism.

There are 5 rows and 4 columns of unit cubes in the base layer.

5 rows of 4 unit cubes each = 20 unit cubes in one layer

There are 4 cubes stacked in the back corner, so there are 4 layers of cubes in the prism.

Imagine repeating a layer of 20 unit cubes four times:

20 unit cubes in 1 layer ∗ 4 layers = 80 unit cubes total

Volume = 80 cubic units

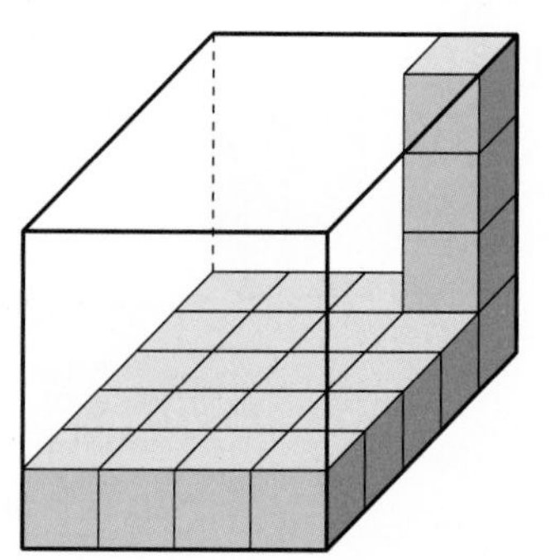

Volume of a Rectangular Prism

You can find the volume of a rectangular prism by thinking about the base and height of the prism. The number of square units in the base corresponds to the number of cubes in one layer. The height of the prism corresponds to the number of layers.

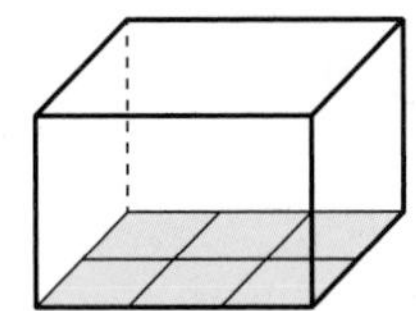

area of the base = 6 sq cm

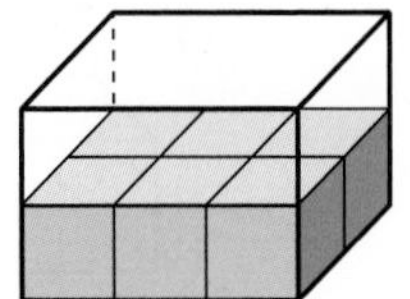

6 cm cubes in 1 layer

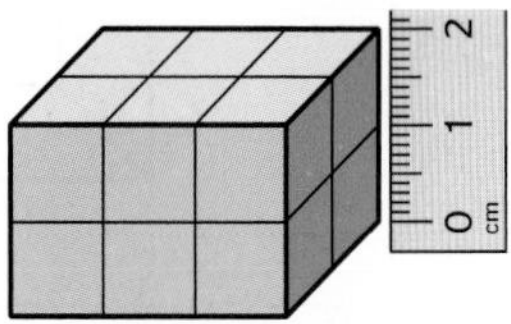

number of layers = 2
height = 2 cm

You can use the following formula to find the volume of the prism:

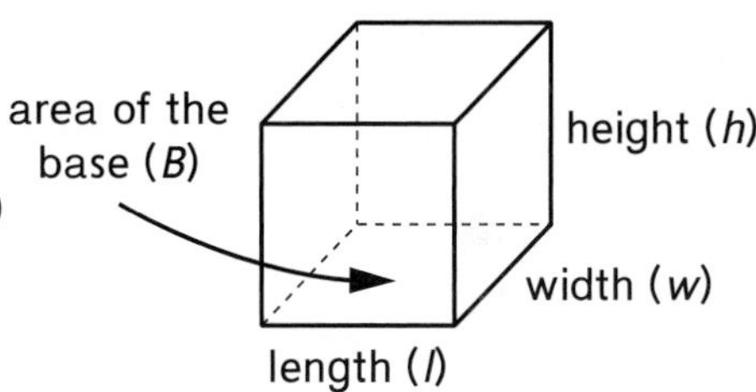

volume = (number of cubes in 1 layer) ∗ (number of layers)
volume = area of the base ∗ height
$V = B * h$

If the area of the base is not shown, you can find the area by using its dimensions. Multiply the length of the base by the width of the base: $A = l * w$. This gives the following formula for the volume of a rectangular prism:

volume = area of the base ∗ height
volume = (length ∗ width) ∗ height
$V = l * w * h$

Both volume formulas, $V = B * h$ and $V = l * w * h$, give the same result.

Example

Find the volume of the rectangular prism.

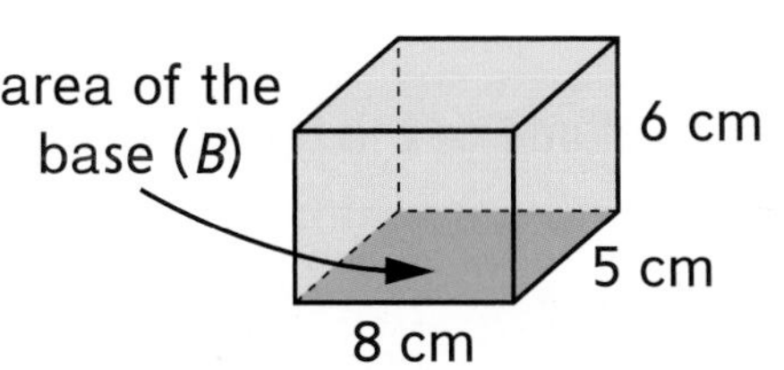

Method 1 Use the formula $V = B * h$. Find the area of the base and multiply by the height of the prism.

- The base is a 5 cm by 8 cm rectangle. The area of the base (B) is 5 cm ∗ 8 cm, or 40 cm^2.
- The height (h) of the prism is 6 cm.
- volume = 40 cm^2 ∗ 6 cm = 240 cm^3

Method 2 Use the formula $V = l * w * h$. Multiply the dimensions of the prism.

- The length (l) of the base is 8 cm.
- The width (w) of the base is 5 cm.
- The height (h) of the prism is 6 cm.
- volume = 8 cm ∗ 5 cm ∗ 6 cm = 240 cm^3

The volume of the rectangular prism is 240 cm^3.

You should use the formula that makes sense for the dimensions you know.

Volume of Figures Composed of Prisms

The figures below are made, or **composed,** of rectangular prisms.

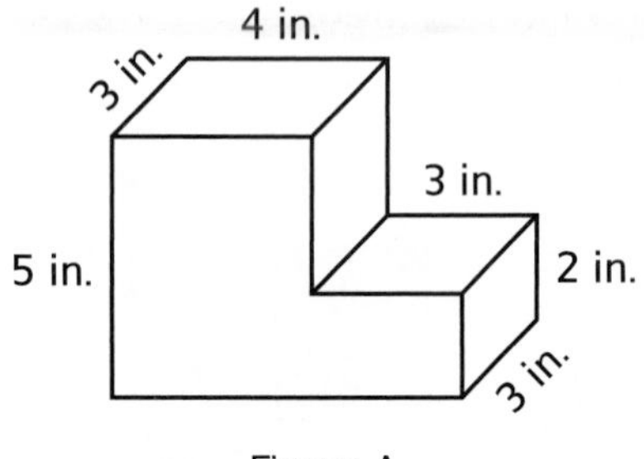

Figure A

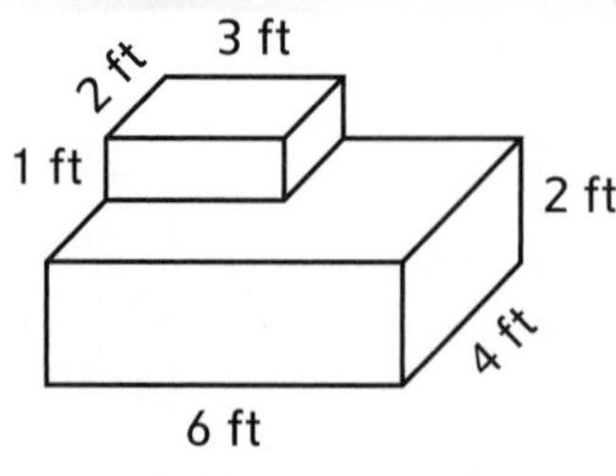

Figure B

You can find the volume of a figure composed of rectangular prisms by decomposing it into rectangular prisms, finding the volume of each prism, and then adding up the smaller volumes.

Example

Find the volume of the figure.

Each cube is 1 cubic centimeter.

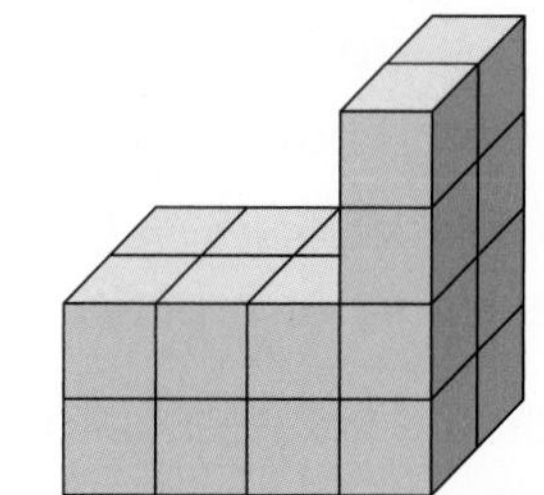

Decompose the figure into rectangular prisms.

Find the volume of each prism. Use the formula $V = l * w * h$.

volume of left prism = 3 cm ∗ 2 cm ∗ 2 cm = 12 cm^3
volume of right prism =1 cm ∗ 2 cm ∗ 4 cm = 8 cm^3

Add the volumes of the two prisms to find the volume of the original figure.

volume of figure = volume of left prism + volume of right prism
= 12 cm^3 + 8 cm^3
= 20 cm^3

The volume of the figure is 20 cubic centimeters.

Check Your Understanding

1. What is another way to find the volume of the figure in the example box.
2. Find the volume of Figure A above.
3. Find the volume of Figure B above.

Check your answers in the Answer Key.

Units of Volume

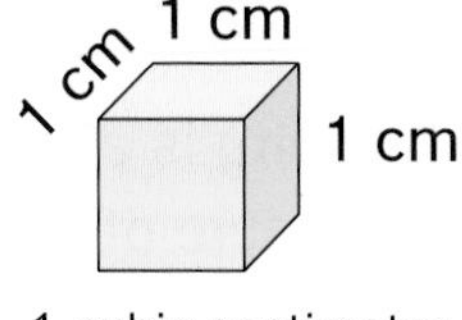

1 cubic centimeter (actual size)

The volume of 3-dimensional objects is measured in **cubic units.** The size of the unit cube used will depend on the size of the object you are measuring.

For smaller objects, such as a tissue box, you might use cubic centimeters (cm^3) or cubic inches ($in.^3$). For larger objects, such as a storage closet, you might use cubic feet (ft^3), cubic yards (yd^3), or cubic meters (m^3).

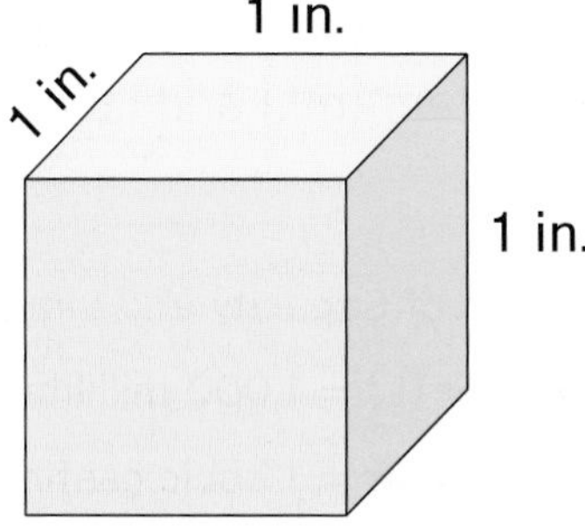

1 cubic inch (actual size)

Each cubic unit is named for the length of its edges. For example, a cube with 1-cm edges is a cubic centimeter, and a cube with 1-inch edges is called a cubic inch. A cubic foot has 1-foot edges, a cubic yard has 1-yard edges, and a cubic meter has 1-meter edges.

You can compare units of volume by thinking about the length of the unit cubes' edges. For example, since 1 inch is longer than 1 centimeter, 1 cubic inch has a greater volume than 1 cubic centimeter.

You can also compare units of volume by imagining a larger unit packed with a smaller unit.

Example

How does a cubic yard compare to a cubic foot?

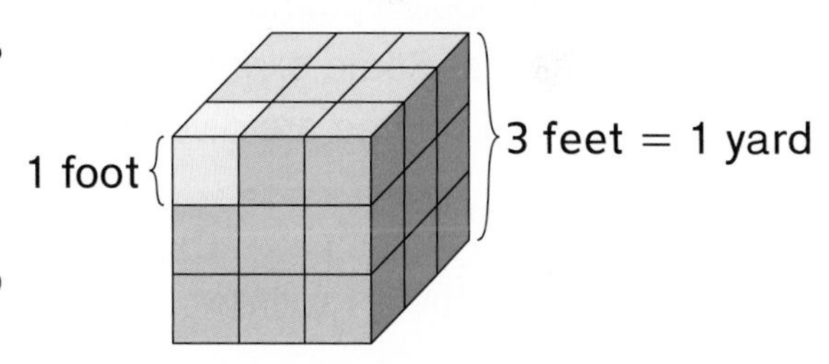

A cubic yard is a cube with 1-yard edges.

A cubic foot is a cube with 1-foot edges.

1 yard = 3 feet, so a cubic yard is equivalent to a 3 ft by 3 ft by 3 ft cube.

The area of the base of a cubic yard is 3 feet $*$ 3 feet = 9 square feet.

The height of a cubic yard is 3 feet.

volume = area of the base $*$ height = 9 square feet $*$ 3 feet = 27 cubic feet

1 cubic yard = 27 cubic feet, so a cubic yard is 27 times as large as 1 cubic foot.

Check Your Understanding

1. Imagine that you are measuring the volume of a pencil box. Which cubic unit would make more sense: cubic centimeters or cubic meters? Why?
2. A family is measuring the volume of the storage space in a moving truck. Should the family measure the volume in cubic inches or cubic yards? Explain your answer.

Check your answers in the Answer Key.

Capacity and Liquid Volume

Capacity is a measure of the amount a container can hold. **Liquid volume** is a measure of the amount of space taken up by a liquid. Granular substances like sand, flour, and sugar can also be measured with units of liquid volume.

Metric Units

In the metric system, units for capacity and liquid volume include milliliters (mL) and liters (L).

Metric System
Units of Capacity and Liquid Volume
1 liter (L) = 1,000 milliliters (mL)
1 milliliter = 1 cubic centimeter (cm^3)

Example

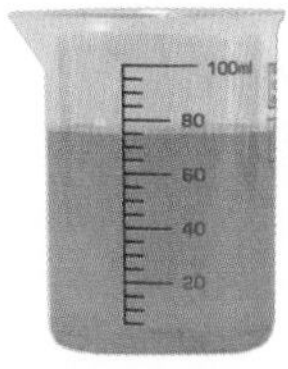

What is the capacity of the beaker? What is the volume of the water in the beaker?

The beaker can hold up to 100 mL of liquid, so its **capacity** is 100 mL.

There are 75 mL of water in the beaker, so the **volume** of liquid is 75 mL.

You can estimate liquid volume by using common objects that you know as personal references. Some examples of personal references for metric units of capacity and liquid volume are given below.

Personal References for Metric Units of Capacity and Liquid Volume	
About 1 milliliter	About 1 liter
• 20 drops of water • the amount of water that could fit in a centimeter cube	• 3 regular-size canned drinks • 1-liter bottle

20 water droplets have a volume of about 1 millileter.

3 regular-size canned drinks have a volume of about 1 liter.

U.S. Customary Units

In the U.S. customary system, units for capacity and liquid volume include cups (c), pints (pt), quarts (qt), and gallons (gal).

U.S. Customary System
Units of Capacity and Liquid Volume
1 gallon (gal) = 4 quarts (qt)
1 quart = 2 pints (pt)
1 pint = 2 cups (c)
1 cup = 8 fluid ounces (fl oz)

Example

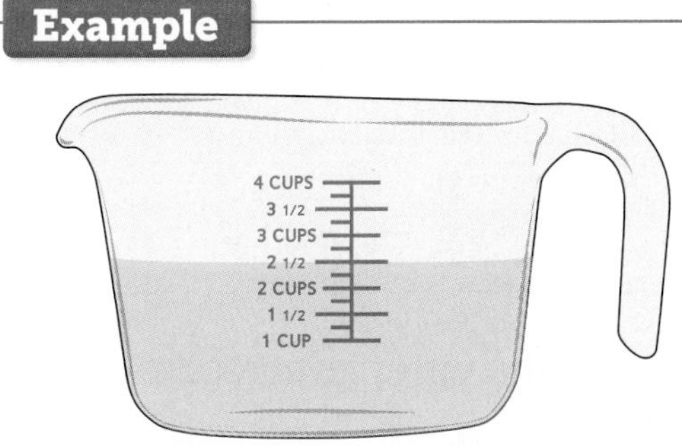

What is the capacity of the measuring cup? What is the volume of the water in the measuring cup?

The measuring cup can hold up to 4 cups of liquid, so its **capacity** is 4 cups.

There are $2\frac{1}{2}$ cups of water in the measuring cup, so the **volume** of liquid is $2\frac{1}{2}$ cups.

Some examples of personal references for U.S. customary units of capacity or liquid volume are given below.

Personal References for U.S. Customary Units of Capacity and Liquid Volume			
About 1 cup	About 1 pint	About 1 quart	About 1 gallon
cup	pint	quart	gallon

Measuring Volume by Displacement

Often it is not practical to use a volume formula to find the volume of an object. For example, many objects, such as rocks or apples, have irregular shapes, so they do not have a clear base or consistent height to measure. One way to find the volume of irregular objects is by using **displacement.** To displace something is to push it aside or move it out of place.

When an object is submerged, or placed underwater, it pushes aside, or **displaces,** some of the water in the container. The amount of water displaced is exactly equal to the volume of the object.

A container that is marked so that the amount of water displaced can be measured is called a *calibrated* container. When an object is submerged in a calibrated container, the volume of the object can be measured accurately.

You can convert from units of liquid volume to cubic units using the fact that 1 milliliter has the same volume as 1 cubic centimeter.

Example

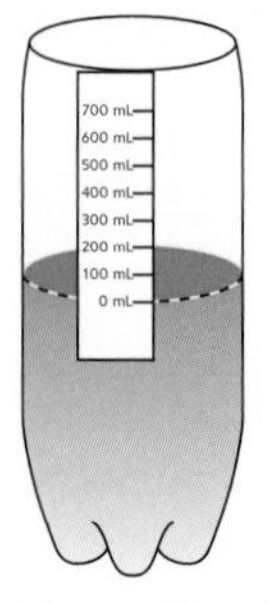

Water level at 0 mL

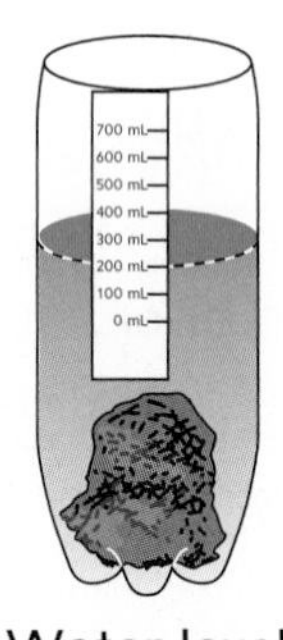

Water level at 200 mL

The bottle is calibrated so that the water level reads 0 mL.

When the rock is submerged in the water, the water level rises to 200 mL.

The rock displaced, or moved, 200 mL of water, so the volume of the rock is 200 mL.

One milliliter of liquid volume is equivalent to 1 cubic centimeter, so 200 mL = 200 cm^3.

The volume of the rock is 200 cm^3.

To measure volume by displacement, you need a container that has enough capacity to hold not only the water you are using, but also the object you are submerging.

Check Your Understanding

1. Denise has calibrated a container so that the water level reads 0 mL. When she submerges an orange, the water level rises to 470 mL. What is the volume of the orange?
2. Kevon poured water into a large measuring cup until it reached the 100 mL line. He added a lime and the water level rose to 160 mL. What is the volume of the lime?

Check your answers in the Answer Key.

Mass and Weight

In everyday life, many people use the words *mass* and *weight* to mean the same thing, but scientifically, there is a difference. **Mass** is a measure of the amount of matter (solid, liquid, and gas) in an object. **Weight** is a measure of the heaviness of an object. Weight depends both on an object's mass and on the pull of gravity.

On the moon, an astronaut's body has the same amount of matter, or mass, as on Earth. However, the astronaut's weight is much less on the moon than on Earth because the moon has less gravitational pull.

133 lb on Earth

22 lb on the moon

Metric Units

The same units can be used to measure both mass and weight. The gram (g) is the basic metric unit for mass and weight. Other units include the kilogram (kg) and the milligram (mg). One tool that can be used for measuring mass is a pan balance.

Metric System
Units of Mass and Weight
1 gram (g) = 1,000 milligrams (mg)
1 kilogram (kg) = 1,000 grams

pan balance

Some examples of personal references for metric units of mass are given below.

Personal References for Metric Units of Mass	
About 1 gram	About 1 kilogram
• dollar bill • large paper clip	• unopened 1-liter bottle of water • baseball bat

A dollar bill has a mass of about 1 gram.

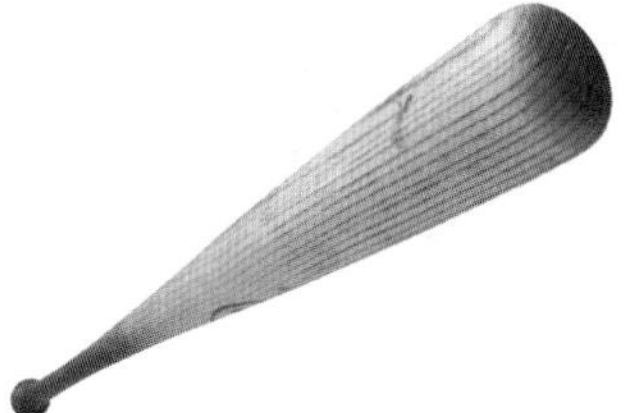

A wooden baseball bat has a mass of about 1 kilogram.

U.S. Customary Units

In everyday life in the United States, we use ounces (oz), pounds (lb), and tons (T) to measure weight. These units can also be used to measure mass, but scientists typically use metric units instead. We can use tools such as weighing scales and spring scales to measure weight.

U.S. Customary System
Units of Weight
1 pound (lb) = 16 ounces (oz)
1 ton (T) = 2,000 pounds

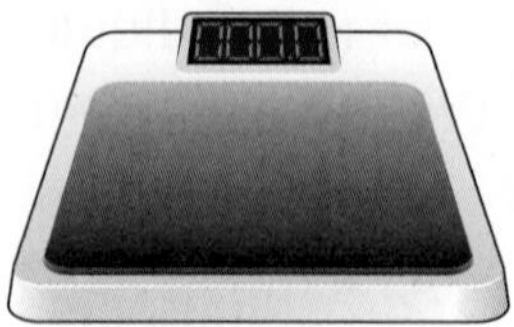

bath scale

Some examples of personal references for U.S. customary units of weight are given below.

Personal References for Weight		
About 1 ounce	About 1 pound	About 1 ton
• a slice of bread • a CD	• a package of 4 sticks of butter • an adult shoe	• a giraffe • a car

A slice of bread weighs about 1 ounce.

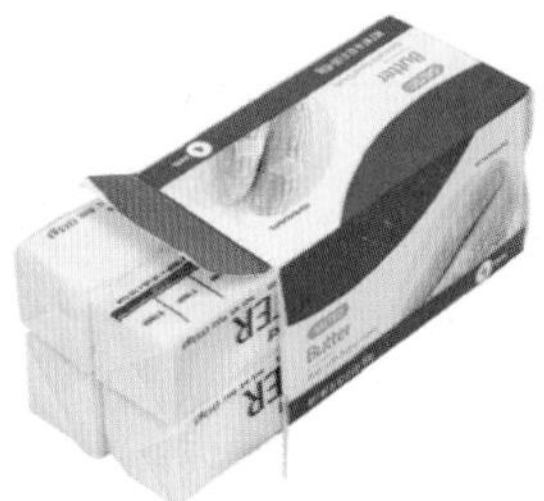

A package of 4 sticks of butter weighs about 1 pound.

A giraffe weighs about 1 ton.

Did You Know?

The abbreviation *lb* is short for the Latin word *libra*. The *libra* was the ancient Roman pound and weighed about 0.72 U.S. customary pound.

Time

We use time to tell when something happens and to measure how long something lasts. Some units of time include seconds (sec), minutes (min), hours (hr), days (d), weeks (wk), months (mo), and years (yr). The table at the right shows how different units of time are related.

Sometimes multiple units are used together to measure a period of time. You can use the relationship between different units to rename measurements.

Units of Time
1 century = 100 years (yr)
1 decade = 10 years
1 year = 12 months (mo)
1 year = 52 weeks (wk) (plus one or two days)
1 year = 365 days (d) (366 days in a leap year)
1 month = 28, 29, 30, or 31 days
1 week = 7 days
1 day = 24 hours (hr)
1 hour = 60 minutes (min)
1 minute = 60 seconds (sec)

Example

Reed is taking a bus to visit his grandparents. The bus takes 1 day and 12 hours to reach his grandparents' town. How many hours is the bus ride?

1 day = 24 hours
1 day and 12 hours = 24 hours + 12 hours = 36 hours

The bus ride is 36 hours long.

Example

A movie is 136 minutes long. How many hours long is the movie?

1 hour = 60 minutes
136 minutes ÷ 60 minutes in one hour = 2 hours, remainder 16 minutes
Check: 2 hours = 2 * 60 minutes = 120 minutes
120 minutes + 16 minutes = 136 minutes

The movie is 2 hours and 16 minutes long.

Check Your Understanding

1. How many years is $3\frac{1}{2}$ decades?
2. Gabby is counting down the days until soccer season begins. The first day of the season is 3 weeks and 2 days away. How many days away is soccer season?

Check your answers in the Answer Key.

Temperature

Temperature is a measure of the hotness or coldness of something. To read a temperature in degrees, you need a reference frame that begins with a zero point and has a number-line scale. The two most commonly used temperature scales, Fahrenheit and Celsius, have different zero points.

Fahrenheit

This scale was invented in the early 1700s by the German physicist G.D. Fahrenheit. On the Fahrenheit scale, pure water freezes at 32°F and boils at 212°F. A saltwater solution freezes at 0°F (the zero point) at sea level. The normal temperature for the human body is 98.6°F. The Fahrenheit scale is used primarily in the United States.

Celsius

This scale was developed in 1742 by the Swedish astronomer Anders Celsius. On the Celsius scale, the zero point (0 degrees Celsius or 0°C) is the freezing point of pure water. Pure water boils at 100°C. The Celsius scale divides the interval between these two points into 100 equal parts. For this reason, it is sometimes called the *centigrade* scale. The normal temperature for the human body is 37°C. The Celsius scale is the standard for most people outside of the United States and for scientists everywhere.

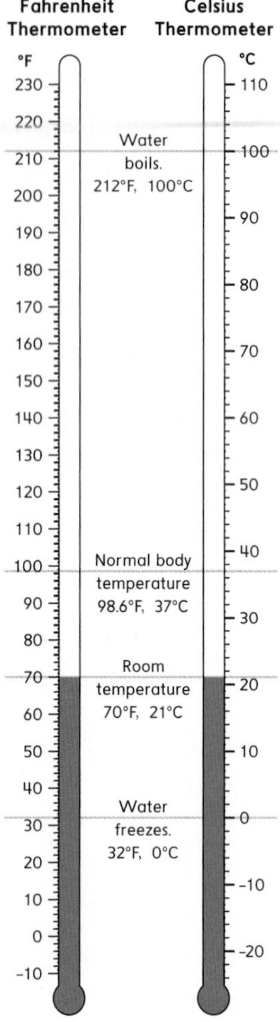

The thermometers show both the Fahrenheit and Celsius scales. Key reference temperatures, such as the boiling and freezing points of water, are indicated. A thermometer reading of 70°F (or about 21°C) is normal room temperature.

A thermometer measures temperature. A common thermometer is a glass tube that contains a liquid. When the temperature goes up, the liquid expands and moves up the tube. When the temperature goes down, the liquid shrinks and moves down the tube.

Here are two formulas for converting from degrees Fahrenheit (°F) to degrees Celsius (°C) and vice versa:

$$F = \frac{9}{5} * C + 32 \text{ and } C = \frac{5}{9} * (F - 32)$$

Example

Find the Celsius equivalent of 82°F.

Use the formula $C = \frac{5}{9} * (F - 32)$ and replace F with 82:

$C = \frac{5}{9} * (82 - 32) = \frac{5}{9} * (50) = 27.78$

The Celsius equivalent of 82°F is about 28°C.

Collecting and Organizing Data

There are different ways to collect information. You can count, measure, ask questions, or observe and describe what you see. The information you collect is called **data.** One count, measurement, or response is called a *data point*. The group or collection of all the data points is called the *data set.*

Once the data have been collected, it helps to organize the information in order to make the data easier to understand. One way to record and organize data is with a **bar graph.** Bar graphs show data in a way that is easy to make comparisons. The title of a bar graph describes the information in the graph. Each bar has a label. Units are given to show how something was counted or measured. When possible, the graph gives the source of the information.

Example

This is a **vertical bar graph.**

- Each bar represents the area in square miles of the lake named below the bar.
- It is easy to compare the lake areas by comparing the lengths of the bars. The area of Lake Superior is about 10,000 square miles larger than Lake Michigan. Lake Huron and Lake Michigan have about the same area.

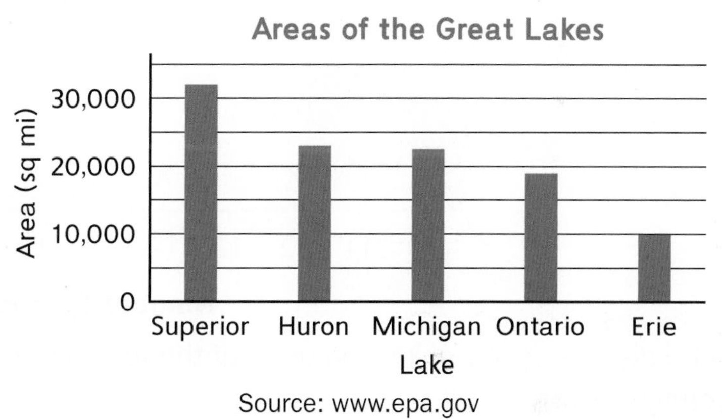

Example

This is a **horizontal bar graph.**

- Each bar represents the number of grams of fat in the food named to the left of the bar.
- It is easy to compare the fat contents of foods by comparing bars. One cup of ice cream contains nearly twice as much fat as one cup of whole milk. One doughnut contains 12 times as much fat as one banana.

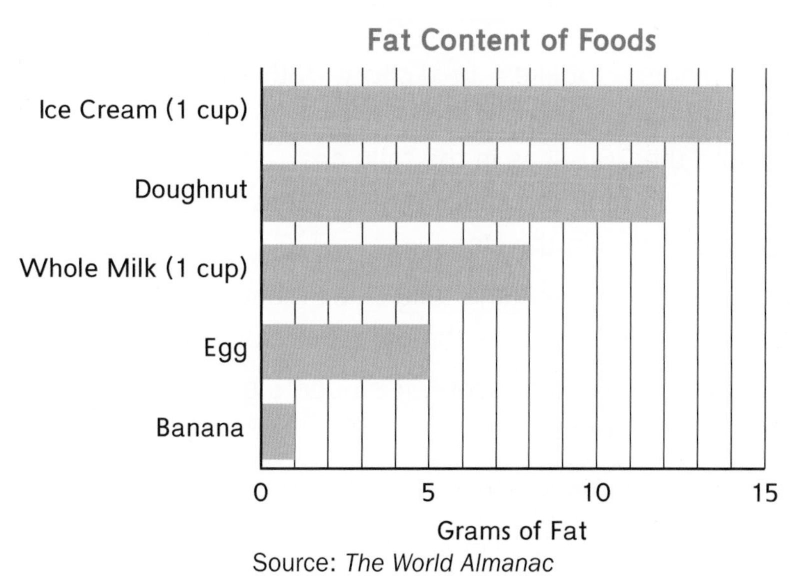

Line Plots

Line plots are another way to record and organize data. A **line plot** uses check marks, Xs, or other marks to show counts. There is one mark for each count.

Example

Ms. Barton's class got the following scores on a 20-word spelling test.

Make a line plot to show the data below.

20	15	18	17	20	12	15	17	19	18	20	16	16
17	14	15	19	18	18	15	10	20	19	18	15	18

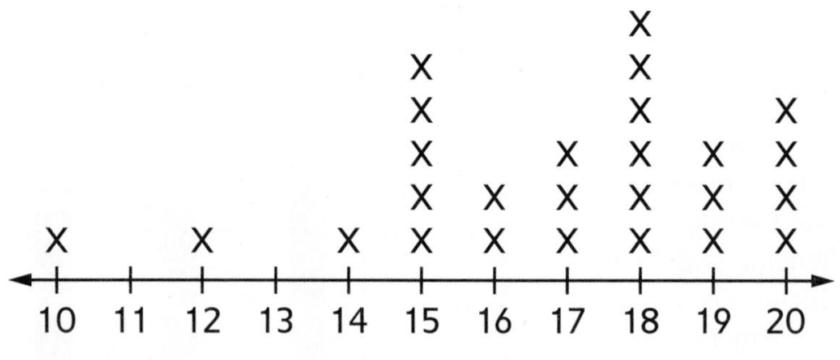

The *data set* is the collection of all of the test scores. Each X represents one *data point,* or one student's score.

There are 5 Xs above 15 in the line plot. This shows that 5 students got a score of 15 on the spelling test. The score of 15 appeared 5 times in the class list of test scores.

A line plot is a lot like a tally chart. Both show one mark for each count. The 26 Xs in the line plot and the 26 tallies in the tally chart show that 26 students took the spelling test. There were 26 total scores in the class list.

Both the line plot and the tally chart make it easier to organize and describe the data. For example:

- 20 was the highest score. 4 students got 20 words correct.
- 10 was the lowest score. 1 student got 10 correct.
- 18 is the score that came up most often.
- Most students got between 15 and 20 correct. 3 students scored lower than 15 correct.

Students' Scores on a 20-Word Spelling Test

Number Correct	Number of Students
10	/
11	
12	/
13	
14	/
15	~~////~~
16	//
17	///
18	~~////~~ /
19	///
20	////

Scaled Line Plots

You can choose the scale of a line plot by looking at a set of data. Identify the smallest (minimum) and largest (maximum) numbers in the data set. These can be the first and last numbers on your scale. Then figure out how to space the rest of your data. The numbers on your scale should have the same interval, or space, between them.

Example

Scientists measured stag beetles in Europe. Here are their measurements (in inches) in order from smallest to largest:

$1\frac{1}{2}$ $1\frac{1}{2}$ $1\frac{1}{2}$ 2 2 2 2 2 2 2 2 2

$2\frac{1}{2}$ $2\frac{1}{2}$ $2\frac{1}{2}$ $2\frac{1}{2}$ $2\frac{1}{2}$ $2\frac{1}{2}$ $2\frac{1}{2}$ 3 3 $3\frac{1}{2}$

Then they drew a line plot to help them see trends in the data.

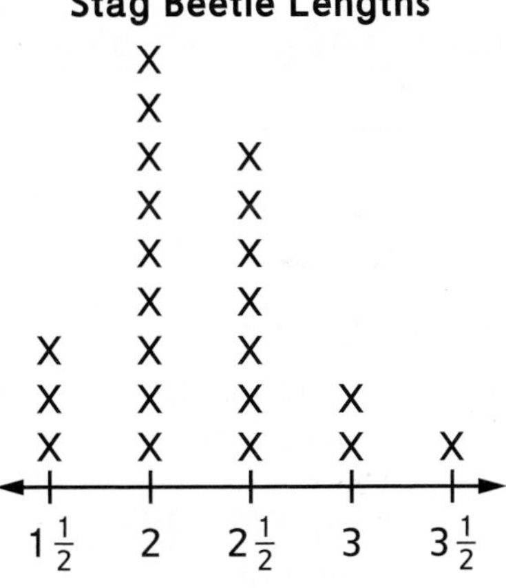

The line plot shows measurement data scaled in $\frac{1}{2}$-inch intervals.

Check Your Understanding

1. Chad planted sunflower seeds. Two weeks later, he measured the heights of the plants to the nearest $\frac{1}{4}$ inch. Here are his results:

   $4\frac{1}{2}$ in. $3\frac{3}{4}$ in. 4 in. $3\frac{3}{4}$ in. $3\frac{3}{4}$ in.

   $3\frac{1}{2}$ in. 4 in. $4\frac{1}{2}$ in. 4 in. $3\frac{3}{4}$ in.

   a. Put the measurements in order from smallest to largest. What is the smallest and largest value in the data set?

   b. Draw a line plot for the data. Which numbers did you use as the first and last numbers on your scale?

   c. What is the size of the interval (or space) between each number on your scale?

   d. Do you have any measurements that do not have Xs? Which one(s)?

Check your answers in the Answer Key.

Line Graphs

Line graphs are used to display information showing patterns, or trends. They can show how something has changed over time.

Some line graphs are called **broken-line graphs.** Line segments connect the points on the graph. Connecting the points makes it easier to see trends.

Line graphs have a horizontal and a vertical scale. Each of these scales is called an **axis** (plural: **axes**). Each axis is labeled to show what is being measured or counted and the units for the measurements or counts.

Broken-Line Graph

Vertical Axis

Horizontal Axis

Joined end to end, the segments look like a broken line.

Data points are recorded on line graphs using two pieces of information: the count or measure from the horizontal axis, and the count or measure from the vertical axis. You can plot points just as you would on a coordinate grid.

Example

This line graph shows the amount of money a school has raised for each raffle ticket sold.

The horizontal axis shows the number of raffle tickets sold. The vertical axis shows the amount of money raised for the number of tickets sold.

Each dot represents the amount of money that was raised when a given number of tickets was sold. For example, the dot above 10 on the horizontal axis shows that $30 was raised when 10 tickets were sold.

The dots are connected with a line to help show patterns. The amount of money raised increases as more tickets are sold.

The line has an arrow on the right end because the amount of money raised will continue to increase as more tickets are sold.

You can use the graph to find information:

- When 40 raffle tickets were sold, the school raised $120.
- When the school had raised $60, twenty raffle tickets had been sold.

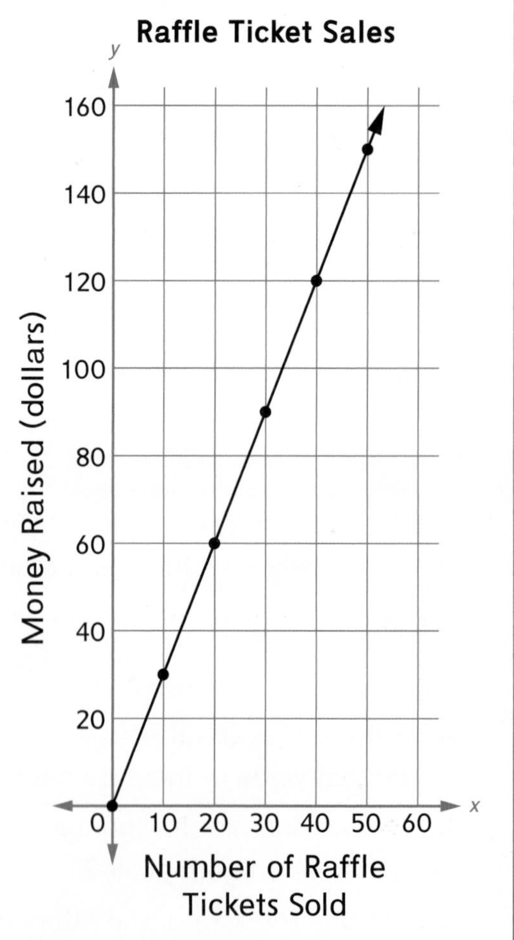

Interpreting Data

Once you have collected, organized, and represented data, you can analyze the data to see what you notice. Here are some questions you might ask when interpreting a data set:

- What is the smallest (minimum) value in the data set?
- What is the largest (maximum) value in the data set?
- What is the difference between the smallest value and the largest value?
- Which value occurs most often in the data set?
- What is the middle value in the data set?

Example

What do you notice about the stag beetle data below?

Stag Beetle Lengths

Length (inches)	$1\frac{1}{2}$	2	$2\frac{1}{2}$	3	$3\frac{1}{2}$
Number of X's	XXX	XXXXXXXXX	XXXXXXX	XX	X

Some of the things you might notice are:

- The shortest stag beetle is $1\frac{1}{2}$ inches. The longest stag beetle is $3\frac{1}{2}$ inches.
- The difference between the measure of the longest and shortest stag beetle can be found by subtracting the shortest length from the longest length:

$$3\tfrac{1}{2} \text{ inches} - 1\tfrac{1}{2} \text{ inches} = 2 \text{ inches}$$

- The most common length for a stag beetle is 2 inches.
- The middle length can be found by putting the data points in order from smallest to largest. The middle length for a stag beetle is 2 inches.

$1\frac{1}{2}$ $1\frac{1}{2}$ $1\frac{1}{2}$ 2 2 2 2 2 2 2 (2 2) $2\frac{1}{2}$ $2\frac{1}{2}$ $2\frac{1}{2}$ $2\frac{1}{2}$ $2\frac{1}{2}$ $2\frac{1}{2}$ $2\frac{1}{2}$ 3 3 $3\frac{1}{2}$

Finding a Typical Value: Evening Out a Data Set

Sometimes you want to find a typical value for a data set. A typical value is a number that could be used to represent a whole set of numbers. One way to find a typical value is to use the value that occurs most often. Another way is to find the middle value. A third way is to even out the data set.

Note Evening out a data set is also referred to as finding the average or mean.

Example

A camp cook wants to know how many pancakes each camper typically eats. Here is his data for one table of three campers: Kayla ate 4, Ryan ate 6, and Ava ate 2 pancakes.

The cook found the total number of pancakes the three campers ate:

$$4 + 6 + 2 = 12 \text{ pancakes}$$

Then he imagined distributing the 12 total pancakes equally. Each camper would have 4 pancakes. When the number of pancakes is evened out, you can say that 4 is the typical number of pancakes that each camper ate.

Redistributing data into equal groups is the same as adding the data points and dividing the sum by the number of data points.
In general, to even out a data set, complete the following:

Step 1 Add all of the data points.
Step 2 Divide the sum by the number of data points.

Example

Willa is setting up beakers of water for a science experiment. When she first pours water into the beakers, each has a different amount of water:

45 mL 60 mL 72 mL 63 mL 47 mL 55 mL

Even out the data above. If the water is distributed equally, how much water will be in each beaker?

Step 1 Add all of the data points: $45 + 60 + 72 + 63 + 47 + 55 = 342$

The sum of all of the data points is 342 milliliters.

Step 2 Divide the sum by the number of data points.

There are 6 data points in the data set. $342 \div 6 = 57$

If the water is distributed equally, there will be 57 mL of water in each beaker.

Volume in the Real World

Many different kinds of work in the real world require estimating or calculating volume.

Heating and Cooling

When people talk about the size of an apartment or building, they often refer to its **area,** or how much floor space it has. This area is often measured in square feet. However, heating and cooling specialists need to know more than the area of a building. They calculate the **volume** of each space in a building because they must heat or cool all of the air that fills the spaces. They consider the length, width, and height of the spaces in order to know how much power and energy it will take to heat or cool the building.

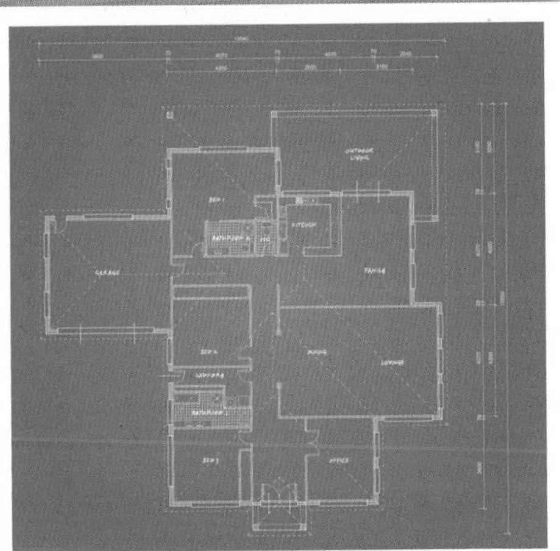

A blueprint can show a home's *footprint*, or the area it takes up on the ground. However, to heat or cool the home, its height matters too. Spaces with very high ceilings will take more energy to heat or cool than spaces with lower ceilings, even if the spaces have the same footprint. A room that is 10 feet long by 12 feet wide has a floor area of 120 square feet. If that room has ceilings that are 8 feet high, the volume of the room is 120 square feet * 8 feet, or 960 cubic feet. But if a room with the same floor area has 10-foot-high ceilings, the volume of the room is 120 square feet * 10 feet, or 1,200 cubic feet.

In the United States, we use British Thermal Units (BTUs) to describe the cooling capability of air conditioners. A small bedroom that has a volume of 10 feet * 10 feet * 8 feet, or 800 cubic feet, needs about 8,400 BTUs per hour to cool it. A large warehouse that is 100 feet * 100 feet * 10 feet, or 100,000 cubic feet, needs about 280,000 BTUs per hour to cool it.

(cl) mstay/Digital Vision Vectors/Getty Images; (cr) Phillip Spears/Getty Images; (bl) robertomm/iStock/Getty Images Plus/Getty Images

Packaging Engineering

A packaging engineer is someone who designs and selects containers and packaging materials for companies that manufacture or ship goods. Packaging engineers designed the boxes, bags, cartons, and bottles you see every day in your home and in stores.

Packaging for Shipping

Packaging engineers must design packages that protect the product and are cost-effective. They consider the advantages and disadvantages of different 3-dimensional packages. How much material will it take to create a particular package? What is the volume of the package, or how much will it hold? Will the product fit inside the package? Packaging engineers use this information to help companies make packages that are economical and practical.

Package engineers think about how well different containers will fit together in larger shipping boxes. Drinks are often packaged into cylinders, even though cylinders pack less efficiently in shipping boxes than rectangular prisms. Notice the wasted space between the cylinders in the top picture. However, making a cylinder requires less material than making a rectangular prism of the same volume. This is why many foods, such as soup and oatmeal, are stored in cylinders, even though cylinders do not ship as efficiently.

Shippers often use packing materials to help items fit snugly in a box. They subtract the volume of the item to be shipped from the total volume of the box to know the volume of the empty space they need to fill with packing materials.

Packaging for Consumers

Packaging engineers must also create packages that appeal to consumers. They think about what consumers want from the packages: Where and how do people want to store and use the item? Does the package need to be stacked when it is stored?

Packaging engineers think about how to package the same volume in the most convenient way. For example, both of these jugs hold one gallon, but the jug on the right is shaped more like a rectangular prism so that it can fit inside a refrigerator door. The liquid capacity of each jug is the same (1 gallon), but the jug on the right may be more appealing to consumers.

Both of these toothpaste containers hold about the same amount of toothpaste, but the one on the right has a base and can stand up on a bathroom counter, which makes it more appealing to some consumers. It requires more material to make, but may result in more sales because of the added convenience.

Earth's Water

There are about 330 million cubic miles (330,000,000 mi^3) of water on Earth, but nearly all of it is salt water. Less than 3% $\left(\frac{3}{100}\right)$, or about 8,400,000 cubic miles, of Earth's water is fresh water, and nearly 70% $\left(\frac{70}{100}\right)$ of that fresh water is frozen in glaciers and icecaps. The rest is located underground, in lakes and rivers, and in the atmosphere.

The Columbia Icefield in Canada is the largest ice mass in North America.

Water resource professionals work to ensure that there is enough fresh water to meet the needs of people living in a particular area. To do so, they must calculate the volume of water available and the volume of water that people typically use. They also make suggestions about how people can reduce the volume of water they use in order to conserve water resources.

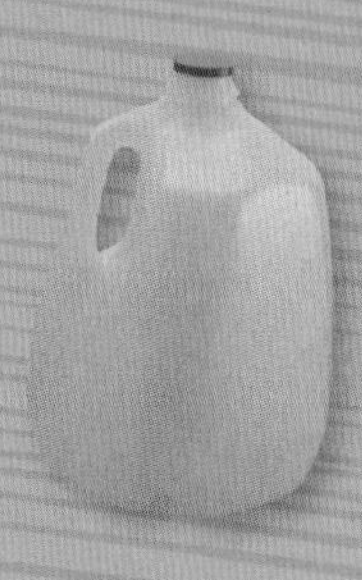

This is one gallon of water. In the United States, each person uses about 100 gallons of water per day for indoor and outdoor household purposes.

Water Use

Water is both a necessity and a luxury. All animals and plants—including people, pets, trees, crops, and livestock—need water to survive. We use water to generate electricity and to fight fires. We use water to wash ourselves and our clothes, fill swimming pools, and water lawns.

Americans use approximately 70 gallons of water per day for use inside the home. On average, we use about 10 gallons of water per day at kitchen and bathroom faucets and 20 gallons per day flushing the toilet.

A hydroelectric power plant

A nuclear power plant

Throughout the world, water is used in the process of creating electricity. In the United States, about 40% $\left(\frac{40}{100}\right)$ of all the fresh water we use helps create electricity. Water is used to create electricity in hydroelectric power plants and in nuclear and coal-fired power plants. That's about 541,300,000 cubic meters of water per day. An Olympic swimming pool, which has standard dimensions of 50 meters by 25 meters by 2 meters, holds about 2,500 cubic meters of water. That means that the water we use to create electric power each day is enough to fill more than 216,000 Olympic swimming pools.

Water Storage

How do water resource professionals know if an area has enough water to meet the needs of the people that live there? They think about how much water people use, and they calculate the **capacity** of the water reservoirs in the area.

Reservoir capacity is measured in cubic meters (m^3). A reservoir's **capacity** is the amount of water it can possibly hold, or its maximum volume. The **volume** of water that a reservoir is holding at any given time depends on the amount of recent snow or rainfall and the amount of water that people have used.

Not all places on Earth have the same capacity for water storage or the same number of people who depend on the water. Water resource professionals calculate the volume of water available to each person in an area by dividing the total amount of available water in cubic meters by the total number of people: water (m^3) / number of people. This is sometimes called "water storage *per capita*," or per person. In North America, water storage per person is about 5,660 cubic meters. In sub-Saharan Africa, water storage per person is about 543 cubic meters due to low levels of rainfall and the lack of large reservoirs to store water. In Asia, water storage per person is about 353 cubic meters because available water reserves must serve an extremely large population.

Water Conservation

Earth's population continues to grow every year, but we still need to share the same amount of available fresh water. Water resource professionals work hard to develop ways for individuals and industries to *conserve* water, or reduce the volume of water that each person uses.

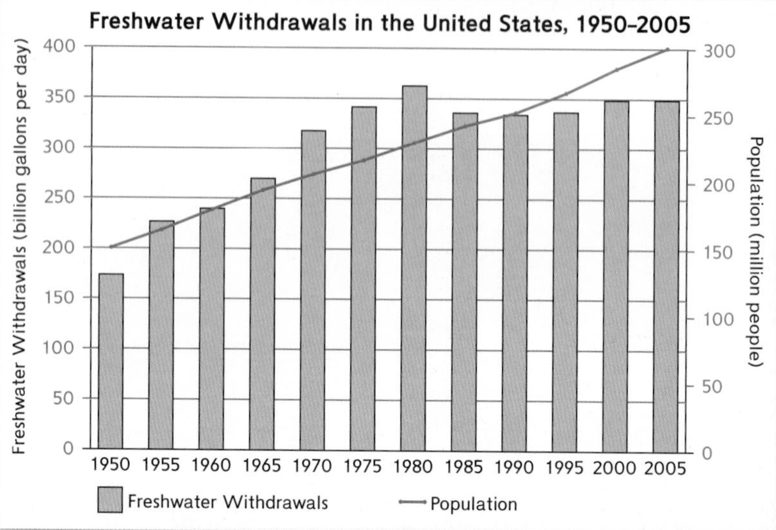

After 1980, water usage in the United States stopped increasing every year. The amount of water that the population of the United States consumes per day has stayed fairly constant from 1985 to today, despite the fact that the population of the United States has continued to grow every year. Because our water usage has not changed, but the population has grown, each person in the United States must be using less water now than they were 30 years ago.

In the last 30 years, many people have made changes to reduce how much water they use every day. These changes include steps such as turning off the water while brushing teeth, taking shorter showers, using less water to wash dishes, and using landscape plants that require less water to survive.

Many new technologies reduce the amount of water required for daily tasks. In 1980, a typical toilet used about 7 gallons of water per flush. New technology now allows high-efficiency toilets to flush using only about 1.6 gallons of water. That reduces the volume of water used in each flush by almost four-fifths. If a family of four replaced a low-efficiency toilet with a high-efficiency model, they would reduce the volume of water they use just flushing the toilet by almost 30,000 gallons every year.

Geometry

Cartesia/Photodisc/Getty Images

Geometry in Our World

The world is filled with geometry. There are angles, segments, lines, and curves everywhere you look. There are 2-dimensional and 3-dimensional shapes of every type.

Many wonderful geometric patterns can be seen in nature. You can find patterns in flowers, spider webs, leaves, seashells, even your own face and body.

The ideas of geometry are also found in the things people create. Think of the games you play. Checkers is played with round pieces. The game board is covered with squares. Baseball and soccer are played with spheres. They are played on fields that are painted with straight and curved lines. The next time you play or watch a game, notice how geometry is important to the way the game is played.

The places we live in are built from plans that use geometry. Buildings almost always have rectangular rooms. Outside walls and roofs often include sections that have triangular shapes. Archways are curved and are often shaped like semicircles (half circles). Staircases may be straight or spiral.

Buildings and rooms are often decorated with beautiful patterns. You see these decorations on doors and windows; on walls, floors, and ceilings; and on railings of staircases.

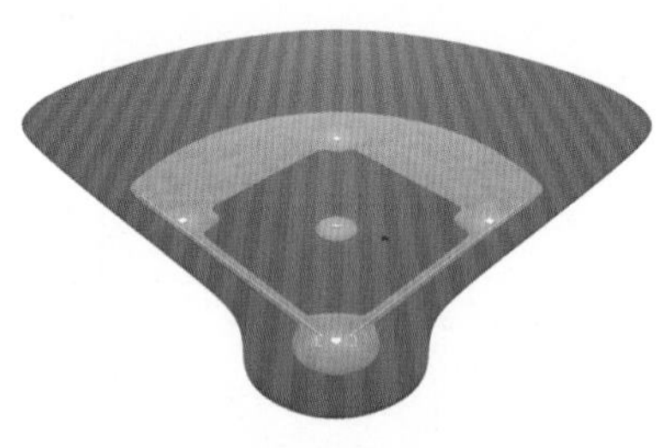

The clothes people wear are often decorated with geometric shapes. So are the things they use every day. Everywhere in the world, people create things using geometric patterns. Examples include quilts, pottery, baskets, and tiles. Some patterns are shown here. Which are your favorites?

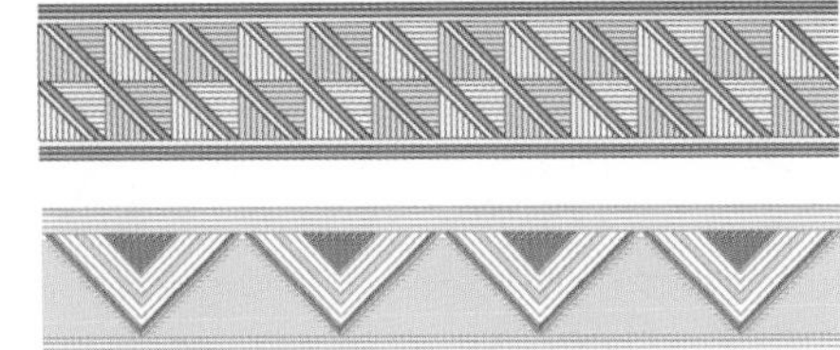

We can also use geometry to provide us with information. Latitude and longitude lines and other coordinate grids used on maps help us locate places. When watching the news or researching on the Internet, you will see graphs that display useful information. Examples include temperature, rainfall, and population growth.

Make a practice of noticing geometric shapes around you. Pay attention to bridges, buildings, and other structures. Look at the ways in which simple shapes such as triangles, rectangles, and circles are combined. Notice interesting designs. Share these with your classmates and your teacher.

In this section, you will learn about lines, angles, relationships between geometric shapes, and coordinate grids. As you learn, try to find examples of geometric figures around you.

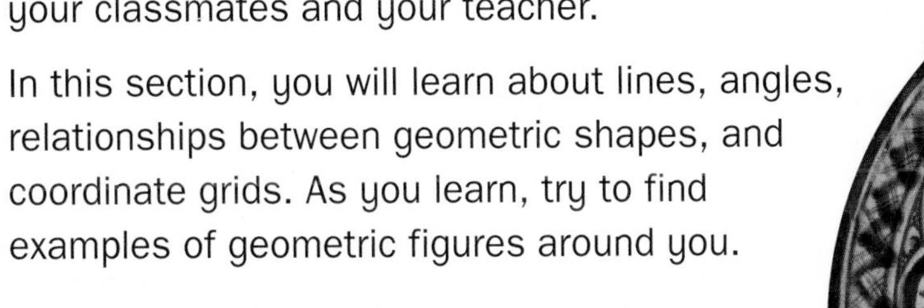

(t)LACMA - Los Angeles County Museum of Art; (c)Peeterv/iStock/360/Getty Images; (b)Map Resources/Alamy

Angles

An **angle** is formed by two rays, two line segments, or a ray and line segment that share the same endpoint.

angle formed by 2 rays

angle formed by 2 segments

angle formed by 1 ray and 1 segment

The endpoint where the rays or segments meet is called the **vertex** of the angle. The rays or segments are called the **sides** of the angle.

S D C sides vertex

Naming Angles

The symbol for an angle is ∠. An angle can be named in two ways:

1. Name the vertex. The angle shown above is angle *S*. Write this as ∠*S*.

2. Name three points: the vertex and one point on each side of the angle. The angle above can be named angle *DSC* (∠*DSC*) or angle *CSD* (∠*CSD*). The vertex must always be listed in the middle, between the points on the sides.

Angle Measures

The size of an angle is the amount of turning about the vertex from one side of the angle to the other. Angles are measured in degrees. A **degree** is a unit of measure for the size of an angle. A full turn about a vertex makes an angle that measures 360 degrees. A **protractor** is a tool used to measure angles. There are two types of protractors: full-circle protractors and half-circle protractors.

full-circle protractor

half-circle protractor

The **degree symbol °** is often used in place of the word *degrees*. The measure of ∠*S* above is 30 degrees or 30°.

Classifying Angles

Angles may be classified according to size.

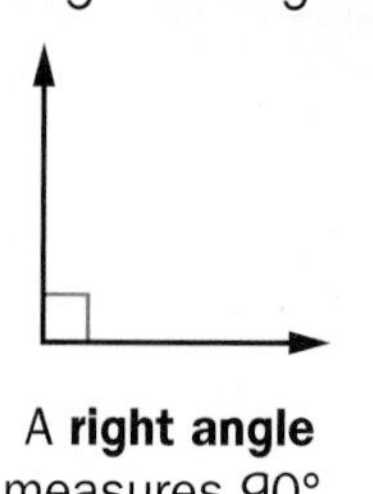

A **right angle** measures 90°.

An **acute angle** measures between 0° and 90°. This angle measures 60°.

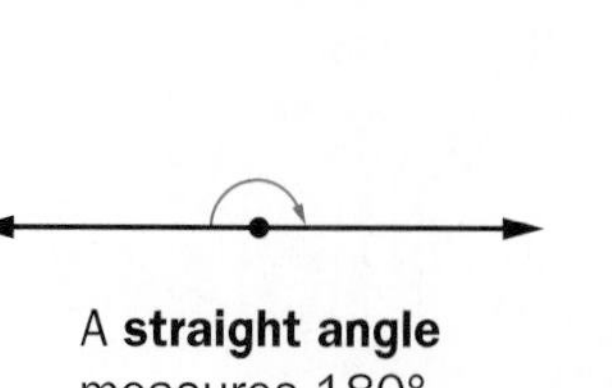

A **straight angle** measures 180°.

An **obtuse angle** measures between 90° and 180°. This angle measures 120°.

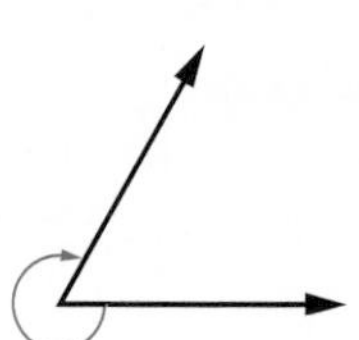

A **reflex angle** measures between 180° and 360°. This angle measures 300°.

Parallel Lines and Segments

Parallel lines are lines on a flat surface that never meet or cross. Think of a straight railroad track that goes on forever. The two rails are parallel lines. The rails never meet or cross, and they are always the same distance apart.

Parallel line segments are parts of lines that are parallel. The top and bottom edges of this page are parallel. If each edge were extended forever in both directions, the lines would remain parallel because they would always be the same distance apart.

The symbol for parallel is a pair of vertical lines $\|$. If $\overline{FE}$ and $\overline{JK}$ are parallel, write $\overline{FE} \| \overline{JK}$. See example.

If lines or segments cross or meet each other, they **intersect.** Lines or segments that intersect and form right angles are called **perpendicular** lines or segments.

The symbol for perpendicular is $\perp$, which looks like an upside-down letter *T*. If $\overleftrightarrow{TS}$ and $\overleftrightarrow{XY}$ are perpendicular, write $\overleftrightarrow{TS} \perp \overleftrightarrow{XY}$. See example below.

Example

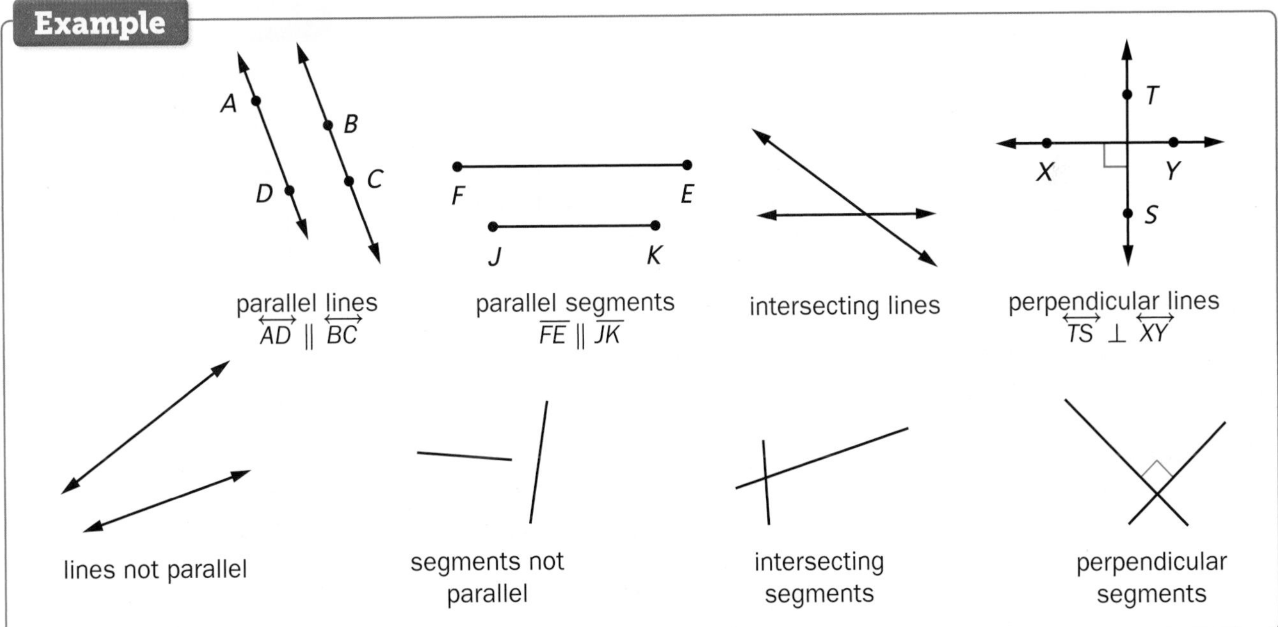

parallel lines $\overleftrightarrow{AD} \| \overleftrightarrow{BC}$

parallel segments $\overline{FE} \| \overline{JK}$

intersecting lines

perpendicular lines $\overleftrightarrow{TS} \perp \overleftrightarrow{XY}$

lines not parallel

segments not parallel

intersecting segments

perpendicular segments

Check Your Understanding

Draw and label the following.

1. parallel line segments *AB* and *EF*
2. a line segment that is perpendicular to both $\overline{AB}$ and $\overline{EF}$

Check your answers in the Answer Key.

Line Segments, Rays, Lines, and Angles

Figure	Symbol	Name and Description
A	A	**point:** a location in space
B, C, endpoints	$\overline{BC}$ or $\overline{CB}$	**line segment:** a straight path between two points called its endpoints
N, M, endpoint	$\overrightarrow{MN}$	**ray:** a straight path that goes on forever in one direction from an endpoint
S, T	$\overleftrightarrow{ST}$ or $\overleftrightarrow{TS}$	**line:** a straight path that goes on forever in both directions
vertex, S, T, P	$\angle T$ or $\angle STP$ or $\angle PTS$	**angle:** two rays or line segments with a common endpoint, called the vertex
A, B, C, D	$\overleftrightarrow{AB} \parallel \overleftrightarrow{CD}$ $\overline{AB} \parallel \overline{CD}$	**parallel lines:** lines that never cross or meet and are everywhere the same distance apart **parallel line segments:** segments that are parts of lines that are parallel
R, E, D, S	none none	**intersecting lines:** lines that cross or meet **intersecting line segments:** segments that cross or meet
B, F, E, C	$\overleftrightarrow{BC} \perp \overleftrightarrow{EF}$ $\overline{BC} \perp \overline{EF}$	**perpendicular lines:** lines that intersect at right angles **perpendicular line segments:** segments that intersect at right angles

The white dividing lines illustrate parallel and perpendicular line segments.

Did You Know?

The use of letters to designate points and lines has been traced back to the Greek mathematician Hippocrates of Chios (about 450 BCE).

Polygons

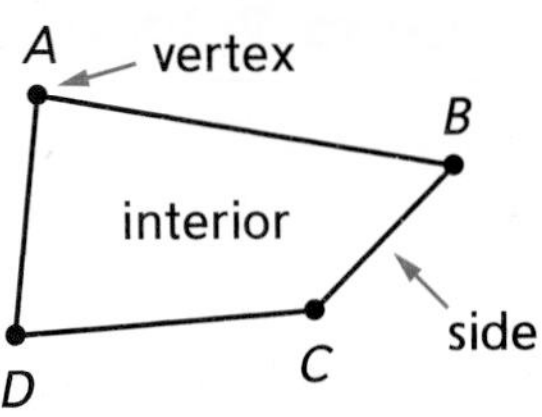

A **polygon** is a flat, 2-dimensional figure made up of line segments called **sides.** A polygon can have any number of sides, as long as it has at least three. The **interior** (inside) of a polygon is not part of the polygon.

- The sides of a polygon are connected end to end and make one closed path.
- The sides of a polygon do not cross.

Each endpoint where two sides meet is called a **vertex.**

Figures That Are Polygons

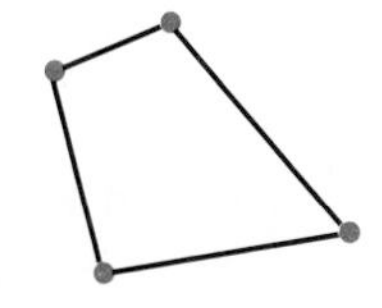
4 sides, 4 vertices

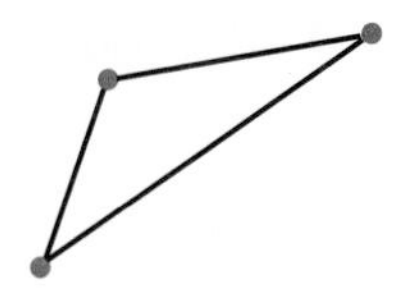
3 sides, 3 vertices

7 sides, 7 vertices

Figures That Are NOT Polygons

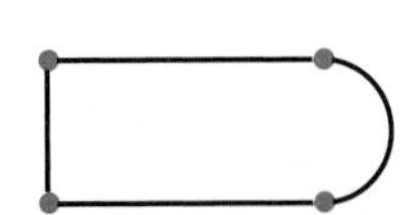
All sides of a polygon must be line segments. Curved lines are not line segments.

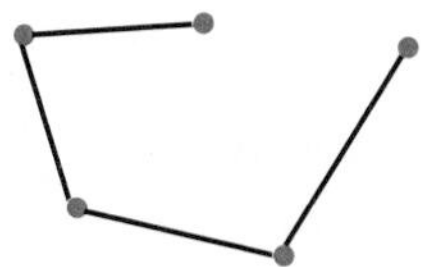
The sides of a polygon must form a closed path.

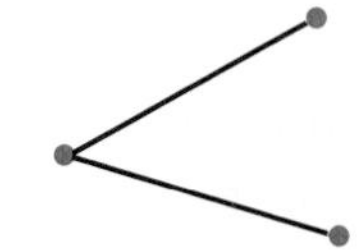
A polygon must have at least 3 sides.

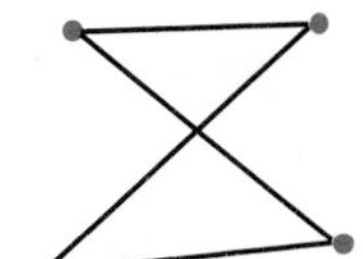
The sides of a polygon must not cross.

The prefix for a polygon's name tells the number of sides it has.

Prefixes

tri- 3

triangle
3 sides

quad- 4

quadrilateral
4 sides

penta- 5

pentagon
5 sides

hexa- 6

hexagon
6 sides

hepta- 7

heptagon
7 sides

octa- 8

octagon
8 sides

nona- 9

nonagon
9 sides

deca- 10

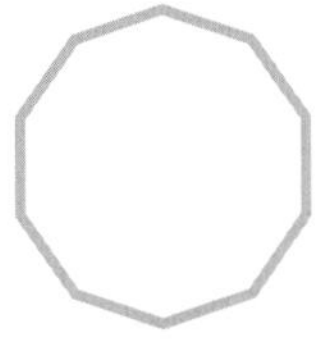
decagon
10 sides

dodeca- 12

dodecagon
12 sides

Convex Polygons

A **convex** polygon is a polygon in which all the sides are pushed outward. The polygons below are all convex.

triangle

quadrilateral

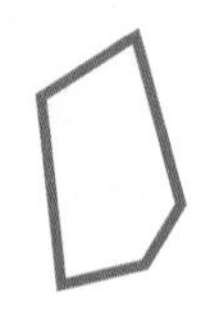
pentagon

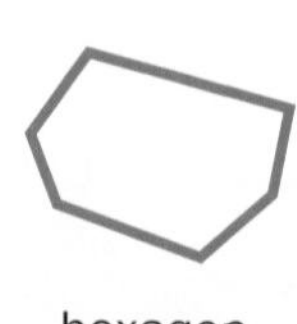
hexagon

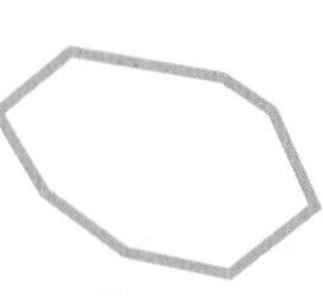
octagon

Concave (Nonconvex) Polygons

A **concave,** or nonconvex, polygon is a polygon in which at least two sides are pushed in. The polygons below are all concave.

quadrilateral

pentagon

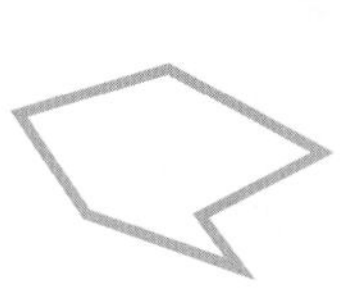
hexagon

octagon

Side or Angle Markings

When the markings on the sides of a polygon are the same, it means the sides are the same length. When the markings on angles of a polygon are the same, it means that the angles have the same measure.

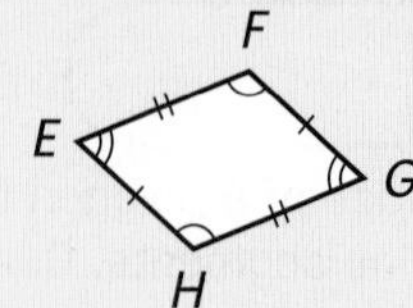

In quadrilateral *EFGH*, side $\overline{EF}$ has the same length as side $\overline{GH}$, and side $\overline{HE}$ has the same length as side $\overline{FG}$. $\angle E$ has the same angle measure as $\angle G$, and $\angle F$ has the same angle measure as $\angle H$.

Regular Polygons

A polygon is a **regular polygon** if the sides all have the same length and the angles inside the figure are all the same size. A regular polygon is always convex. The polygons below are all regular.

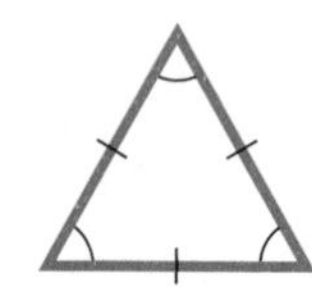
equilateral triangle

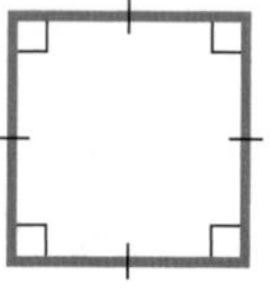
square

regular pentagon

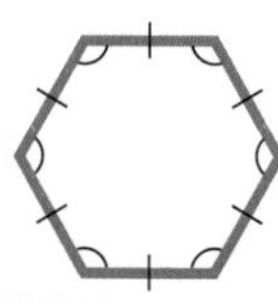
regular hexagon

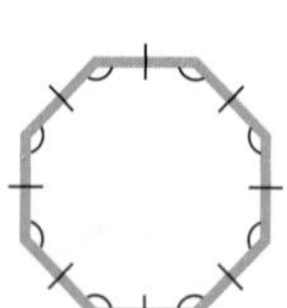
regular octagon

regular nonagon

Check Your Understanding

1. What is the name of a polygon that has
   a. 6 sides? **b.** 4 sides? **c.** 8 sides?
2. **a.** Draw a convex heptagon. **b.** Draw a concave decagon.

Check your answers in the Answer Key.

Triangles

Triangles have fewer sides and angles than any other polygon. The prefix *tri-* means *three*. All triangles have three vertices, three sides, and three angles.

For the triangle shown here:

- The vertices are the points *B*, *C*, and *A*.
- The sides are $\overline{BC}$, $\overline{BA}$, and $\overline{CA}$.
- The angles are $\angle B$, $\angle C$, and $\angle A$.

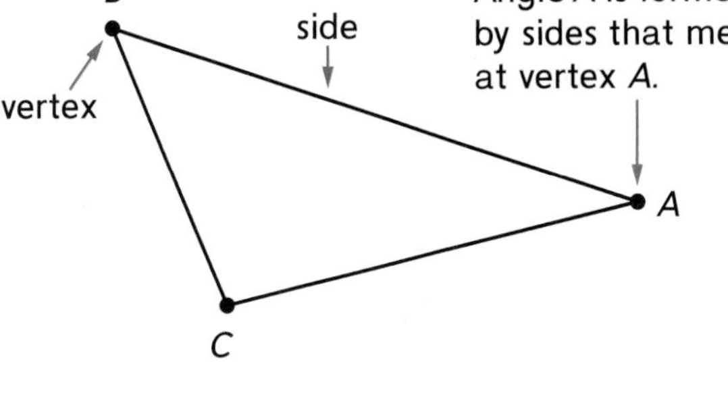

The symbol for triangle is △. Triangles have three-letter names. You name a triangle by listing each letter name for the vertices in order. The triangle above has six possible names: △*BCA*, △*BAC*, △*CAB*, △*CBA*, △*ABC*, and △*ACB*.

Triangles may be classified according to the length of their sides and angle measures.

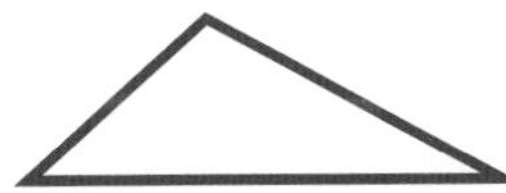

A **scalene triangle** is a triangle with sides that all have different lengths and angles that all have different measures.

An **isosceles triangle** is a triangle that has at least two sides that have the same length and two angles that have the same measure.

An **equilateral triangle** is a triangle with sides that all have the same lengths and angles that all have the same measure.

A **right triangle** is a triangle that contains one right angle (an angle that measures 90°). In a right triangle, the sides that form the right angle are perpendicular to each other. Right triangles have many different shapes and sizes.

Is an equilateral triangle also an isosceles triangle? Why or why not?

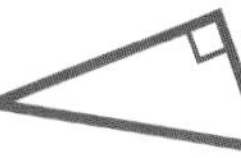

Some right triangles are scalene triangles, and some right triangles are isosceles triangles. But a right triangle cannot be an equilateral triangle because the side opposite the right angle is always longer than each of the other sides.

Check Your Understanding

1. Draw and label an equilateral triangle named △*DEF*. Write the five other possible names for this triangle.
2. Draw an isosceles triangle.
3. Draw a right scalene triangle.

Check your answers in the Answer Key.

Hierarchy

A **hierarchy** is a classification system that shows how you can sort a group of related objects or ideas. It is often represented by a diagram with the most general **category** of objects at the top. As you move down a hierarchy, each **subcategory** is more specific than the categories above it. That is, the objects in a subcategory have all of the attributes of the objects or ideas in the categories above the subcategory, as well as at least one additional attribute.

Example

Justin's class assignment is to create a hierarchy of pets.

He creates a list of some different types of pets:

cat	dog	fish	bird
snake	turtle	hamster	lizard

Justin thinks of attributes to classify the pets in his list and uses them to create a hierarchy:

- pets with no legs
- pets with at least 2 legs
- pets with 4 legs

In this hierarchy, all of the pets on Justin's list fit in the category of pets. So "pets" is the most general group and is shown at the top of Justin's hierarchy.

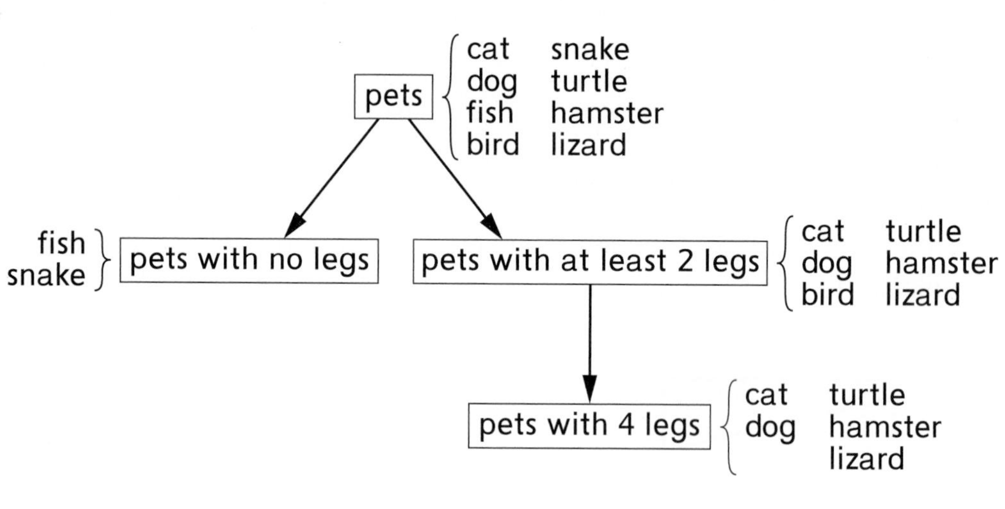

The groups below a category are defined by the types of attributes they share. These groups, or subcategories, get more specific as you move down a hierarchy. For example:

- "Pets with at least 2 legs" is a subcategory of "pets."
- "Pets with 4 legs" is a subcategory of "pets with at least 2 legs" *and* "pets."

All of the pets in the subcategory "pets with 4 legs" also share the attributes of the higher subcategory "pets with at least 2 legs." For example, it is true that a cat has at least 2 legs *and* that a cat has 4 legs.

Example

Justin's hierarchy on the previous page can also be represented using a Venn diagram. This diagram shows the same relationships as the hierarchy on the previous page.

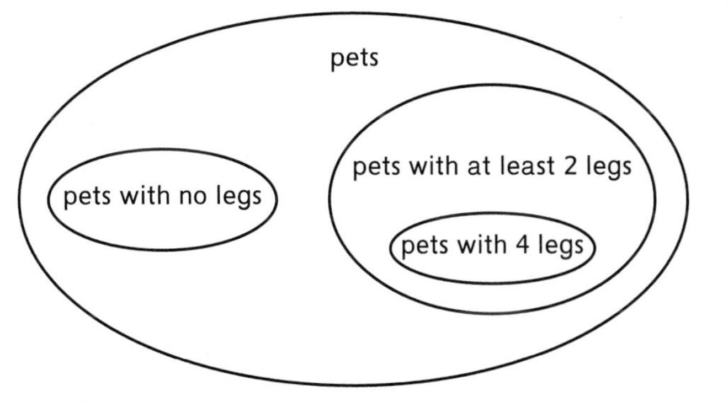

Check Your Understanding

Think of other attributes you can use to sort Justin's list of pets. Use your attributes to create a different hierarchy.

Check your answers in the Answer Key.

Triangle Hierarchy

In mathematics, a hierarchy can be used to classify shapes.

This triangle hierarchy shows how triangles can be classified into categories and subcategories.

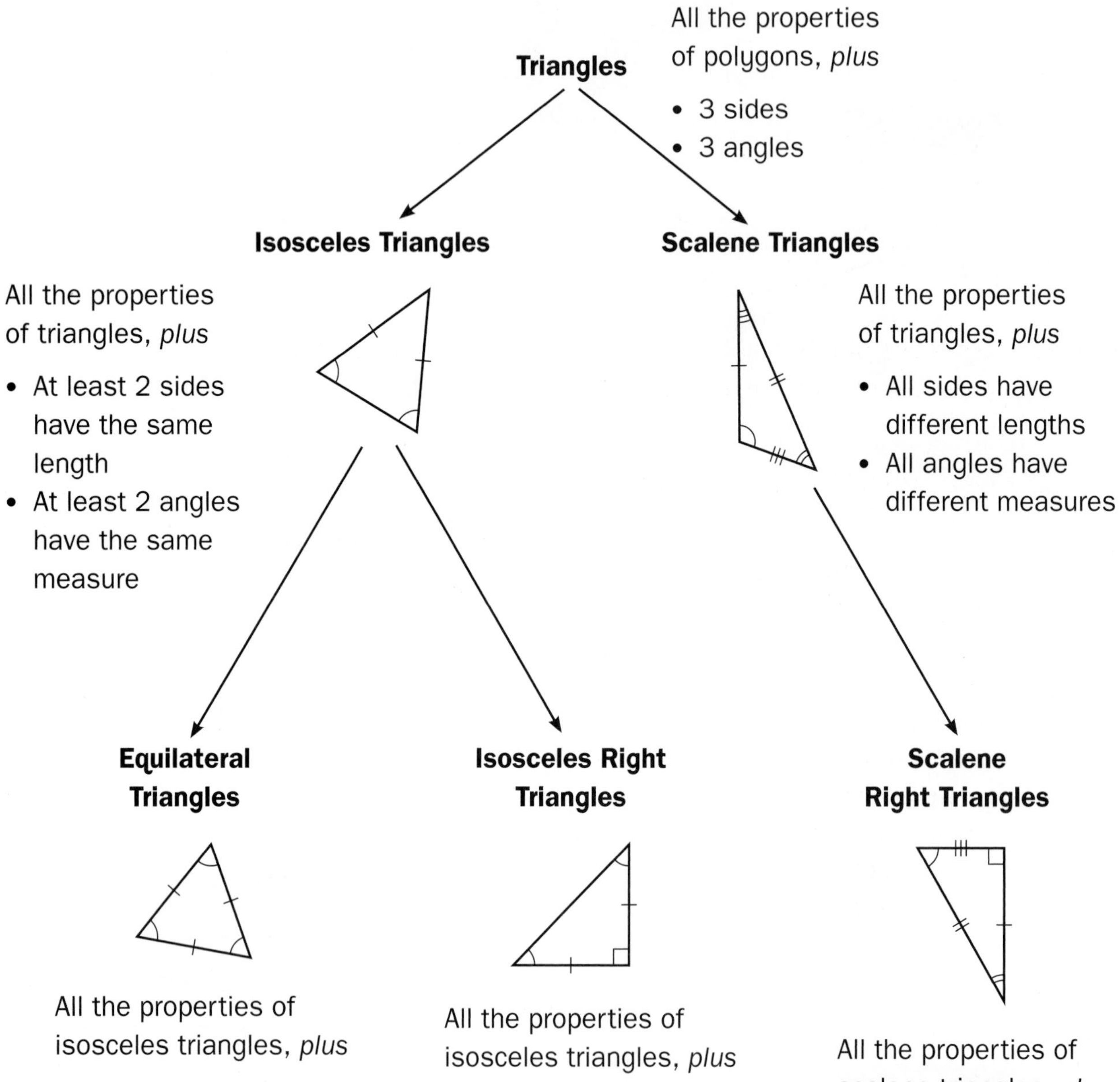

All the properties of isosceles triangles, *plus*

- All sides have the same length
- All angles have the same measure

All the properties of isosceles triangles, *plus*

- 1 right angle

All the properties of scalene triangles, *plus*

- 1 right angle

Mathematical hierarchies can also be represented using a Venn diagram.

This diagram shows the same relationships as the triangle hierarchy on the previous page.

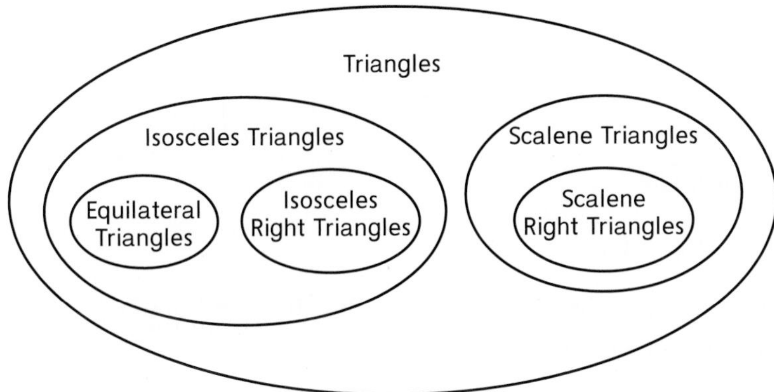

Both representations of this triangle hierarchy give a lot of information about the properties of triangles and the relationships among triangles. For example:

- Isosceles triangles and scalene triangles are both subcategories of triangles.
- Equilateral triangles and isosceles right triangles are both subcategories of isosceles triangles *and* triangles.
- Scalene right triangles is a subcategory of both scalene triangles *and* triangles.

Sometimes there is confusion about how to read a hierarchy. For example, it may appear that isosceles triangles are either equilateral triangles or isosceles right triangles. However, the isosceles triangle shown below does not have the additional properties or attributes of equilateral triangles or isosceles right triangles.

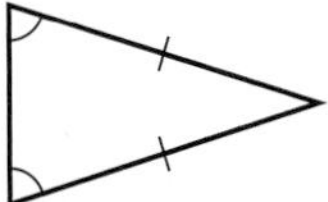

This triangle belongs in the subcategory of "isosceles triangle," but it is NOT part of the subcategories of "equilateral triangle" or "isosceles right triangle."

Quadrilaterals

A **quadrilateral** is a polygon with four sides. The prefix *quad-* means *four.* All quadrilaterals have four vertices and four sides. Another name for quadrilateral is **quadrangle.**

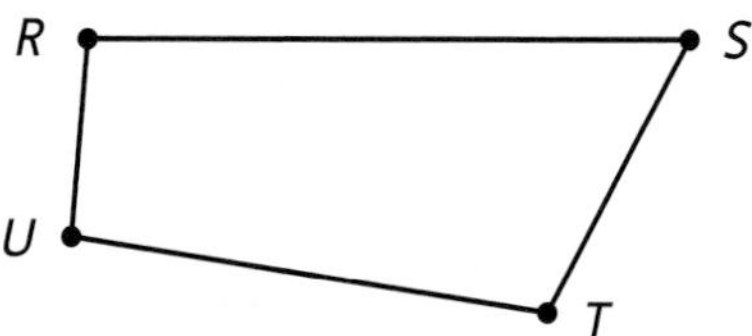

For the quadrilateral shown here:

- The vertices are R, S, T, and U.
- The sides are $\overline{RS}$, $\overline{ST}$, $\overline{TU}$, and $\overline{UR}$.
- The angles are $\angle R$, $\angle S$, $\angle T$, and $\angle U$.

A quadrilateral is named by listing in order the letter names for the vertices. The quadrilateral above has eight possible names:

RSTU, *RUTS*, *STUR*, *SRUT*, *TURS*, *TSRU*, *URST*, *UTSR*

Some quadrilaterals have *at least* one pair of parallel sides. These quadrilaterals are called **trapezoids.**

Reminder: Two sides are parallel if they never meet, no matter how far they are extended in either direction.

Figures That Are Trapezoids

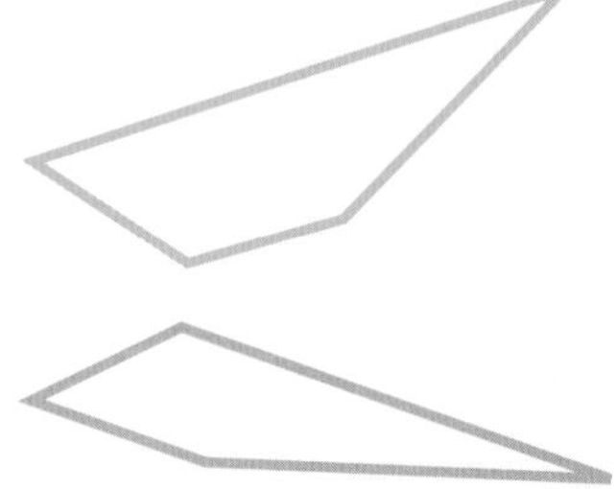

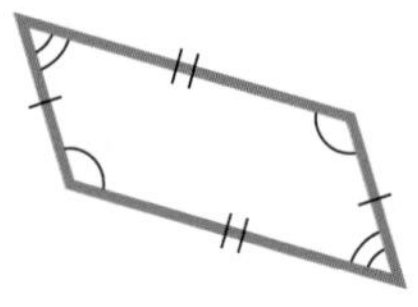

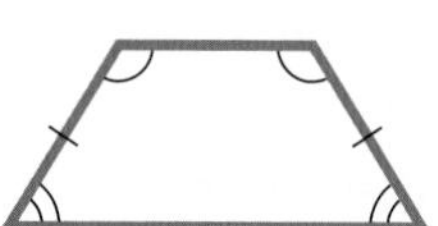

Each figure has *at least* one pair of parallel sides.

Figures That Are NOT Trapezoids

No parallel sides

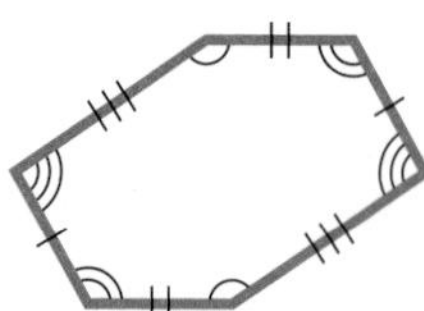

Not a quadrilateral

No parallel sides

Did You Know?

In mathematics, there are times when concepts or ideas are defined differently. In *Everyday Mathematics,* a **trapezoid** is defined as a quadrilateral that has *at least* one pair of parallel sides. Using this definition, parallelograms are also trapezoids. Other sources define a trapezoid as a quadrilateral with *exactly* one pair of parallel sides. Using this definition, parallelograms are not trapezoids since parallelograms have more than one pair of parallel sides.

Quadrilateral Hierarchy

This quadrilateral hierarchy shows how quadrilaterals are classified.

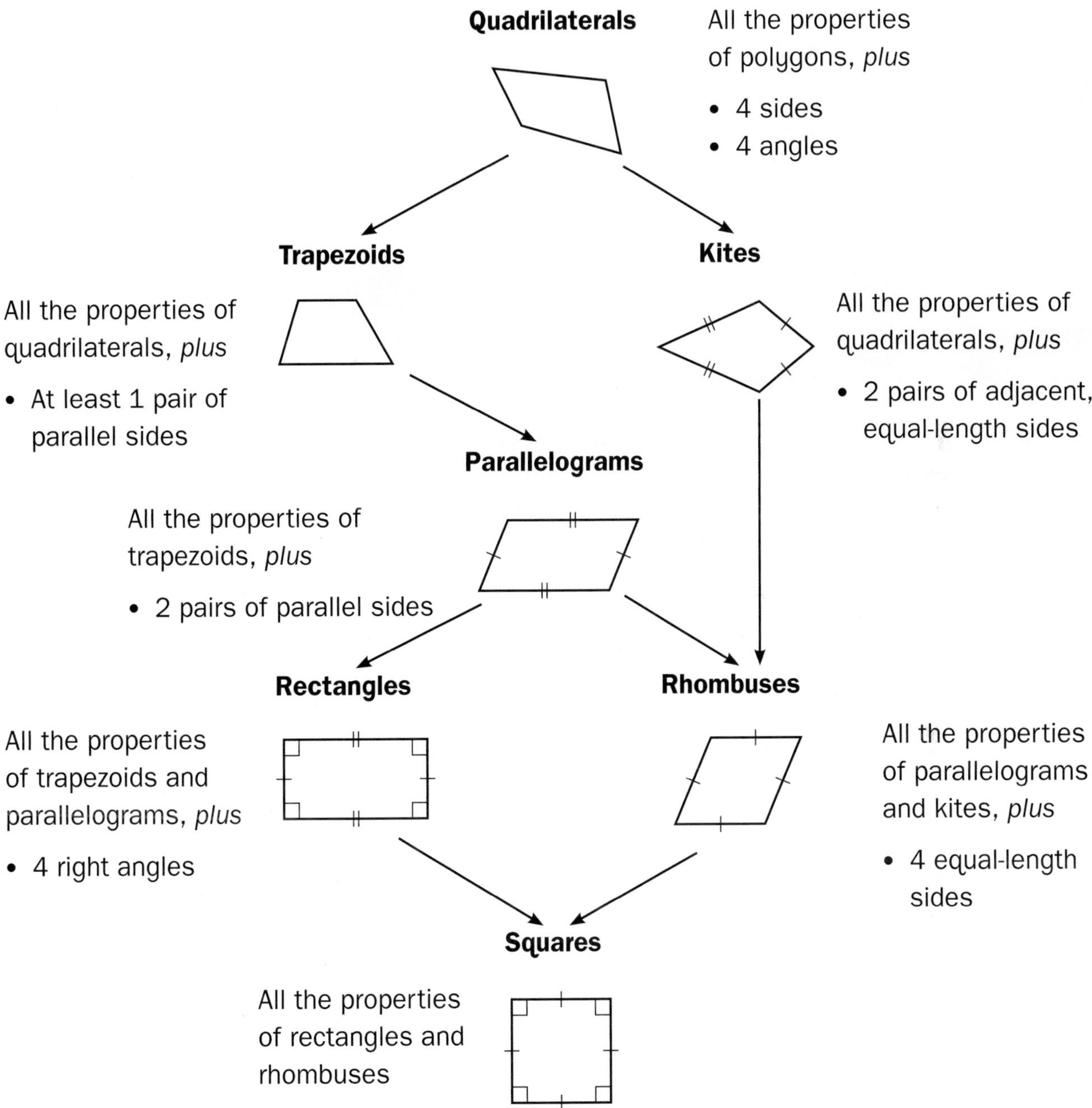

This quadrilateral hierarchy gives a lot of information about the properties of quadrilaterals and the relationships among quadrilaterals. For example:

- Kites are a subcategory of quadrilaterals, so kites have all of the attributes of quadrilaterals.
- Rhombuses are a subcategory of parallelograms, kites, trapezoids, and quadrilaterals, so rhombuses have all of the attributes of parallelograms, kites, trapezoids, and quadrilaterals. So a rhombus is also a parallelogram, kite, trapezoid, and quadrilateral.

Geometric Solids

Polygons and circles are flat, **2-dimensional** figures. The surfaces they enclose take up a certain amount of area, but they do not have any thickness and do not take up any volume.

Three-dimensional shapes have length, width, *and* thickness. They take up volume. Boxes, chairs, books, cans, and balls are all examples of 3-dimensional shapes.

A **geometric solid** is the surface or surfaces that surround a 3-dimensional shape. The surfaces of a geometric solid may be flat, curved, or both. Despite its name, a geometric solid is hollow; it does not include the points within its interior.

- A flat surface of a solid is called a **face.**
- A curved surface of a solid does not have a special name.

Examples

Describe the surfaces of each geometric solid.

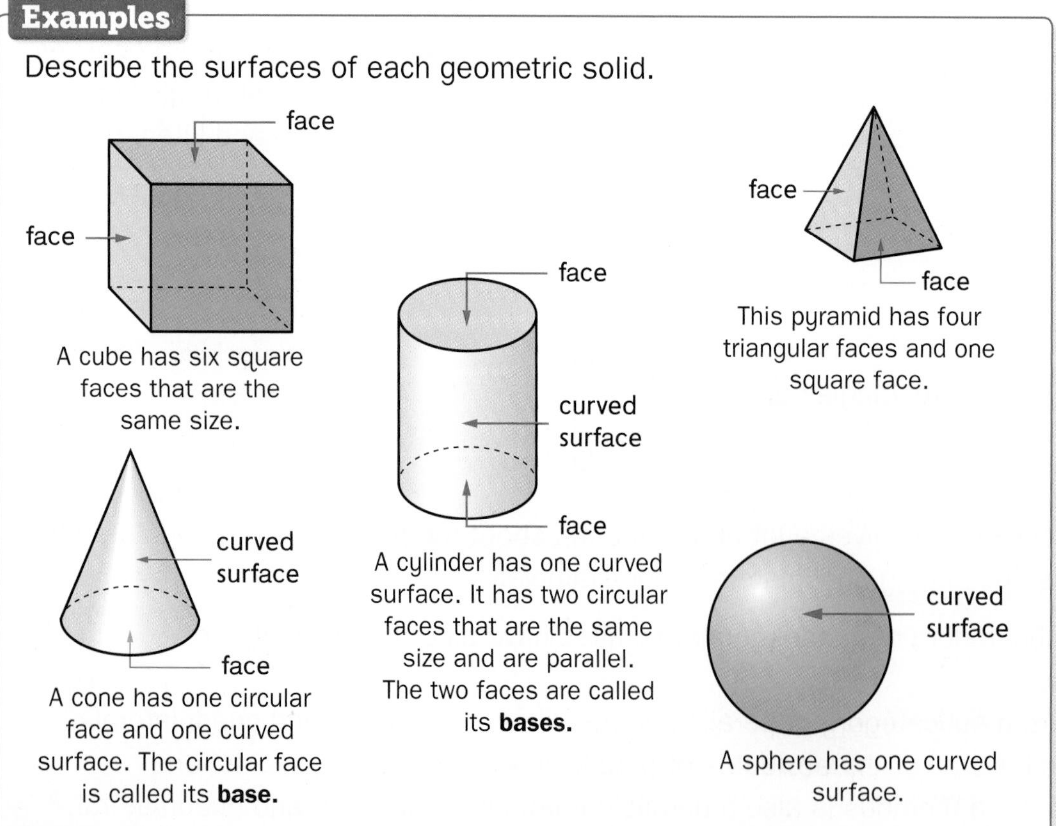

A cube has six square faces that are the same size.

This pyramid has four triangular faces and one square face.

A cylinder has one curved surface. It has two circular faces that are the same size and are parallel. The two faces are called its **bases.**

A cone has one circular face and one curved surface. The circular face is called its **base.**

A sphere has one curved surface.

The **edges** of a geometric solid are the line segments or curves where surfaces meet.

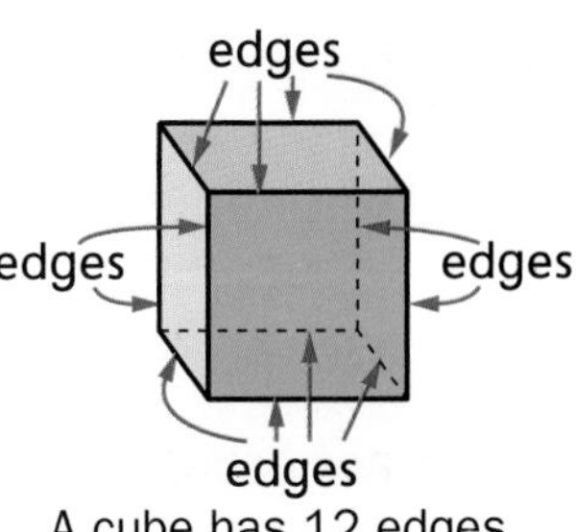

A cube has 12 edges.

A corner of a geometric solid is called a **vertex** (plural *vertices*).

A vertex is usually a point at which edges meet. The vertex of a cone is an isolated corner completely separated from the edge of the cone.

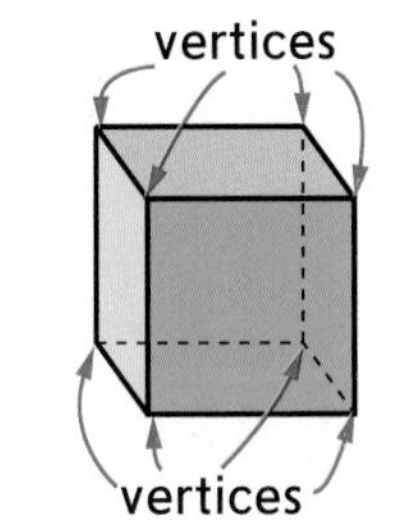

A cube has 8 vertices.

A cone has one edge and one vertex. The vertex opposite the circular base is called the **apex.**

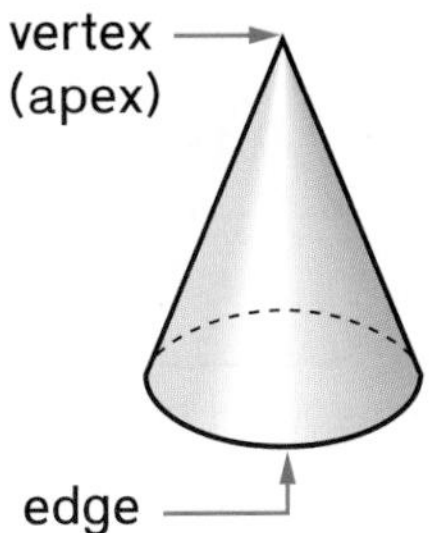

The pyramid shown here has eight edges and five vertices. The vertex opposite the rectangular base is called the **apex.**

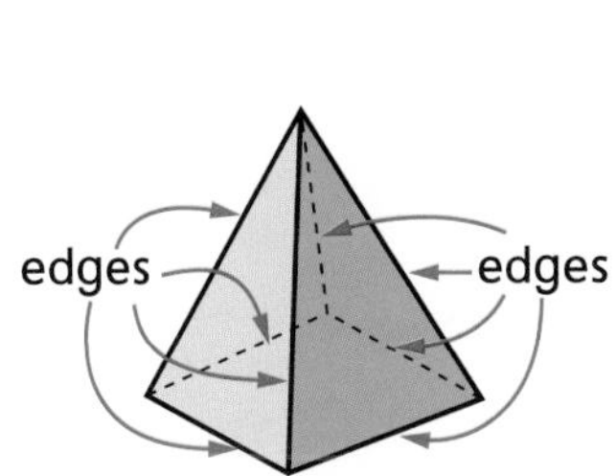

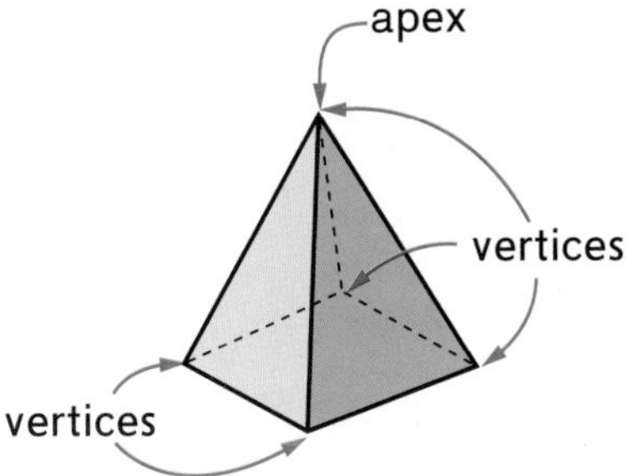

A cylinder has two edges. It has no vertices.

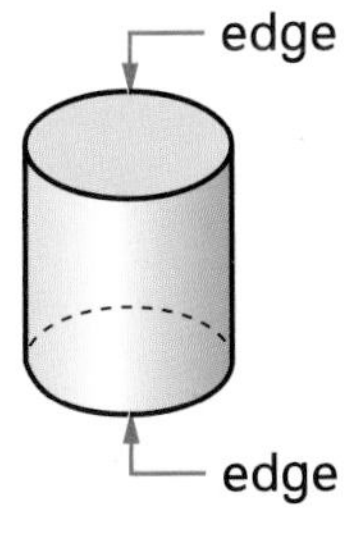

A sphere has no edges and no vertices.

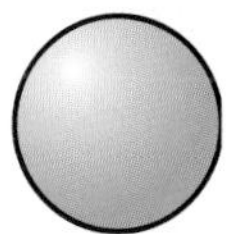

Did You Know?

Euler's Theorem is a formula that tells how the number of faces, edges, and vertices in a polyhedron are related. Let *F, E, V* denote the number of faces, edges, and vertices of a polyhedron. Then $F + V - E = 2$. (Polyhedrons are defined on the next page.)

Check Your Understanding

1. a. How are cylinders and cones alike? **b.** How do they differ?

2. a. How are pyramids and cones alike? **b.** How do they differ?

Check your answers in the Answer Key.

Polyhedrons

A **polyhedron** is a geometric solid whose surfaces are all formed by polygons. These surfaces are the faces of the polyhedron. A polyhedron does not have any curved surfaces.

Two important groups of polyhedrons are shown below. These are **pyramids** and **prisms.**

Pyramids

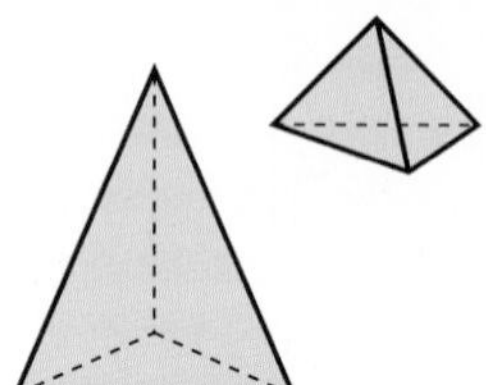

triangular pyramids

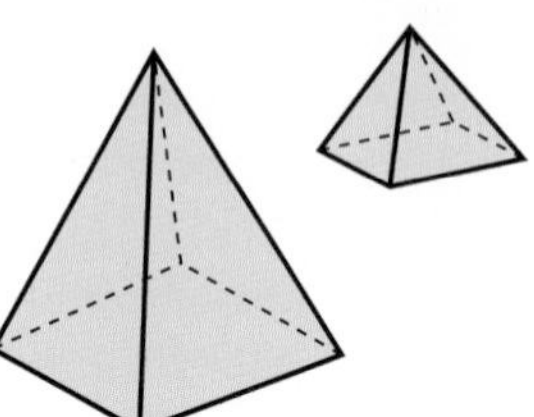

rectangular pyramids

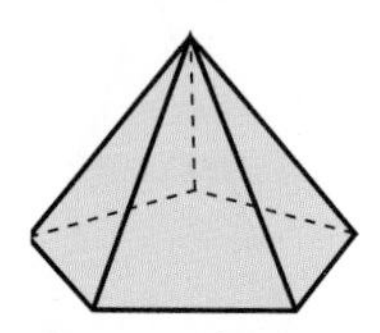

pentagonal pyramid

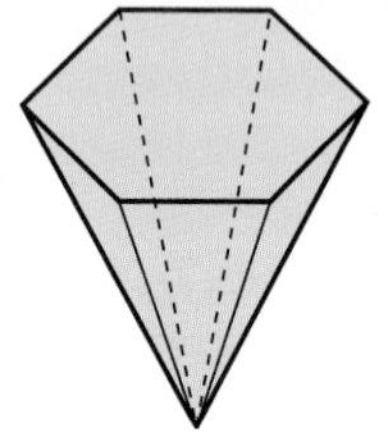

hexagonal pyramid

Prisms

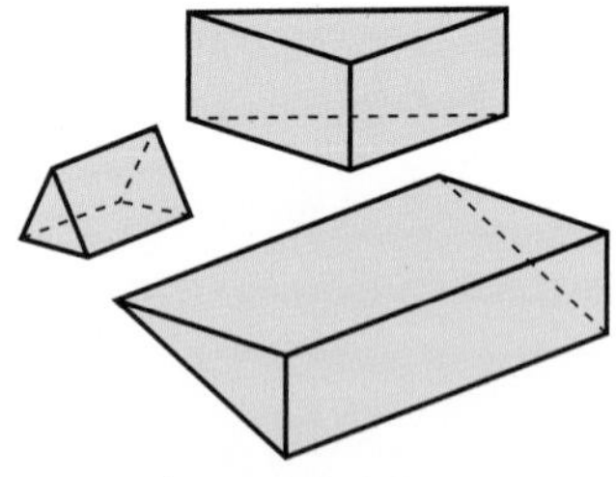

triangular prisms

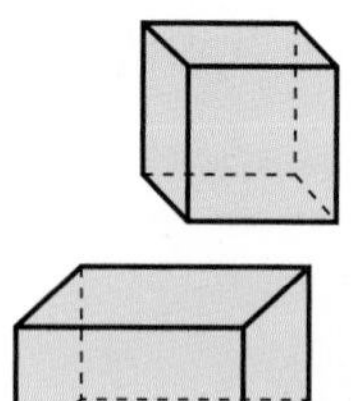

rectangular prisms

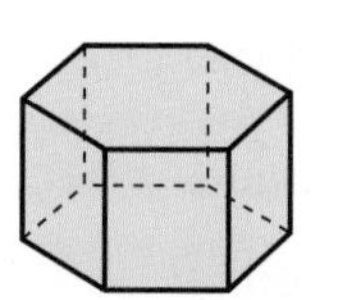

hexagonal prism

Many polyhedrons are not pyramids or prisms. Some examples are shown below.

Polyhedrons That Are NOT Pyramids or Prisms

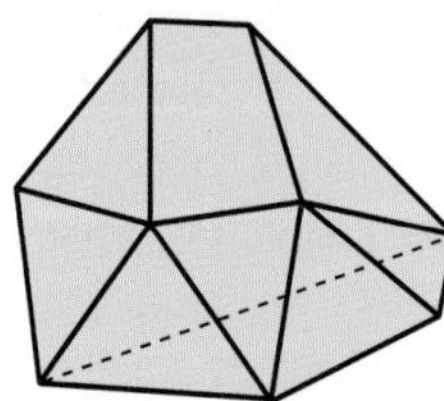

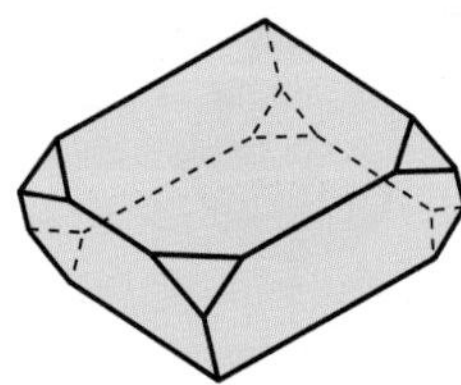

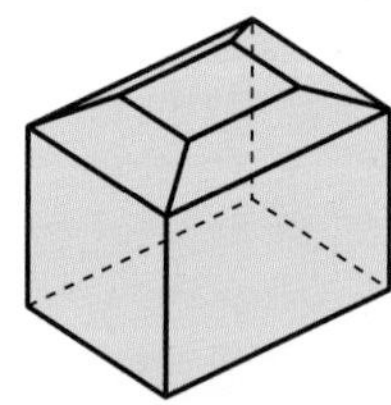

Check Your Understanding

1. a. How many faces does a rectangular pyramid have?

b. How many faces have a rectangular shape?

2. a. How many faces does a triangular prism have?

b. How many faces have a triangular shape?

Check your answers in the Answer Key.

Prisms

All of the geometric solids below are **prisms.**

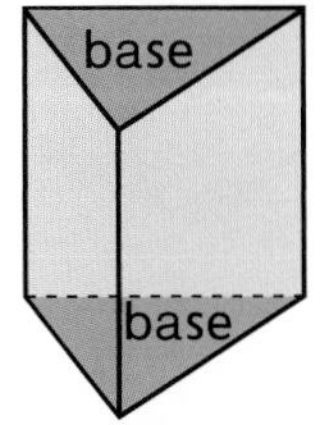

triangular prism

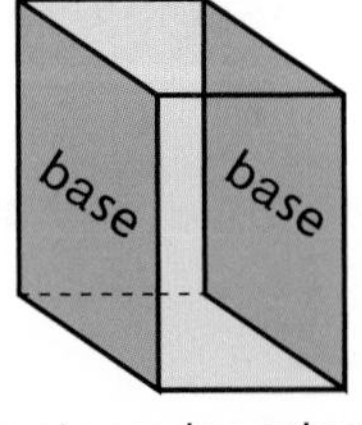

rectangular prism

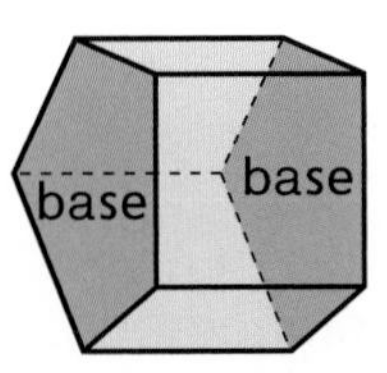

pentagonal prism

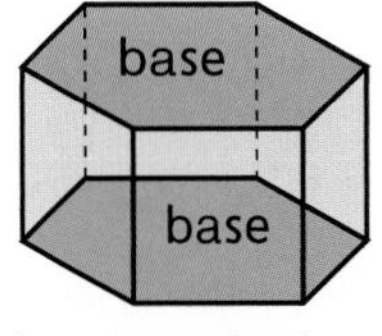

hexagonal prism

The two shaded faces of each prism are called **bases.**

- The bases have the same size and shape.
- The bases are parallel. This means that the bases will never meet, no matter how far they are extended.
- The other faces connect the bases and are all shaped like parallelograms.

Note Notice that the edges connecting the bases of a prism are parallel to each other.

The shape of its bases is used to name a prism. If the bases are triangular shapes, it is called a **triangular prism.** If the bases are rectangular shapes, it is called a **rectangular prism.** Rectangular prisms have three possible pairs of bases.

The number of faces, edges, and vertices that a prism has depends on the shape of the base.

Example

The triangular prism shown here has five faces—three rectangular faces and two triangular bases. It has nine edges and six vertices.

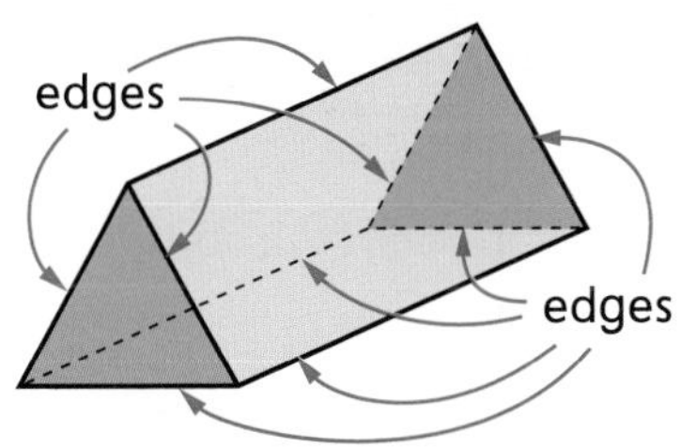

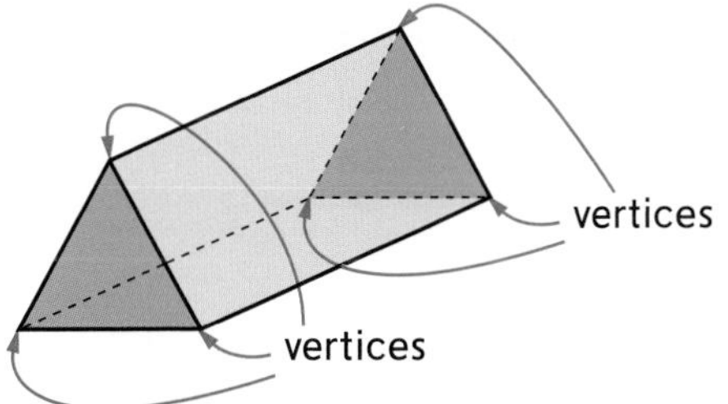

Check Your Understanding

1. **a.** How many faces does a hexagonal prism have?
   b. How many edges?
   c. How many vertices?
2. What is the name of a prism that has ten faces?

Check your answers in the Answer Key.

Did You Know?

A *rhombohedron* is a six-sided prism whose faces are all parallelograms. Every rectangular prism is also a rhombohedron.

Regular Polyhedrons

A polyhedron is **regular** if:

- it is convex.
- each face is a regular polygon.
- the faces all have the same size and shape.

There are only five kinds of regular polyhedrons.

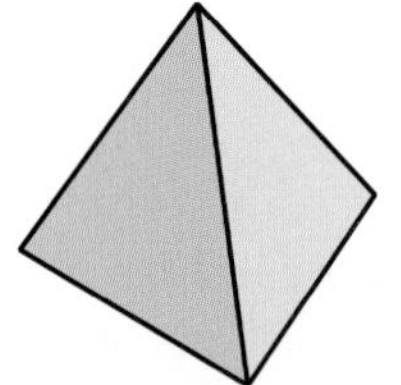

regular tetrahedron

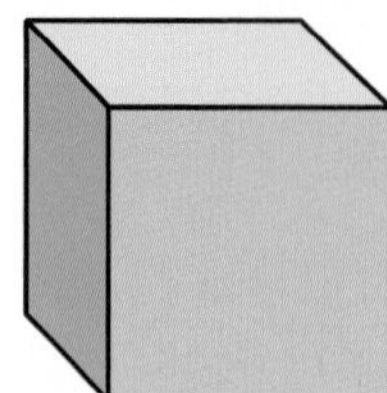

cube

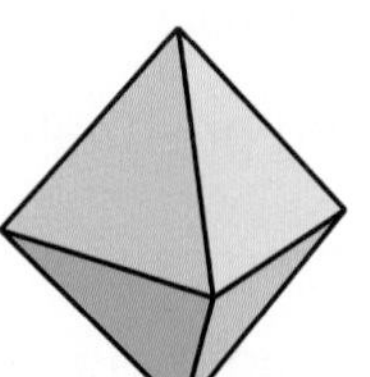

regular octahedron

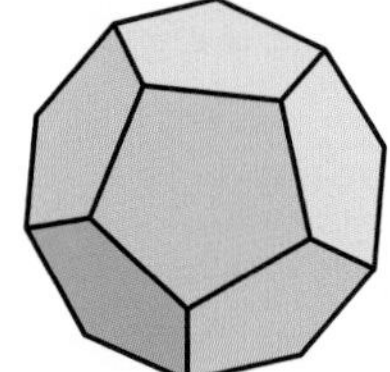

regular dodecahedron

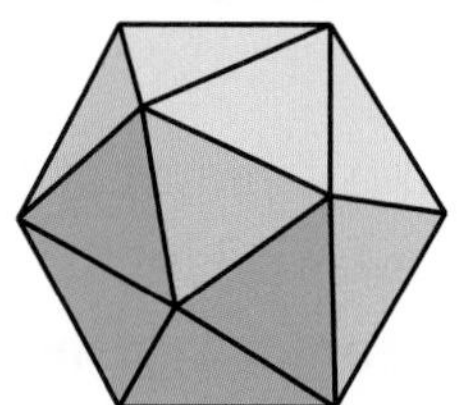

regular icosahedron

Name	Shape of Face	Number of Faces
regular tetrahedron	equilateral triangle	4
cube	square	6
regular octahedron	equilateral triangle	8
regular dodecahedron	pentagon	12
regular icosahedron	equilateral triangle	20

Did You Know?

The ancient Greeks proved that there are exactly five regular polyhedrons. They were described by Plato (427–327 BCE) and are often called the *Platonic solids.*

Plato associated the tetrahedron with fire, the cube with earth, the octahedron with air, the dodecahedron with heavenly bodies, and the icosahedron with water.

Check Your Understanding

1. Which regular polyhedrons have faces formed by equilateral triangles?

2. a. How many edges does a rectangular polyhedron have?
b. How many vertices?

3. a. How are regular tetrahedrons and regular octahedrons alike?
b. How are they different?

Check your answers in the Answer Key.

Coordinate Grid

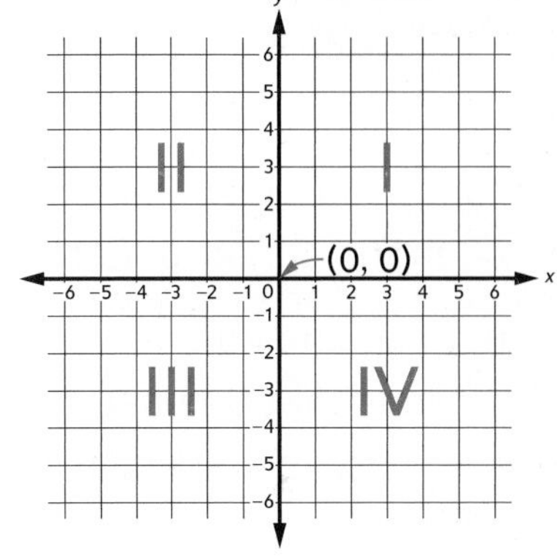

The ordered pair (0, 0) names the origin.

A **coordinate grid** is used to name points on a flat surface. It is made up of two number lines, called **axes,** that are perpendicular, or at right angles to each other. The horizontal number line is the ***x*-axis** and the vertical number line is the ***y*-axis.** The axes intersect at their 0 points to form four quadrants. The point where the axes meet is called the **origin.**

Every point on a coordinate grid can be named by an **ordered pair.** The numbers that make up an ordered pair are called the **coordinates** of the point. The first coordinate, called the ***x*-coordinate,** gives the position of the point along the horizontal axis. The second coordinate, called the ***y*-coordinate,** gives the position of the point along the vertical axis.

On the grid at the right, the ordered pair (7, 5) names point *A*. The numbers 7 and 5 are the coordinates of point *A*.

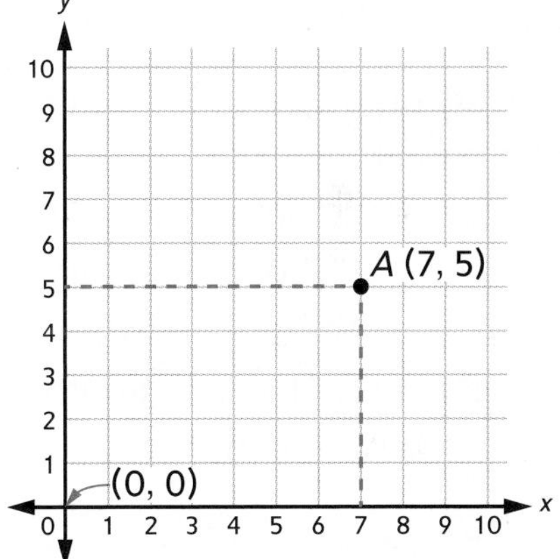

You can plot ordered pairs following the steps below. To plot the ordered pair (6, 8), see the grid to the right.

Step 1 Locate 6 on the *x*-axis. Draw a vertical line.

Step 2 Locate 8 on the *y*-axis. Draw a horizontal line.

Step 3 Point *B* (6, 8) is located at the intersection of the two lines.

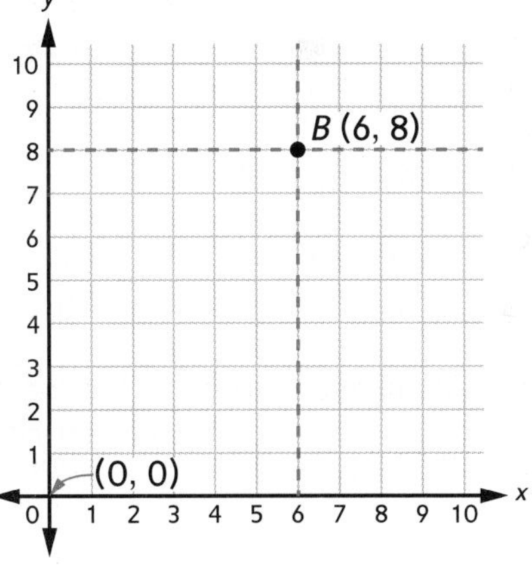

Another way to plot the ordered pair (6, 8) is to start at the origin (0, 0) and move 6 units to the right along the *x*-axis to (6, 0). From (6, 0), move up 8 units along a line parallel to the *y*-axis to (6, 8).

The order of the numbers in an ordered pair is important. The coordinate (6, 8) does not name the same point as the coordinate (8, 6). The point (8, 6) is located 8 units from the origin in the horizontal direction and 6 units from the origin in the vertical direction. It would appear to the right of the point (6, 8) and lower on the grid.

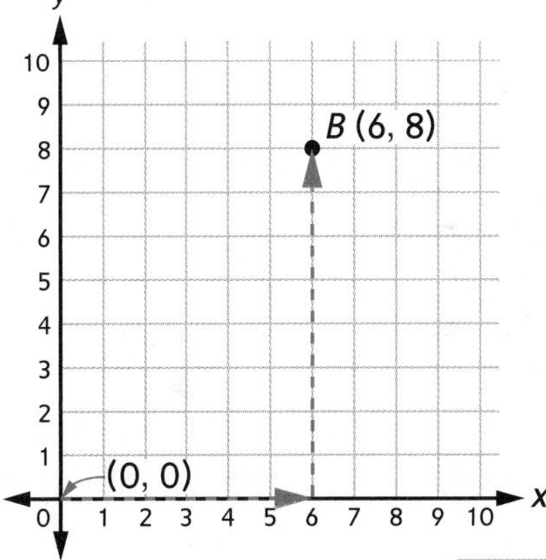

Latitude and Longitude

Earth is almost a perfect **sphere.** All points on Earth are about the same distance from its center. Earth rotates on an **axis,** which is an imaginary line through the center of Earth connecting the North Pole and the South Pole.

Reference lines are drawn on globes and maps to make places easier to find. Lines that go east and west around Earth are called **lines of latitude.** The equator is a special line of latitude. Every point on the equator is the same distance from the North Pole and the South Pole. The lines of latitude are often called **parallels** because each one is a circle that is parallel to the equator.

The **latitude** of a place is measured in **degrees.** The symbol for degrees is (°). Lines north of the equator are labeled °N (degrees north); lines south of the equator are labeled °S (degrees south). The number of degrees tells how far north or south of the equator a place is. The area north of the equator is the Northern Hemisphere. The area south of the equator is the Southern Hemisphere.

Did You Know?

In order to locate places more accurately, each degree is divided in 60 minutes. One minute equals $\frac{1}{60}$ degree. The symbol for minutes is (′). For example, 31°23′N means $31\frac{23}{60}$ degrees north.

Examples

The latitude of the North Pole is 90°N.
The latitude of the South Pole is 90°S.

The poles are the points farthest north and farthest south on Earth.

The latitude of Cairo, Egypt, is 30°N. We say that Cairo is 30 degrees north of the equator.

The latitude of Durban, South Africa, is 30°S. Durban is in the Southern Hemisphere.

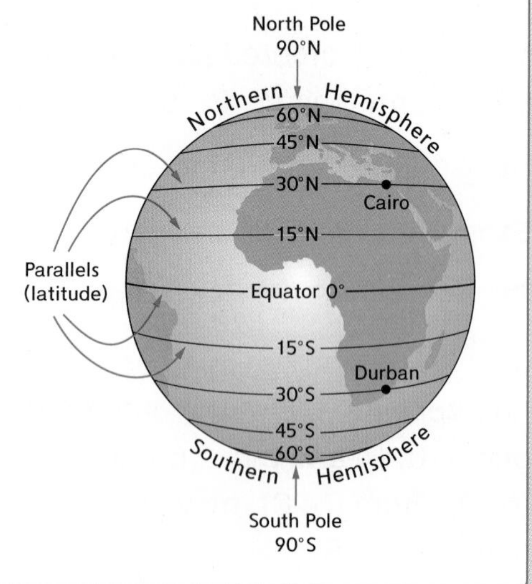

A second set of lines runs from north to south. These are semicircles (half-circles) that connect the poles. They are called **lines of longitude** or **meridians.** The meridians are not parallel, since they meet at the poles.

The **prime meridian** is the special meridian labeled 0°. The prime meridian passes through Greenwich, England (near London). Another special meridian is the international date line. This meridian is labeled 180° and is exactly opposite the prime meridian on the other side of the world.

The **longitude** of a place is measured in degrees. Lines west of the prime meridian are labeled °W. Lines east of the prime meridian are labeled °E. The number of degrees tells how far west or east of the prime meridian a place is. The area west of the prime meridian is called the Western Hemisphere. The area east of the prime meridian is called the Eastern Hemisphere.

Examples

The longitude of Greenwich, England, is 0° because it lies on the prime meridian.

The longitude of Durban, South Africa, is 30°E. Durban is in the Eastern Hemisphere.

The longitude of Gambia (a small country in Africa) is 15°W. We say that Gambia is 15 degrees west of the prime meridian.

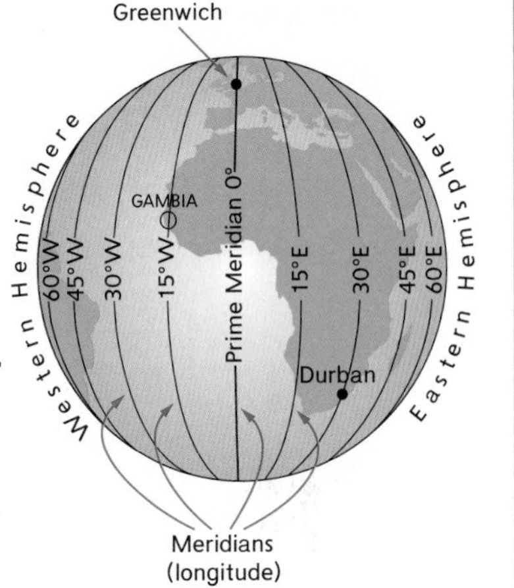

When lines of both latitude and longitude are shown on a globe or map, they form a pattern of crossing lines called a *grid*. The grid can help you locate places on the map. Any place on the map can be located by naming its latitude and longitude.

Examples

This map may be used to find approximate latitude and longitude for the cities shown. For example, Denver, Colorado, is about 40° North and 105° West.

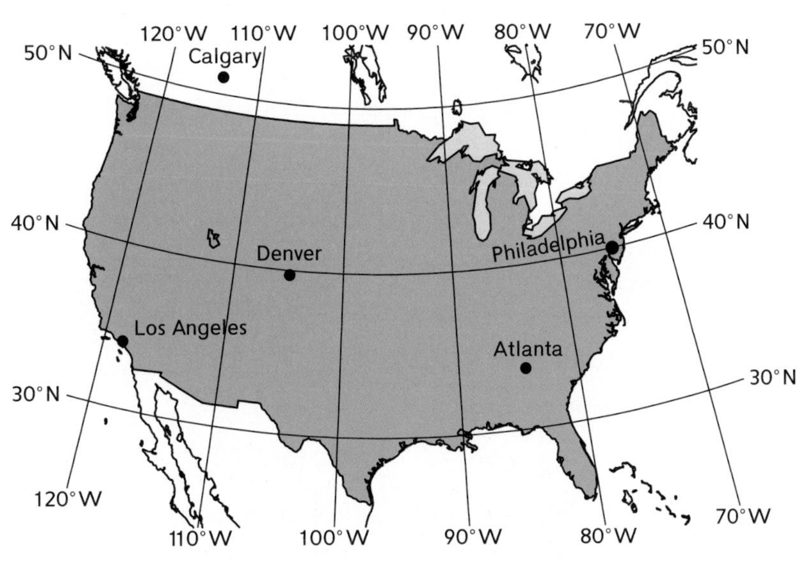

Congruent Figures

Figures that have the same shape and size are **congruent figures.** Figures are congruent if they would match exactly if you could place one figure on top of the other. You can use a coordinate grid to determine whether figures are congruent.

Example

Line segments are congruent if they have the same length.

$\overline{EF}$ and $\overline{CD}$ are both 3 units long. They have the same shape and the same length. These line segments are congruent.

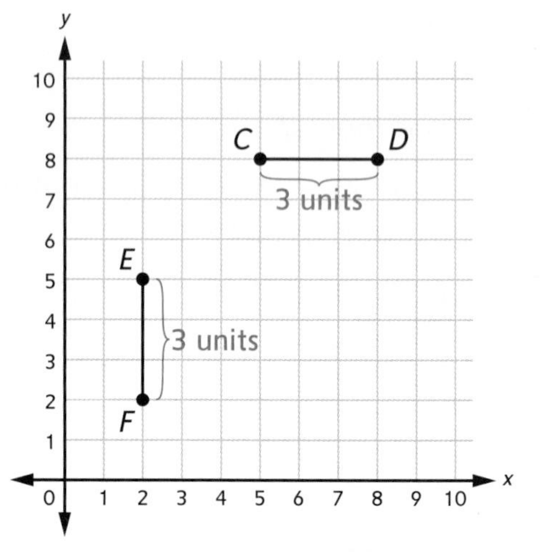

Example

Polygons are congruent if all of their sides have the same length and all of their angles have the same measure.

Quadrilaterals *WXYZ* and *KLMN* both have two sides that are 3 units long and two sides that are 4 units long. All the angles of both quadrilaterals measure 90°. If you placed one figure on top of the other, they would match exactly. These figures are congruent.

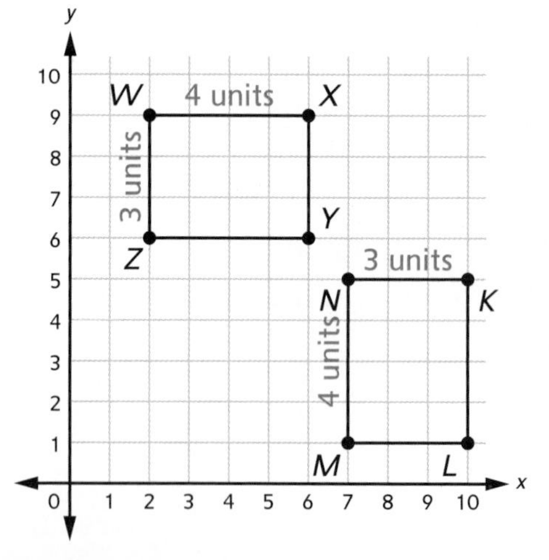

Check Your Understanding

Besides using a coordinate grid, which of these methods could you use to make a congruent copy of this square?

a. Use a copy machine to copy the square.

b. Use tracing paper and trace the square.

c. Measure the sides with a ruler, then draw the sides at right angles to each other using a protractor.

Check your answers in the Answer Key.

Similar Figures

Figures that have exactly the same shape are called **similar figures.** Usually, one figure is an enlargement or a reduction of the other. The size-change factor tells the amount of enlargement or reduction. Congruent figures are similar because they have the same shape. The size-change factor for congruent figures is 1X because they have the same size.

Examples

If a copy machine is used to copy a picture, the copy will be similar to the original.

Original Copy
Exact copy
Copy machine set to 100%.
Size-change factor is 1X.

Original Copy
Enlargement
Copy machine set to 200%.
Size-change factor is 2X.

Original Copy
Reduction
Copy machine set to 50%.
Size-change factor is $\frac{1}{2}$X.

Example

Similar figures can be shown on a coordinate grid by applying the same rule to all the *x*- and *y*-values of each coordinate in the original figure.

For example, a size-change factor of $\frac{1}{3}$X means that you divide the *x*- and *y*-values of each coordinate in the original figure by 3. The new image will be one-third as large as the original figure. Quadrilaterals *ABCD* and *MNOP* are similar. The smaller quadrilateral (new figure) was created by applying a rule.

Quadrilateral *ABCD* (original figure)	Quadrilateral *MNOP* (new figure) Rule: Divide each coordinate of the original ordered pair by 3
A (9, 9)	*M* (3, 3)
B (3, 6)	*N* (1, 2)
C (6, 3)	*O* (2, 1)
D (12, 6)	*P* (4, 2)

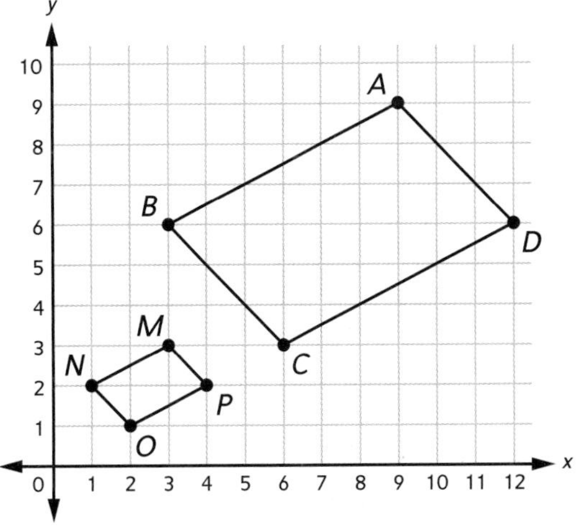

Reflections, Translations, and Rotations

In geometry, a figure can be moved from one place to another. Three different ways to move a figure are shown below.

- A **reflection** moves a figure by "flipping" it over a line.
- A **translation** moves a figure by "sliding" it to a new location.
- A **rotation** moves a figure by "turning" it around a point.

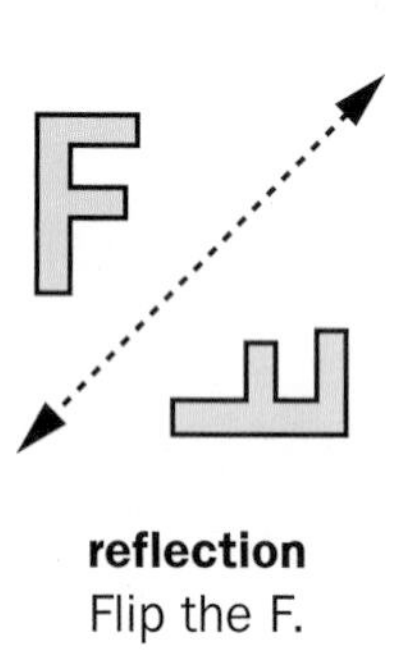

reflection
Flip the F.

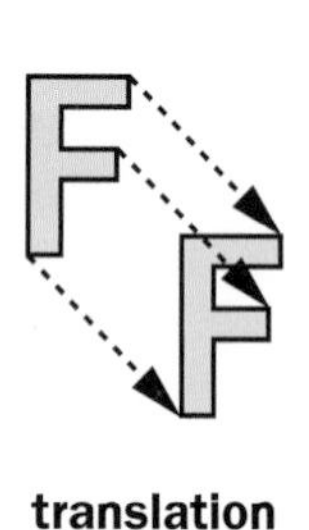

translation
Slide the F.

rotation
Turn the F.

An approximate reflection is shown here. The line of reflection is the water's edge, along the bank.

The original figure, before it has been moved, is called the **preimage.** The new figure produced by the move is called the **image.**

Each point of the preimage is moved to a new point of the image called its *matching point.* A point and its matching point are also called *corresponding points.*

For each of the moves shown above, the image has the same size and shape as the preimage. The image and preimage are congruent shapes.

Reflections

A reflection is a "flipping" motion of a figure. The line that the figure is flipped over is called the **line of reflection.** The preimage and the image are on opposite sides of the line of reflection.

For any reflection:

- The preimage and the image have the same size and shape.
- The preimage and the image are reversed.
- Each point and its matching point are the same distance from the line of reflection.

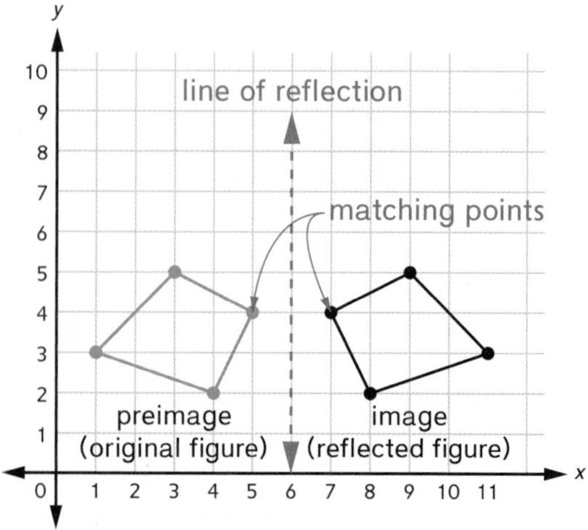

These matching points each have a distance of 1 unit from the line of reflection.

Lissa Harrison

Translations

A translation is a "sliding" motion of a figure. Each point of the figure slides the same distance in the same direction. Imagine the letter T drawn on a coordinate grid.

- If each point of the letter T slides 6 units to the right, the result is a *horizontal translation*.
- If each point of the letter T slides 7 units upward, the result is a *vertical translation*.
- Suppose that each point of the letter T slides 6 units to the right, then 7 units upward. The result is a *diagonal translation*.

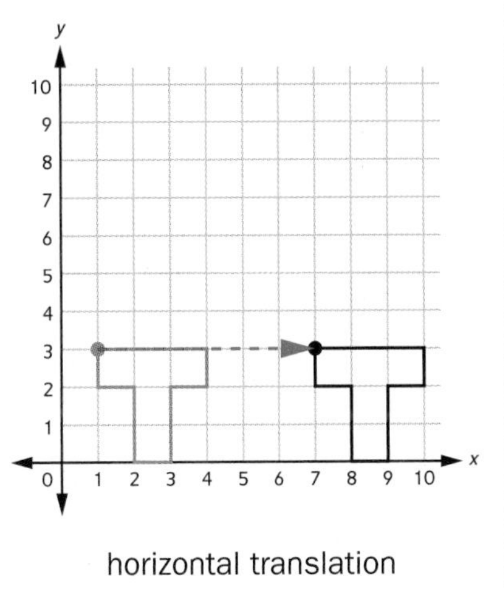

horizontal translation

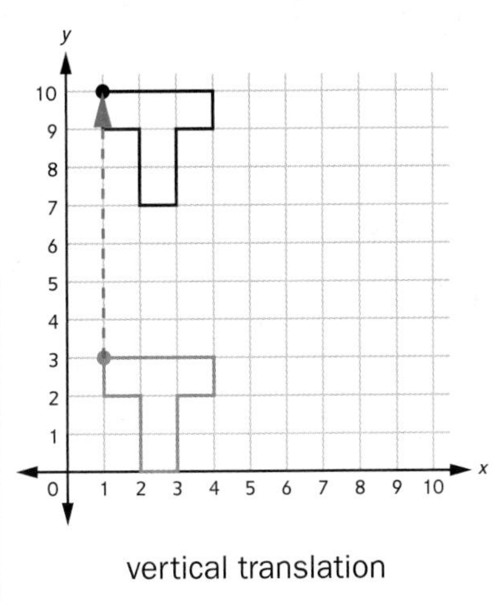

vertical translation

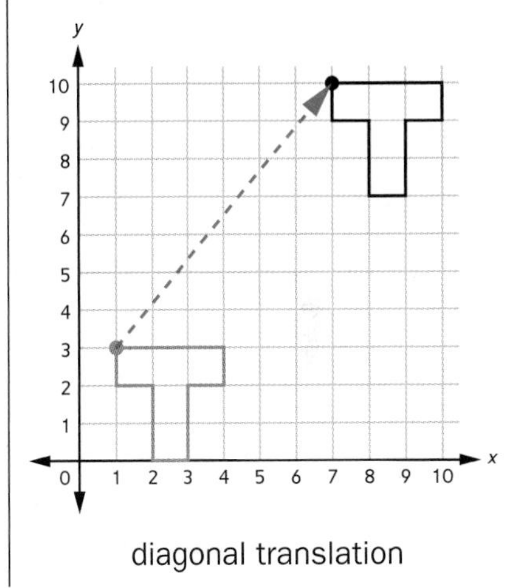

diagonal translation

Rotations

When a figure is rotated, it is turned a certain number of degrees around a particular point.

A figure can be rotated *clockwise* (the direction in which clock hands move). The figure can also be rotated *counterclockwise* (the opposite direction to the way clock hands move).

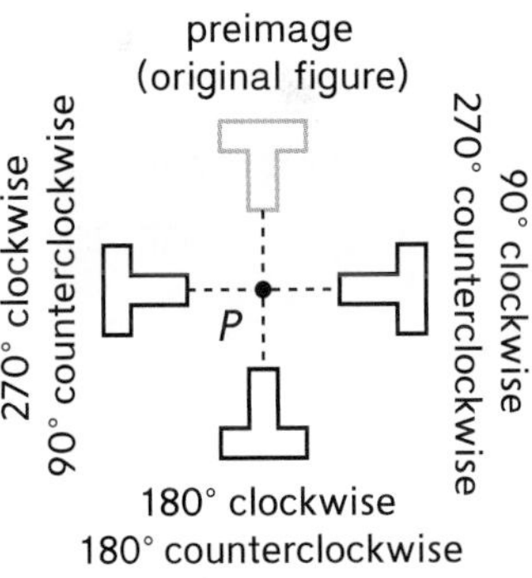

Line Symmetry

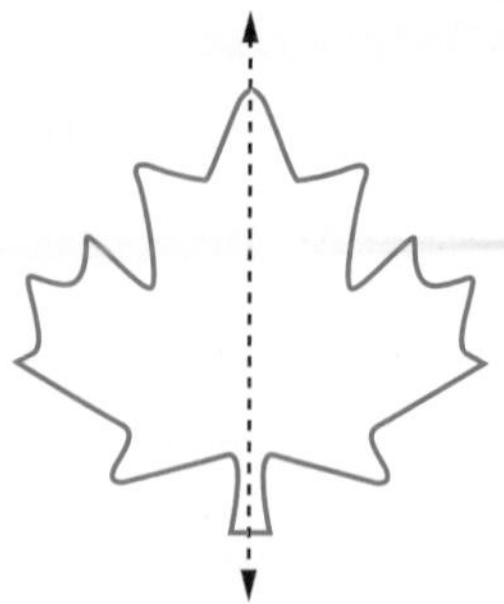
line of symmetry

A dashed line is drawn through the figure at the right. The line divides the figure into two parts. Both parts look alike but are facing in opposite directions.

The figure is **symmetric about a line.** The dashed line is called a **line of symmetry** for the figure.

An easy way to check whether a figure has *line symmetry* is to fold it in half. If the two halves match exactly, the figure is line symmetric. The fold line is the line of symmetry.

Examples

The letters M, A, C, and H are symmetric. The lines of symmetry are drawn for each letter.

The letter H has two lines of symmetry. If you fold along either line, the two halves match exactly.

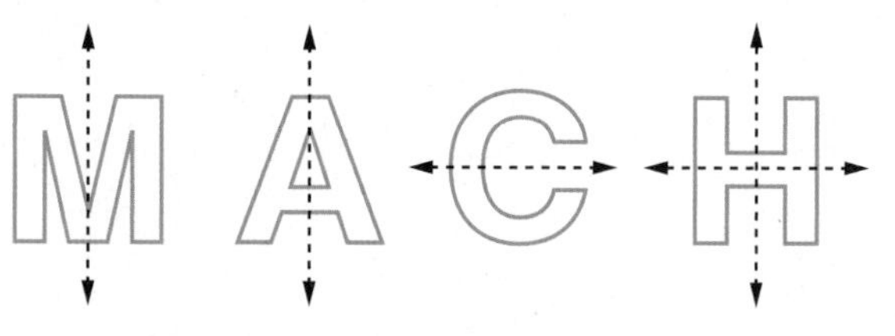

The figures below are all line symmetric. The line of symmetry is drawn for each figure. If there is more than one line of symmetry, they are all drawn.

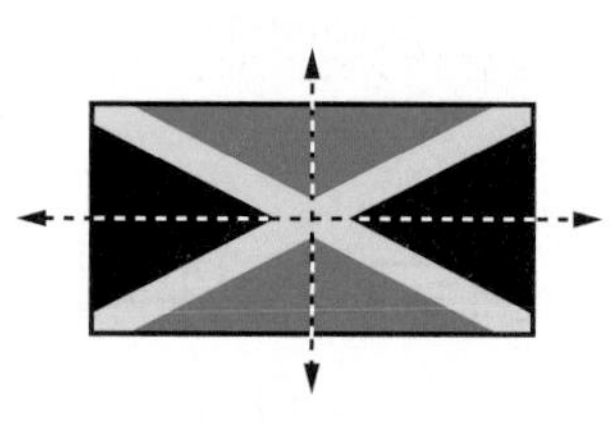
flag of Jamaica

butterfly

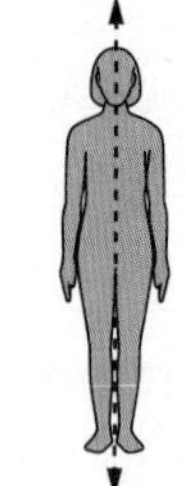
human body

Polynesian Navigation

Polynesia is a vast area in the Pacific Ocean that contains more than 1,000 islands. Polynesians throughout history have been great explorers. Archeological evidence indicates that Polynesian ancestors were migrating from Southeast Asia to Samoa as far back as 800 BCE. By around 400 CE, Polynesians had crossed thousands of miles of uncharted ocean to reach Hawaii. There is now evidence that the Polynesians may have even reached South America.

Because of their knowledge of the sea and their impressive ocean voyages, Polynesians have been compared to Vikings. The Vikings were the first Europeans to reach North America, some 500 years before Columbus. However, Polynesian journeys covered a much greater area of ocean over a much longer period of time.

Ancient Wayfinders

The Polynesians made their long ocean voyages in huge double-hulled canoes like this one. The navigators of these voyages were called *wayfinders* because they used the positions of celestial bodies (natural objects visible in the sky such as stars and planets), ocean currents, waves, winds, and the habits of birds and other animals to find their way from one island to the next.

Using wayfinding methods of navigation, the Polynesians were able to settle every habitable island in a huge triangle, with New Zealand on the southwest vertex, Easter Island on the southeast, and Hawaii on the north.

Recreating the Voyages

People have often wondered how the Polynesians were able to find tiny islands in the vast Pacific Ocean without maps or mechanical instruments like a compass. In 1976, the members of the Polynesian Voyaging Society, based in Hawaii, set out to find the answer. They built the Hōkūle'a, a replica of a Polynesian double-hulled voyaging canoe, and used traditional wayfinding methods to sail it several times from Hawaii (21.3°N, 157.8°W) to Tahiti (17.7°S, 149.4°W), a distance of more than 2,600 miles.

This is a drawing of the Hōkūle'a. The legend at the right shows its length, beam (width), sail area, and displacement (weight). Supplies and sleeping quarters are located below the ship's 40-foot deck. The canoe's two hulls stabilize it in rough seas.

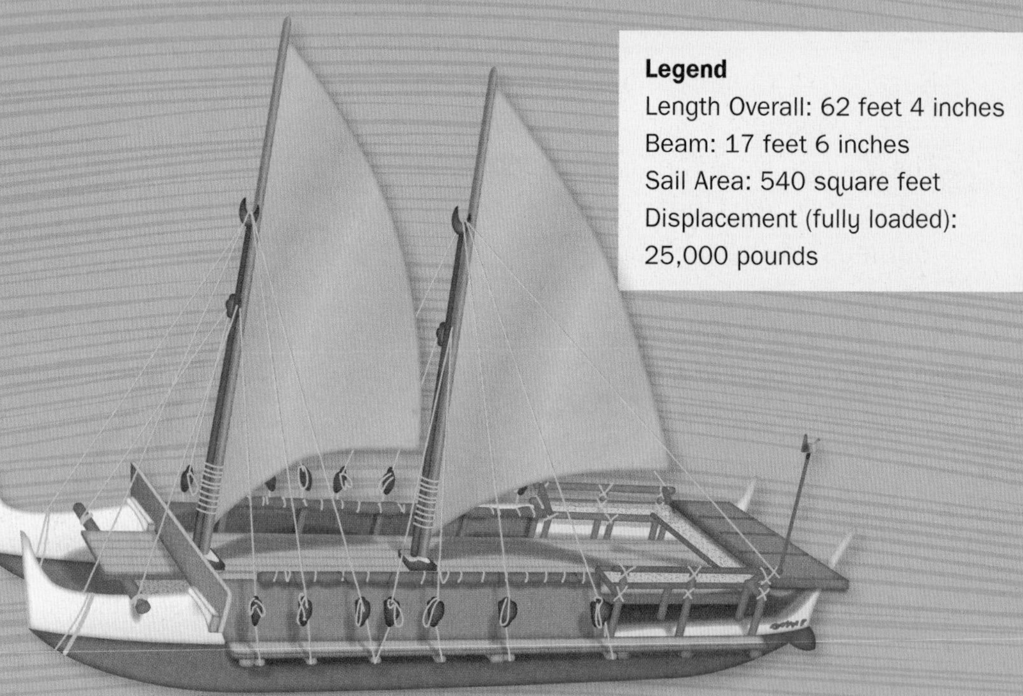

Legend
Length Overall: 62 feet 4 inches
Beam: 17 feet 6 inches
Sail Area: 540 square feet
Displacement (fully loaded): 25,000 pounds

Like the ancient Polynesian wayfinders, navigators of the Hōkūle'a used many clues to set their course. For example, the exact position on the horizon where the sun rises and sets changes every day following a consistent pattern. The navigators learned this pattern and used it to guide their canoes.

Design Pics/Dan Sherwood

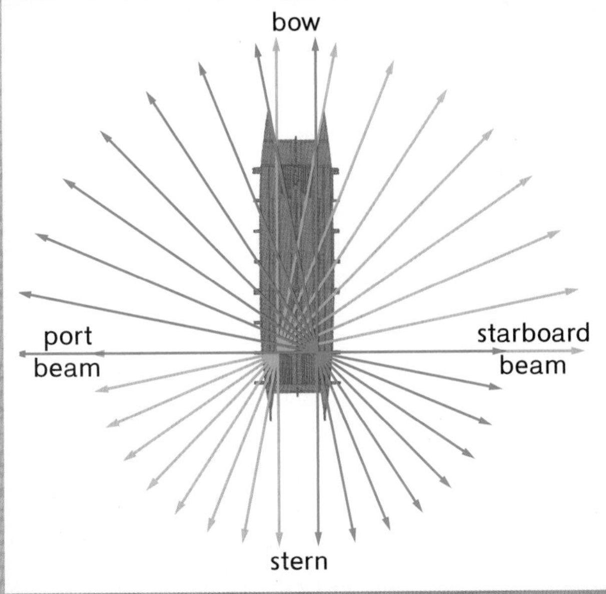

To hold a steady course on the Hōkūle'a, wayfinders aligned the rising or setting of the sun to marks on the railings of the canoe. There are eight marks on each side of the canoe. Each mark is paired with a single point located on the opposite side at the stern (back) of the canoe. A wayfinder standing at the stern can use the alignment of the markings and the sun to set a course in one of 32 different directions.

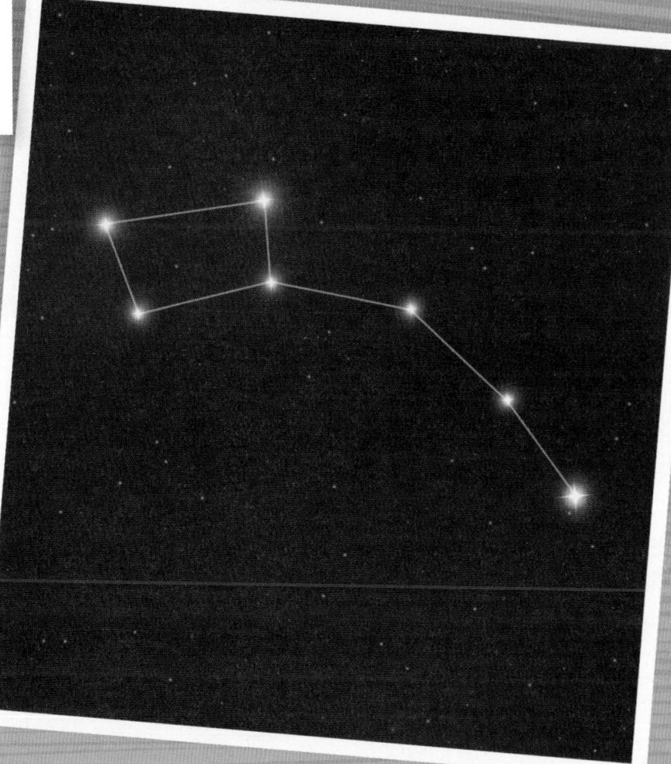

At night, stars served as a compass. Wayfinders knew about 150 stars by name and memorized their paths across the night sky. The rising and setting positions of most stars can be used to tell direction. While sailing in the Northern Hemisphere, one important exception is Polaris, the North Star, which doesn't rise or set like the other stars. It appears to be stationary and always points to the north.

Wayfinders also memorized the cycles of the moon. They knew that the moon rises and sets at different points on the horizon during the 29.5 days it takes for the moon to complete one phase-change cycle (from a new moon, when it looks dark, to a full moon, and back to a new moon). Tracking the location of the rising and setting moon is one way that wayfinders were able to guide their canoes with remarkable accuracy.

Winds, Currents, and Swells

During the middle of the day and on cloudy days and nights when celestial bodies cannot be seen, wayfinders often relied on winds, currents, and swells to estimate the canoe's speed and direction.

When the wind blows, it creates waves and swells. Swells are waves created by strong storms that can travel great distances. A swell will continue to move across a vast ocean like the Pacific long after the storm has ended.

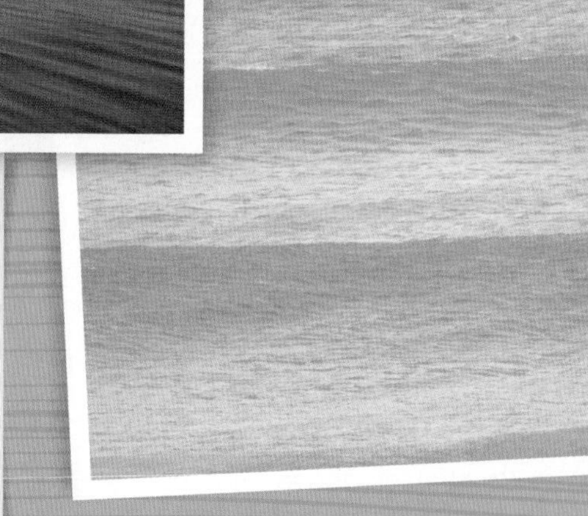

Winds and currents can be fairly predictable. For instance, when sailing between the latitudes 9°N and 25°N, the wind generally comes from the east at 10 to 20 miles per hour. The current generally flows to the west at about 0.5 miles per hour. Knowing the canoe can travel about 120 miles per day in these conditions, wayfinders could estimate the distance traveled and the approximate position of the canoe.

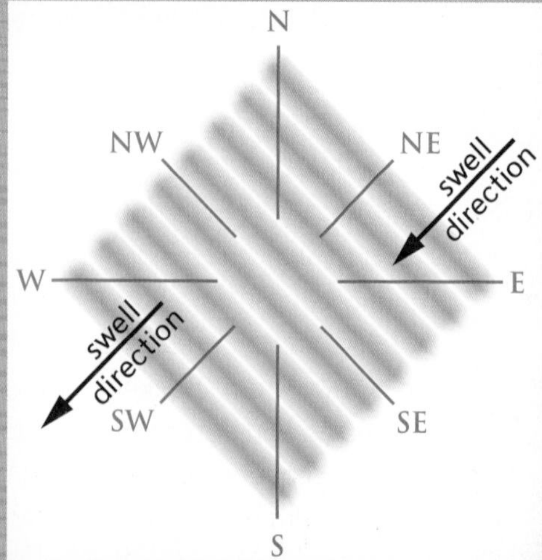

Swells are more predictable than waves. For example, if a swell comes from the northeast, it will always travel to the southwest. Wayfinders used ocean swells to orient their canoes. By observing the motion caused by swells, they could stay on course, even on a cloudy night. An experienced wayfinder could recognize and use as many as four or five different swells at once.

Seamarks

When traveling on land, we often use landmarks to find our way. On the sea, wayfinders used mid-ocean clues, or seamarks, to determine their canoe's daily position. These seamarks included schools of fish, flocks of birds, and ocean swells. Wayfinders used hundreds of seamarks that were passed on from generation to generation.

Navigators on the Hōkūle'a used a school of porpoises as a seamark on their voyages from Hawaii to Tahiti. Finding this seamark indicated that they reached a point around the latitude 9°N.

Seabirds can be helpful in locating land because some of them go out to sea in the morning to feed on fish and return to land at night to rest. In the morning, a wayfinder might sail in the direction that the birds are coming from to find land. In late afternoon, they might follow the birds as they return to land.

The white, or fairy, tern is a reliable indicator of nearby land. These birds venture up to 120 miles from their island nests. In general, a sighting of a large group of terns is a more reliable sign of land than just one or two birds.

A noddy tern indicates that land is nearby. These birds will fly about 40 miles from their nests in search of food.

Modern Navigation

With the modern Global Positioning System (GPS), we no longer need to look to the stars to find our way. GPS apps available on smartphones and other devices allow you to not only determine where you are, but to also plot a course to where you want to go.

GPS uses a network of about 30 satellites orbiting Earth. The system was originally developed for military use, but now anyone with a GPS device can pinpoint his or her exact location by receiving radio signals broadcast from these satellites.

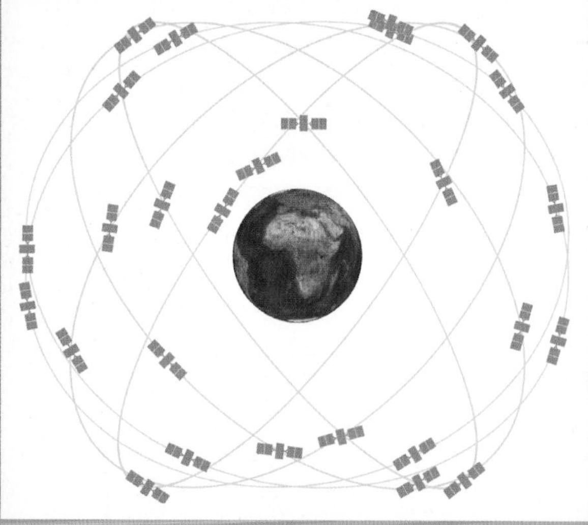

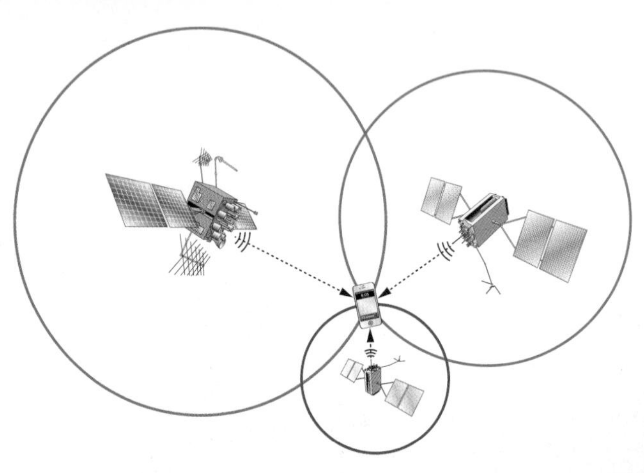

Each satellite broadcasts its position and time. GPS devices use this information and the amount of time it takes to receive a signal to calculate the exact distance to a given satellite. When this information has been received from at least three satellites, the exact location of the device can be determined by trilateration, a process illustrated in the diagram above.

Navigating in the Pacific using the methods of the Polynesian wayfinders is never exact. However, the crews of the Hōkūle‘a have shown that it can be done successfully.

With today's easy access to GPS, why might it still be worthwhile for people to learn and use ancient navigation methods like those of the Polynesian wayfinders?

Games

P. Ughetto/PhotoAlto

Games

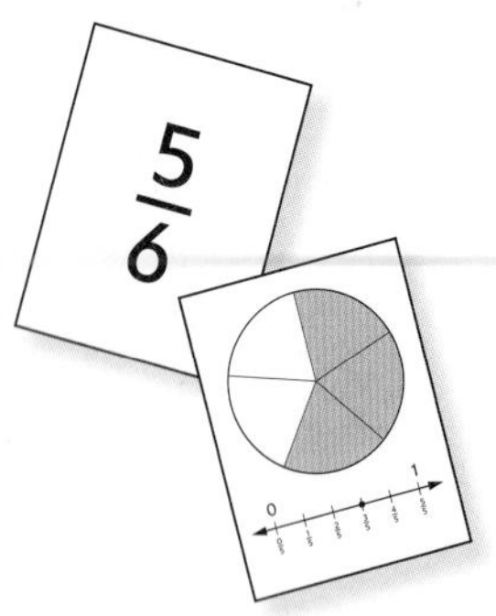

Fraction Cards

Throughout the year, you will play games that help you practice important math skills. Playing mathematics games gives you a chance to practice math skills in a different way. We hope that you will play often and have fun.

In this section of your *Student Reference Book*, you will find the directions for many games. The numbers in most games are generated randomly. This means that when you play games over and over, it is unlikely that you will repeat the same problems.

Many students have created their own variations of these games to make them more interesting. We encourage you to do this too.

Number Cards

Materials

The materials for each game are different and may include cards, dice, coins, counters, and calculators. Many games use the set of fraction cards found in the back of your *Math Journal 1* or in your eToolkit. For some games you will have to make a gameboard, a score sheet, or a set of cards that are not number cards. These instructions are included with the game directions. More complicated gameboards and card decks are available from your teacher.

Number Cards. You need a deck of number cards for many of the games. You can use an Everything Math Deck, a deck of regular playing cards, or make your own deck out of index cards. An Everything Math Deck is part of the eToolkit.

An Everything Math Deck includes 54 cards. There are 4 cards each for the numbers 0–10. And there is 1 card for each of the numbers 11–20.

You can also use a deck of regular playing cards after making a few changes. A deck of playing cards includes 54 cards (52 regular cards, plus 2 jokers). To create a deck of number cards, use a permanent marker to mark the cards in the following ways:

- Mark each of the 4 aces with the number 1.
- Mark each of the 4 queens with the number 0.
- Mark the 4 jacks and 4 kings with the numbers 11 through 18.
- Mark the 2 jokers with the numbers 19 and 20.

Baseball Multiplication (1 to 10 Facts)

Materials
- ☐ 1 *Baseball Multiplication* Game Mat (*Math Masters,* p. G3)
- ☐ number cards 1–10 (4 of each)
- ☐ 4 counters
- ☐ 1 calculator or 1 multiplication/division table

Players 2 teams of one or more players each

Skill Multiplying with automaticity

Object of the Game To score more runs in a 3-inning game.

2nd base
3rd base
1st base
home

Directions

1. Shuffle the cards and place the deck number-side down on the table.
2. Teams take turns being the pitcher and the batter. The rules are similar to the rules of baseball, but this game lasts only 3 innings.
3. The batter puts a counter on home plate. The pitcher draws 2 cards. The batter multiplies the numbers on the cards and gives the answer. The pitcher checks the answer and may use a calculator to do so.
4. If the answer is correct, the batter looks up the product in the Hitting Table to the right. If it is a hit, the batter moves all counters on the field the number of bases shown in the table. The pitcher tallies each out on the scoreboard.
5. An incorrect answer is a strike and another pitch is thrown (2 more cards are drawn). Three strikes make an out.
6. A run is scored each time a counter crosses home plate. The batter tallies each run scored on the scoreboard.
7. After each hit or out, the batter puts a counter on home plate. The batting and pitching teams switch roles after the batting team has made 3 outs. The inning is over when both teams have made 3 outs. Shuffle all cards and replace the deck between innings or when all cards have been used.

The team with more runs at the end of 3 innings wins the game. If the game is tied at the end of 3 innings, play continues into extra innings until one team wins.

Scoreboard					
Inning		1	2	3	Total
Team 1	outs				
	runs				
Team 2	outs				
	runs				

Hitting Table 1 to 10 Facts	
1 to 21	Out
24 to 45	Single (1 base)
48 to 70	Double (2 bases)
72 to 81	Triple (3 bases)
90 to 100	Home Run (4 bases)

Comstock Images/Alamy Images

Build-It

Materials ☐ 1 set of fraction cards

☐ 1 *Build-It* Gameboard for each player (*Math Masters,* p. G15)

Players 2

Skill Comparing and ordering fractions

Object of the Game To be the first player to arrange 5 fraction cards in order from smallest to largest.

Directions

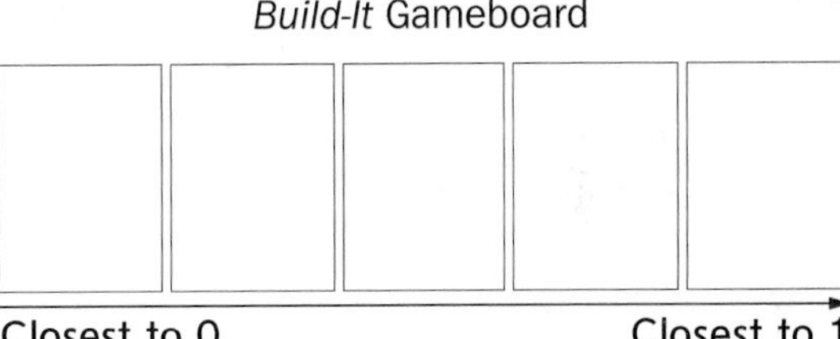

1. Shuffle the fraction cards. Deal 1 card number-side up on each of the 5 spaces on each *Build-It* gameboard. Do not change the order of the cards.
2. Put the remaining cards number-side down for a draw pile. Turn the top card over and place it number-side up in a discard pile.
3. Players take turns. When it is your turn:
   - Take either the top card from the draw pile or the top card from the discard pile.
   - Decide whether to keep this card or put it on the discard pile.
   - If you keep the card, it must replace 1 of the 5 cards on your gameboard. Put the replaced card on the discard pile.

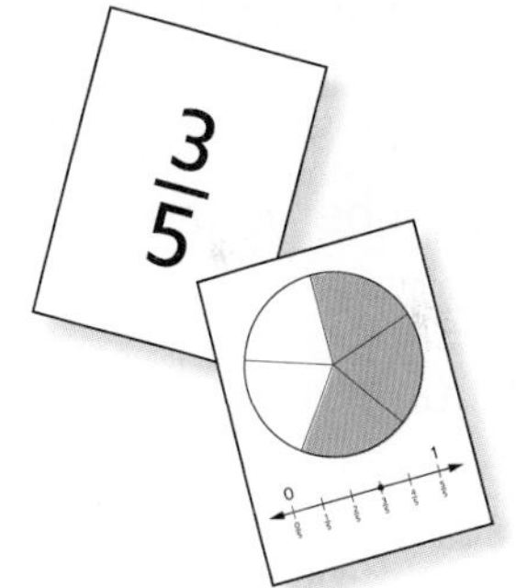

4. If all the cards in the draw pile are used, shuffle the discard pile. Place the cards number-side down in a draw pile. Turn over the top card to start a new discard pile.
5. When you think your 5 cards are in order from the smallest to the largest, say, "Built it!" and turn over your cards to check the pictures on the back. If the cards are in order, with no equivalent fractions, you are the winner. If not, flip them back over and continue until someone wins.

Buzz Games

Materials	none
Players	5–10
Skill	Finding multiples of a number and common multiples of two numbers
Object of the Game	To correctly say "BUZZ," "BIZZ," "BIZZ-BUZZ," or the next number when it is your turn.

Buzz

Directions

1. Players sit in a circle and choose a leader. The leader names any whole number from 3 to 9. This number is the BUZZ number. The leader also chooses the STOP number. The STOP number should be at least 30.
2. The player to the left of the leader begins the game by saying "one." Play continues clockwise with each player saying either the next whole number or "BUZZ."
3. A player must say "BUZZ" instead of the next number if:
   - The number is the BUZZ number or a multiple of the BUZZ number; or
   - The number contains the BUZZ number as one of its digits.
4. If a player makes an error, the next player starts with 1.
5. Play continues until the STOP number is reached.
6. For the next round, the player to the right of the leader becomes the new leader.

Example

The BUZZ number is 4. Play should proceed as follows: 1, 2, 3, BUZZ, 5, 6, 7, BUZZ, 9, 10, 11, BUZZ, 13, BUZZ, 15, and so on.

Bizz-Buzz

Directions

Bizz-Buzz is played like *Buzz*, except the leader names 2 numbers: a BUZZ number and a BIZZ number.

Players say:

1. "BUZZ" if the number is a multiple of the BUZZ number.
2. "BIZZ" if the number is a multiple of the BIZZ number.
3. "BIZZ-BUZZ" if the number is a multiple of both the BUZZ number and the BIZZ number.

Example

The BUZZ number is 6, and the BIZZ number is 3. Play should proceed as follows: 1, 2, BIZZ, 4, 5, BIZZ-BUZZ, 7, 8, BIZZ, 10, 11, BIZZ-BUZZ, 13, 14, BIZZ, 16, and so on. The numbers 6 and 12 are replaced by "BIZZ-BUZZ" since 6 and 12 are multiples of both 3 and 6.

Decimal Domination

Materials
- ☐ number cards 0–9 (4 of each)
- ☐ 2 counters per player
- ☐ 1 coin
- ☐ calculator (optional)

Players 2

Skill Predicting decimal products; multiplying decimals

Object of the Game To get the most points.

Directions

1. Shuffle the cards and place them number-side down in a pile.
2. Players take turns flipping a coin. Heads means to create the largest possible product. Tails means to create the smallest possible product.
3. After the coin flip, each player draws 4 cards and uses them to create 2 numbers, at least one of which is a decimal. Players use counters to indicate decimal points.

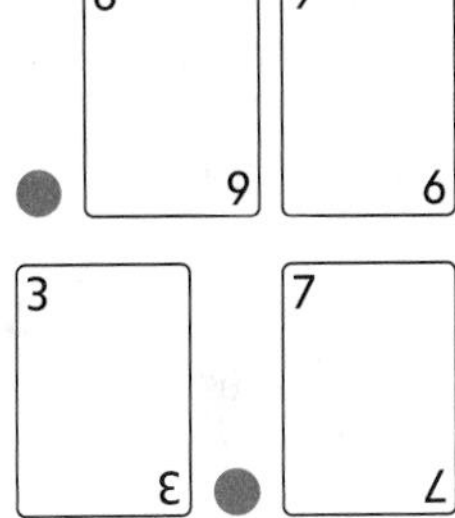

4. Players calculate their product. The player whose product is larger or smaller (depending on the coin flip) gets a point.
5. Players may use a calculator to check their partner's products.
6. Play continues until one player reaches 5 points. Players may have to reshuffle used cards and add them back into the deck.

Example

Naomi flips a coin that lands on tails, so the goal is to make the smallest product. Naomi and her partner, Alex, each draw 4 cards. Naomi draws 4, 1, 0, and 8. Alex draws 6, 9, 7, and 3. Naomi makes the problem .14 ∗ .08 and calculates the product to be 0.0112. Alex makes the problem .69 ∗ 3.7 and calculates the product to be 2.553. Naomi has the smaller product, so she gets a point.

Decimal Top-It

Materials	☐ number cards 0–9 (4 of each) ☐ 1 six-sided die ☐ 1 Decimal Place-Value Mat for every 2 players (*Math Masters,* p. TA25) ☐ Thousandths Grids (optional) (*Math Masters,* p. TA23)
Players	2 to 4
Skill	Comparing decimals
Object of the Game	To make the largest decimal numbers and have the lowest total score.

Directions

1. Shuffle the cards and place the deck number-side down on the table.
2. Players take turns rolling the die to determine the number of digits in their decimal number. If you roll:
   - 1 or 4: Make a 1-digit decimal.
   - 2 or 5: Make a 2-digit decimal.
   - 3 or 6: Make a 3-digit decimal.
3. After rolling the die, each player draws the number of cards that corresponds to the number of digits in his or her decimal number. Each player then chooses how to place the cards on his or her row of the game mat.
4. At the end of each round, players read their numbers aloud and compare them to the other players' numbers. You can use the thousandths grids to help compare numbers. The player with the largest number for the round scores 1 point. The player with the next-largest number scores 2 points, and so on.
5. Play 5 rounds for a game. Shuffle the deck between each round. The player with the *smallest* total number of points at the end of the 5 rounds wins the game.

Example

Phil and Claire played *Decimal Top-It*. Phil rolled a 2, so he used 2 cards to make a 2-digit decimal. Claire rolled a 6, so she used 3 cards to make a 3-digit decimal.

	Ones	.	Tenths	Hundredths	Thousandths
Phil	0	.	3	5	
Claire	0	.	6	4	2

Claire's number is larger than Phil's number. So Claire scores 1 point for this round, and Phil scores 2 points.

Variation

Decimal Top-It (Decimal Subtraction): Steps 1–3 remain the same. At the end of each round, players read their numbers aloud and compare them to the other players' numbers. The player with the smallest number scores 0 points. Each other player's score is the difference between his or her own number and the smallest number. (Subtract the smaller number from the larger number.) At the end of 4 rounds, players find their total scores. The player with the *largest* total score wins the game.

Example

If Phil and Claire played this variation, Claire still wins. Since 0.642 – 0.35 = 0.292, Claire scores 0.292 for the round. Phil has the smaller number, so he scores 0 points.

Decimal Top-It: Addition or Subtraction

Materials
- ☐ number cards 0–9 (4 of each)
- ☐ 2 counters per player
- ☐ calculator (optional)

Players 2

Skill Adding, subtracting, and comparing decimals

Object of the Game To collect more cards.

Decimal Top-It: Addition

Directions

1. Shuffle the deck and place it number-side down on the table.
2. Each player turns over 6 cards and, using the counter as a decimal point, forms 2 numbers with digits in the ones, tenths, and hundredths places. Players may put their cards in any order.
3. Each player finds the sum of his or her numbers. Players then compare their sums. The player with the larger sum takes all the cards. Players can check their answers with a calculator.
4. The game ends when there are not enough cards left for each player to have another turn. The player with the most cards wins.

Decimal Top-It: Subtraction

Directions

This game is played just like *Decimal Top-It: Addition*, except players find the difference between their numbers in Step 3. The player with the larger difference takes all the cards.

Example

Tony and Melissa are playing *Decimal Top-It: Addition.*

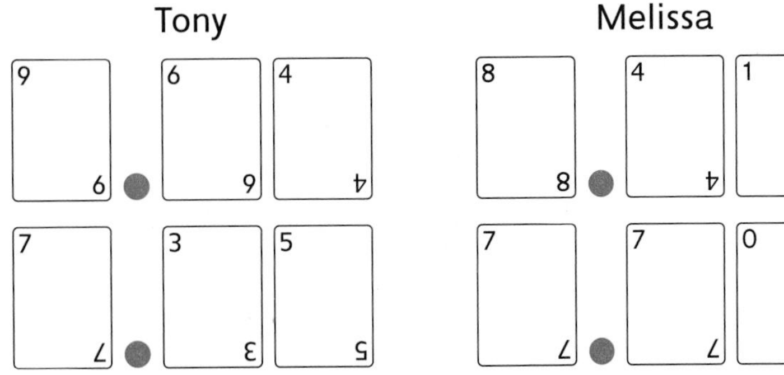

Tony turns over these cards: 4, 6, 3, 5, 9, 7. He makes the decimals 9.64 and 7.35.

Melissa turns over these cards: 7, 4, 1, 0, 7, 8. She makes the decimals 8.41 and 7.70.

Tony adds:

$$
\begin{array}{r}
9.64 \\
+\ 7.35 \\
\hline
16.00 \\
0.90 \\
0.09 \\
\hline
16.99
\end{array}
$$

Melissa adds:

	8.	4	1
+	7.	7	0
	15.	11	1
	16.	1	1

Tony takes all 12 cards because 16.99 is larger than 16.11.

Division Arrays

Materials ☐ number cards 6–20 (1 of each)

☐ 1 six-sided die

☐ 40 counters

Players 2 to 4

Skill Dividing to find equal shares

Object of the Game To have the highest total score.

Directions

1. Shuffle the cards and place the deck number-side down on the table.
2. Players take turns. When it is your turn, draw 2 cards and add the numbers together. Take as many counters as the sum of the 2 cards. You will use the counters to make an array.
   - Roll the die. The number on the die is the number of equal rows you must have in your array.
   - Make an array with the counters.
   - Your score is the number of counters in one complete row. If there are no leftover counters, your score is double the number of counters in one row.
3. Players keep track of their scores. The player with the highest total score at the end of 5 rounds wins.

Example

David draws a 14 card and a 12 card. $14 + 12 = 26$. He takes 26 counters. He rolls a 4 and makes an array with 4 rows by putting 6 counters in each row. Two counters are left over.

David scores 6 because there are 6 counters in each row.

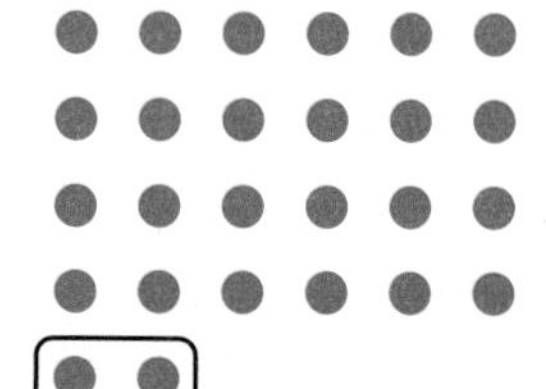

Example

Maria draws a 15 card and an 18 card. $15 + 18 = 33$. She takes 33 counters. She rolls a 3 and makes an array with 3 rows by putting 11 counters in each row.

Maria scores $11 * 2 = 22$ because there are 11 counters in each row, with none left over.

Division Dash

Materials ☐ number cards 1–9 (4 of each)
☐ 1 *Division Dash* Record Sheet for each player (*Math Masters*, p. G13)

Cards Drawn	Division Problem	Quotient	Score
Example: 6, 4, 5	640 / 50 = ?	12	12

Players 1 or 2

Skill Dividing 3-digit numbers by 2-digit numbers

Object of the Game To reach a score of 100.

Directions

1. Shuffle the cards and place the deck number-side down on the table.
2. Each player follows the instructions below:
   - Turn over 2 cards and lay them on the table to make a 2-digit number. Players may put their cards in any order. Multiply the number by 10. The result is your *dividend*, the number you are dividing.
   - Turn over another card and lay it on the table. Multiply the number by 10. The result is your *divisor*, the number you are dividing by.
   - Divide your dividend (the 3-digit number) by your divisor (the 2-digit number) and record the result. This result is your *quotient*. Ignore any remainders. Calculate mentally or on paper.
   - Add your quotient to your previous score and record your new score. If this is your first turn, your previous score was 0.
3. Players repeat Step 2 until one player's score is 100 or more. The first player to reach at least 100 wins. If there is only one player, the object of the game is to reach 100 in as few turns as possible.

Example

Turn 1: Brandon draws 6, 4, and then 5. He divides 640 by 50. Quotient = 12. The remainder is ignored. His score is 12 + 0 = 12.

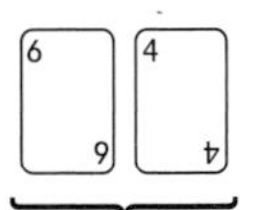

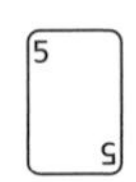

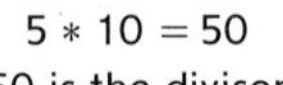

64 ∗ 10 = 640
640 is the dividend.

5 ∗ 10 = 50
50 is the divisor.

Turn 2: Brandon then draws 8, 2, and then 1. He divides 820 by 10. Quotient = 82. His score is 82 + 12 = 94.

Turn 3: Brandon then draws 5, 7, and then 8. He divides 750 by 80. Quotient = 9. The remainder is ignored. His score is 9 + 94 = 103.

Brandon has reached 100 in 3 turns and the game ends.

Doggone Decimal

Materials	☐ number cards 0–9 (4 of each)
	☐ 4 index cards labeled 0.1, 1, 10, and 100
	☐ 2 counters per player (to use as decimal points)
	☐ 1 calculator for each player
Players	2
Skill	Estimating products of whole numbers and decimals
Object of the Game	To collect more number cards.

Directions

1. One player shuffles the number cards and deals 4 cards to each player.
2. The other player shuffles the index cards, places them number-side down, and turns over the top card. The number that appears (0.1, 1, 10, or 100) is the *target number*.
3. Using 4 number cards and 2 counters as decimal points, each player forms 2 numbers. Each number must have 2 digits and a decimal point.
   - Players try to form 2 numbers whose product is as close as possible to the target number.
   - The decimal point can go anywhere in a number—for example:

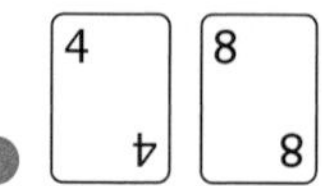
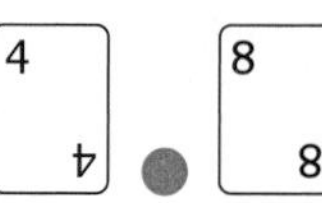
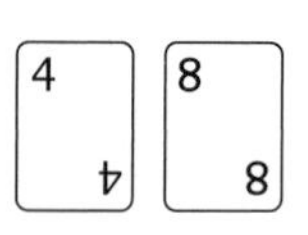

4. Each player computes the product of his or her numbers using a calculator.
5. The player whose product is closer to the target number takes all 8 number cards.
6. Four new number cards are dealt to each player and a new target number is turned over. Repeat Steps 3–5 using the new target number.
7. The game ends when all the target numbers have been used.
8. The player with more number cards wins the game. In the case of a tie, reshuffle the index cards and turn over a new target number. Play one tie-breaking round.

Example

The target number is 10.

Brianna is dealt 1, 4, 8, and 8. She forms the numbers 8.8 and 1.4.
Evelyn is dealt 2, 3, 6, and 9. She forms the numbers 2.6 and 3.9.

Brianna's product is 12.32 and Evelyn's is 10.14.
Evelyn's product is closer to 10. She wins the round and takes all 8 cards.

Exponent Ball

Materials
- ☐ 1 *Exponent Ball* Gameboard (*Math Masters*, p. G28)
- ☐ number cards 1–4 (4 of each, 2 black and 2 red or blue)
- ☐ 2 dice
- ☐ 1 counter

Players 2

Skill Multiplying and dividing by powers of 10; comparing numbers

Object of the Game To score more points in 4 turns.

Exponent Ball Gameboard

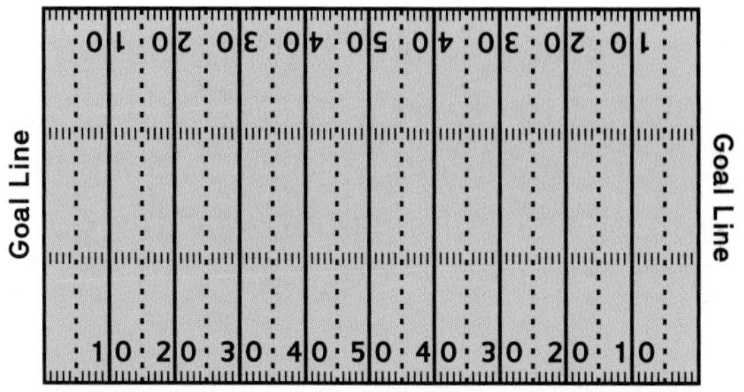

Table 1: Runs	
Value of Expression	**Move Ball**
0.0001 to 0.00099	Backward 15 yards
0.001 to 0.0099	Forward 10 yards
0.01 to 99	Forward 20 yards
100 to 3,999	Forward 30 yards
4,000 to 39,999	Forward 40 yards
40,000 and above	Forward 50 yards

Directions

1. The game is similar to U.S. football. Players take turns. A turn consists of 4 chances to advance the ball to the opposite goal line and score. Player 1 first places the ball (the counter) on one of the 20-yard lines. Player 1 must move 80 yards forward to score.

2. The first 3 chances must be runs on the ground. To *run*:
   - Roll both dice and form a decimal with a digit in the ones place and a digit in the tenths place. This is your starting number.
   - Draw a card. This is the exponent for the power of 10.
   - If the card is black, multiply your starting number by the power of 10. If the card is red or blue, divide your starting number by the power of 10. Write an expression to show how to multiply or divide your number. Then find the value of the expression. Use the value of the expression and Table 1 on the gameboard to find how far to move the ball forward or backward.
   - When all of the cards have been drawn, shuffle them to use again.

See the next page for an example.

Note If a backward move should carry the ball behind the goal line, the ball (counter) is put on the goal line.

Example

Sally rolls a 3 and a 5. Her starting number is 3.5. She draws a red 2, so she needs to divide her starting number by 10^2. She calculates the value of the expression $3.5 \div 10^2$ and gets 0.035. According to Table 1, Sally moves the ball 20 yards forward toward the goal because $0.01 < 0.035 < 99$.

Exponent Ball Gameboard

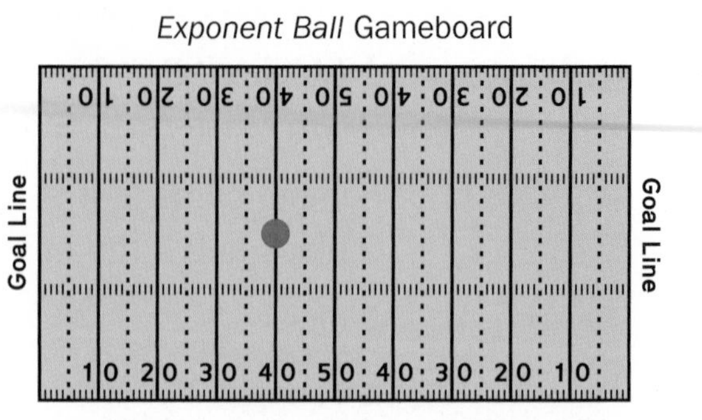

3. If Player 1 does not score in the first 3 chances, he or she may choose to run or kick on the fourth chance. To *kick*:
   - Roll one die and multiply the number shown by 10. The result is the distance the ball travels (see Table 2 on the gameboard).

Table 2: Kicks

Value of Roll	Move Ball
1	Forward 10 yards
2	Forward 20 yards
3	Forward 30 yards
4	Forward 40 yards
5	Forward 50 yards
6	Forward 60 yards

4. If the ball reaches the goal line on a run, the player scores 7 points. If the ball reaches the goal line on a kick, the player scores 3 points. If a player scores, the other player puts the ball on the 20-yard line closest to him or her, and begins his or her turn.

5. If the ball does not reach the goal line in 4 chances, that player's turn ends. The other player starts from where the ball stopped and moves toward the opposite goal line.

6. A game consists of 4 turns for each player. The player with more points at the end of the game wins.

Fraction Capture

Materials ☐ 1 *Fraction Capture* Gameboard (*Math Masters,* p. G19)

☐ 1 *Fraction Capture* Record Sheet for each player (*Math Masters,* p. G20)

☐ 2 six-sided dice

Players 2

Skill Adding fractions

Object of the Game To get the higher score.

Directions

1. Player 1 rolls two dice, makes a fraction using both numbers, and records the fraction on the *Fraction Capture* Record Sheet.
2. Player 1 initials sections of one or more gameboard squares to capture fractions that add up to the fraction formed in Step 1. Note that Player 1 must initial at least two sections, but can initial more than 2; the sections can be in the same gameboard square or different squares; and once a section is initialed, that section cannot be used again.
3. Player 1 records a fraction addition expression on the record sheet to show the fractions captured. Player 2 checks the sum of Player 1's captured fractions.
4. Players take turns. If a player can't find a combination of uncaptured fractions that add up to the fraction formed from the dice, the player's turn is over.
5. Play continues until each player has had 5 turns.
6. A player claims a gameboard square when that player has initialed sections making up more than $\frac{1}{2}$ of the square. If each player has initialed $\frac{1}{2}$ of the square, no one claims that square.
7. Each player gets 1 point for each gameboard square claimed and 1 point for each addition expression that includes fractions with unlike denominators. The player with the most points wins.

Example

Karli rolls a 5 and a 6 and forms the fraction $\frac{5}{6}$. She initials two $\frac{1}{3}$ sections in one gameboard square and one $\frac{1}{6}$ section in another gameboard square. She records $\frac{5}{6}$ and $\frac{1}{3} + \frac{1}{3} + \frac{1}{6}$ on her record sheet.

Devin checks her answer and agrees that $\frac{1}{3} + \frac{1}{3} + \frac{1}{6} = \frac{5}{6}$. Now it is Devin's turn.

Fraction Of

Materials ☐ 1 set of *Fraction Of* Fraction Cards (Set 1) (*Math Journal 1*, Activity Sheet 14)

☐ 1 set of *Fraction Of* Whole Cards (*Math Journal 1*, Activity Sheet 15)

☐ 1 *Fraction Of* Gameboard and Record Sheet for each player (*Math Masters*, p. G24)

Players 2

Skill Solving fraction-of problems

Object of the Game To get the higher score.

Directions

1. Shuffle the decks separately. Place both decks number-side down on the table.
2. Players take turns. On your turn, draw 1 card from each deck. Place the cards on your gameboard to create a fraction-of problem.
   - The fraction card shows what fraction of the whole you must find.
   - The whole card offers 3 possible choices. Choose a whole that will result in a fraction-of problem with a whole-number answer. There may be more than one choice.
   - Solve the fraction-of problem and set the 2 cards aside. The answer to the problem is your score for the round.

Example

Player 1 draws $\frac{1}{10}$ and 28 35 30.

$\frac{1}{10}$ of 28 will *not* result in a whole-number answer.

$\frac{1}{10}$ of 28 is between 2 and 3.

$\frac{1}{10}$ of 35 will *not* result in a whole-number answer.

$\frac{1}{10}$ of 35 is between 3 and 4.

$\frac{1}{10}$ of 30 *will* result in a whole-number answer.

$\frac{1}{10}$ of 30 is 3.

Player 1 chooses 30 as the whole. Player 1's score for the round is 3.

Example

Player 2 draws $\frac{1}{2}$ and [12 / 30 / 25].

Player 2 could choose 12 or 30 as the whole.

Player 2 chooses 30 because it will give more points than 12, finds $\frac{1}{2}$ of 30, and earns 15 points.

3. Play continues until all of the *Fraction Of* Whole Cards have been used. The player with more points wins.

Variation

For a more challenging version of the game, add 1 set of *Fraction Of* Fraction Cards (Set 2) (*Math Journal 2*, Activity Sheet 17) to the deck of *Fraction Of* Fraction Cards.

Fraction Card   of   WHOLE (Choose one.) Whole Card

Round	Fraction-Of Problem	Solution (Points)
Sample	$\frac{1}{5}$ of 25	5
1		
2		
3		
4		
5		
6		
7		
8		
	Total Score	

Fraction Of Gameboard and Record Sheet

Fraction Spin

Materials	☐ 1 *Fraction Spin* Record Sheet (*Math Masters*, p. G16) ☐ 1 *Fraction Spin* Spinner (*Math Masters*, p. G17) ☐ 1 large paper clip
Players	2
Skill	Estimating sums and differences of fractions
Object of the Game	To complete 10 true number sentences.

Directions

1. Each player writes his or her name in one of the boxes on the record sheet.
2. Players take turns. When it is your turn:
   - Anchor the paper clip to the spinner with the point of your pencil. Spin.
   - Write the fraction you spin on any one of the blanks below your name.
3. The first player to complete 10 true number sentences is the winner.

Example

Ella spun $1\frac{7}{8}$ and $\frac{2}{3}$ on her first two turns, and filled in 2 blank spaces.

On her next turn, Ella spins $1\frac{3}{4}$. She has two choices:

- Choice 1: She can write $1\frac{3}{4}$ in a sentence where there are two blanks. **OR**,
- Choice 2: She can use $1\frac{3}{4}$ to form the true number sentence $1\frac{3}{4} - \frac{2}{3} > \frac{1}{2}$.

Ella cannot use $1\frac{3}{4}$ in the first sentence because $1\frac{7}{8} + 1\frac{3}{4}$ is not < 3.

Ella	Ella	Ella
Name	Name	Name
$1\frac{7}{8}$ + ____ < 3	$1\frac{7}{8}$ + ____ < 3	$1\frac{7}{8}$ + ____ < 3
____ + ____ > 3	____ + ____ > 3	____ + ____ > 3
____ – ____ < 1	$1\frac{3}{4}$ – ____ < 1	____ – ____ < 1
____ – $\frac{2}{3}$ > $\frac{1}{2}$	____ – $\frac{2}{3}$ > $\frac{1}{2}$	$1\frac{3}{4}$ – $\frac{2}{3}$ > $\frac{1}{2}$
	Choice 1	Choice 2

Fraction Top-It: Addition

Materials	☐ 1 set of fraction cards
Players	2
Skill	Adding and comparing fractions
Object of the Game	To collect more cards.

Directions

1. Shuffle the deck and place it number-side down on the table.
2. Each player turns over 2 cards and figures out the sum of the fractions. The player with the larger sum takes all the cards. In case of a tie for the larger sum, each player turns over 2 more cards and says the sum of the fractions. The player with the larger sum takes all the cards from both plays.
3. The player with more cards after 10 rounds wins.

Variation

Use only cards with denominators of 2, 3, 4, 6, and 12 for an easier version of the game.

Fraction/Whole Number Top-It

Materials	☐ number cards 1–10 (4 of each)
	☐ 1 set of fraction cards
Players	2 to 4
Skill	Multiplying whole numbers and fractions
Object of the Game	To collect the most cards.

Directions

1. Shuffle the number cards and fraction cards. Place them number-side down in 2 separate piles on the table.
2. Each player turns over 1 number card and 1 fraction card. Each player calculates the product of his or her whole number and fraction and calls it out as a mixed number or fraction less than one. The player with the largest product takes all the cards.
3. In case of a tie for the largest product, each tied player repeats Step 2. The player with the largest product takes all the cards from both plays.
4. The game ends after 10 rounds. The player with the most cards wins.

Example

Amy turns over a 3 and a $\frac{3}{5}$.

Roger turns over a 7 and a $\frac{2}{8}$.

Amy's product is $3 * \frac{3}{5} = \frac{9}{5}$.

Roger's product is $7 * \frac{2}{8} = \frac{14}{8}$.

$\frac{9}{5} = 1\frac{4}{5}$ $\quad \frac{14}{8} = \frac{7}{4} = 1\frac{3}{4}$

Amy's product is larger, so she takes all of the cards.

Amy: 3, $\frac{3}{5}$ — Roger: 7, $\frac{2}{8}$

Variation

Each player turns over 2 fraction cards. Each player calculates the product of his or her fractions, and the player with the largest product takes all the cards.

Hidden Treasure

Materials ☐ 1 sheet of *Hidden Treasure* Gameboards for each player (*Math Masters,* p. G26)

☐ 2 pencils

☐ 1 red pen or crayon

Players 2

Skill Plotting ordered pairs; developing a search strategy

Object of the Game To find the other player's hidden point on a coordinate grid.

Directions

1. Each player uses 2 grids. Players sit so they cannot see what the other is writing.
2. Each player secretly marks a point on his or her Grid 1. Use the red pen or crayon. These are the "hidden" points.
3. Player 1 guesses the location of Player 2's hidden point by naming an ordered pair. To name the ordered pair (1, 2), say "1 comma 2."
4. If Player 2's hidden point is at that location, Player 1 wins.
5. If the hidden point is not at that location, Player 2 marks the guess in pencil on his or her Grid 1. Player 2 counts the fewest number of "square sides" needed to travel from the hidden point to the guessed point and tells it to Player 1. Repeat Steps 3–5 with Player 2 guessing and Player 1 answering.
6. Play continues until one player finds the other's hidden point.

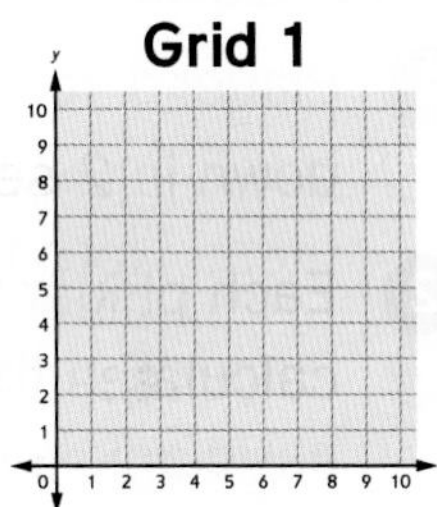

Hide your point here.

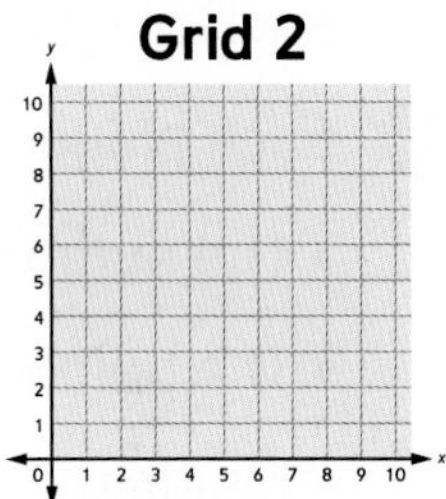

Guess the other player's point here.

Example

Player 1 marks a hidden point at (2, 5).

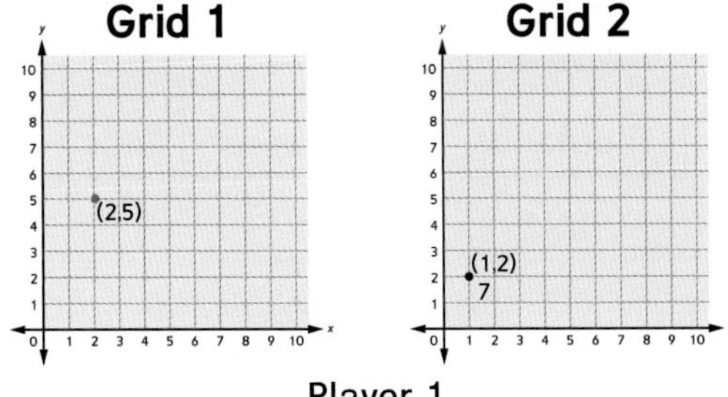

Player 1

Player 2 marks a hidden point at (3, 7).

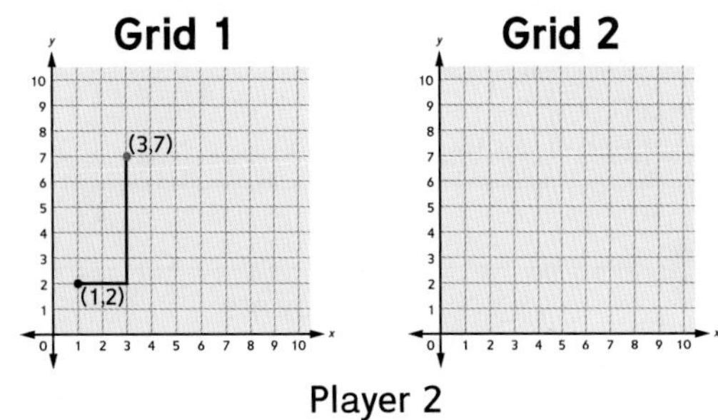

Player 2

- Player 1 guesses that Player 2's hidden point is at (1, 2) and marks it on Grid 2 in pencil.
- Player 2 marks the point (1, 2) in pencil on Grid 1 and tells Player 1 that (1, 2) is 7 units (square sides) away from the hidden point.
- Player 1 writes "7" next to the point (1, 2) on his or her Grid 2. Player 1's turn is over, and Player 2 makes a guess.

High-Number Toss

Materials ☐ 1 six-sided die

☐ 1 *High-Number Toss* Record Sheet (*Math Masters*, p. G10)

Players 2

Skill Understanding place value; using exponents for powers of 10

Object of the Game To make the largest numbers possible.

Directions

1. Player 1 rolls the die and writes the number on any of his or her 4 short blank lines. It does not have to be the first blank—it can be any of them. Note that the last blank line is in the exponent position.
2. Player 2 rolls the die and writes the number on any of his or her blank lines.
3. Players take turns rolling the die and writing the number until each player has a number on each of the 4 lines.
4. Each player then uses the 4 numbers to build a number.
   - The numbers on the first 3 blanks represent a 3-digit number.
   - The number on the last blank is an exponent for a power of 10.
   - Multiply the 3-digit number and the power of 10. Write your number in standard notation on the long line.
5. Players read their numbers. Compare the numbers and write a relation symbol (<, >, or =) between the numbers. The player with the larger number wins the round. In the case of a tie, replay the round.

Note If you don't have a die, you can use a deck of number cards. Use all cards with the numbers 1–6. Instead of rolling the die, draw the top card from the number-side down deck.

Hundred Millions	Ten Millions	Millions	,	Hundred Thousands	Ten Thousands	Thousands	,	Hundreds	Tens	Ones

Example

First three digits — Powers of 10

Player 1: 1 3 2 $* 10^6 =$ 132,000,000

Player 2: 3 5 6 $* 10^4 =$ 3,560,000

Player 1 has the larger number and wins the round.

Multiplication Bull's Eye

Materials ☐ number cards 0–9 (4 of each)

☐ 1 six-sided die

Players 2

Skill Estimating products of 2- and 3-digit numbers

Object of the Game To score more points.

Directions

Number on Die	Target Range of Product
1	500 or less
2	501 - 1,000
3	1,001 - 3,000
4	3,001 - 5,000
5	5,001 - 7,000
6	more than 7,000

1. Shuffle the deck and place it number-side down on the table.
2. Players take turns. When it is your turn:
   - Roll the die. Look up the target range of the product in the table at the right.
   - Take 4 cards from the top of the deck.
   - Use the cards to try to form 2 numbers whose product falls within the target range. Do not use a calculator or pencil and paper when forming your numbers.
   - Multiply the 2 numbers on paper using U.S. traditional multiplication to determine whether the product falls within the target range. If it does, you have hit the bull's eye and score 1 point. If it doesn't, you score 0 points.
   - Sometimes it is impossible to form 2 numbers whose product falls within the target range. If this happens, you score 0 points for that turn.
3. The game ends when each player has had 5 turns. The player with more points wins.

Example

Tom rolls a 3, so the target range of the product is from 1,001 to 3,000. He turns over the cards 5, 7, 9, and 2.

Tom uses estimation to try to form 2 numbers whose product falls within the target range—for example, 97 and 25.

He then finds the product using U.S. traditional multiplication:

```
    1
    3
   97
 * 25
  485
+1940
 2425
```

Since the product is between 1,001 and 3,000, Tom has hit the bull's eye and scores 1 point.

Some other possible winning products from the 5, 7, 2, and 9 cards are:

25 * 79, 27 * 59, 9 * 257, and 2 * 579.

Multiplication Wrestling

Materials ☐ number cards 0–9 (4 of each)

☐ 1 *Multiplication Wrestling* Record Sheet for each player (*Math Masters*, p. G12)

Players 2

Skill Multiplying 2-digit numbers

Object of the Game To get the larger product of two 2-digit numbers.

Directions

1. Shuffle the deck and place it number-side down on the table.
2. Each player draws 4 cards and forms two 2-digit numbers. Players should form their numbers so that their product is as large as possible.
3. Each player creates 2 "wrestling teams" by writing each of their numbers as a sum of 10s and 1s.
4. Each player's 2 teams wrestle. Each member of the first team (for example, 70 and 5) is multiplied by each member of the second team (for example, 80 and 4). Then the 4 products are added.
5. The player with the larger product wins the round and receives 1 point.
6. To begin a new round, each player draws 4 new cards to form 2 new numbers. The player with the highest score at the end of 3 rounds is the winner.

Example

Player 1:

Draws 4, 5, 7, and 8.

Forms 75 and 84.

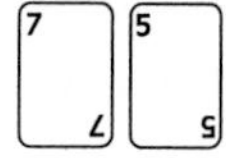

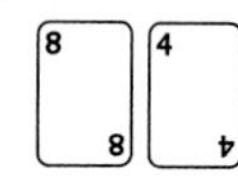

75 * 84

Team 1		Team 2
(70 + 5)	*	(80 + 4)

Products: 70 * 80 = 5,600
70 * 4 = 280
5 * 80 = 400
5 * 4 = 20

Total (add 4 products)
5,000
1,200
+ 100
6,300

75 * 84 = 6,300

Player 2:

Draws 1, 4, 9, and 6.

Forms 64 and 91.

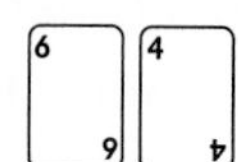

64 * 91

Team 1		Team 2
(60 + 4)	*	(90 + 1)

Products: 60 * 90 = 5,400
60 * 1 = 60
4 * 90 = 360
4 * 1 = 4

Total (add 4 products)
5,000
700
120
+ 4
5,824

64 * 91 = 5,824

6,300 is greater than 5,824, so Player 1 gets 1 point.

Name That Number

Materials ☐ 1 set of number cards

Players 2 or 3

Skill Naming numbers with expressions that contain grouping symbols

Object of the Game To collect the most cards.

Directions

1. Shuffle the deck and deal 5 cards to each player. Place the remaining cards number-side down on the table between the players. Turn over the top card and place it beside the deck. This is the target number for the round.
2. Players try to match the target number by adding, subtracting, multiplying, or dividing the numbers on as many of their cards as possible. A card may only be used once.
3. Players write their solutions on a sheet of paper using grouping symbols as needed.

   When players have written their best solutions:

   - Each player sets aside the cards he or she used to match the target number.
   - Each player replaces the cards he or she set aside by drawing new cards from the top of the deck.
   - The old target number is placed on the bottom of the deck.
   - A new target number is turned over, and another round is played.
4. Play continues until there are not enough cards left to replace all of the players' cards. The player who has set aside the most cards wins the game.

Example

Target number: 16

Player 1's cards: 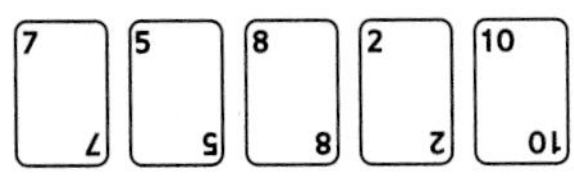

Some possible solutions:

10 + 8 − 2 = 16 (3 cards used)

10 + (7 ∗ 2) − 8 = 16 (4 cards used)

{10 / (5 ∗ 2)} + 8 + 7 = 16 (all 5 cards used)

{(8 + 7) / 5} ∗ 2 + 10 = 16 (all 5 cards used)

Number Top-It

Materials	☐ number cards 0–9 (4 of each)
	☐ 1 *Number Top-It* Mat (*Math Masters,* pp. G7–G8)
	☐ 1 *Number Top-It* Record Sheet (*Math Masters,* p. G9)
Players	2
Skill	Understanding place value for whole numbers
Object of the Game	To make the larger 7-digit number.

Directions

1. Shuffle the cards and place the deck number-side down on the table.
2. Each player uses one row of boxes on the *Number Top-It* Mat. In each round, players take turns turning over the top card from the deck and placing it number-side up on any one of their empty boxes. Each player takes a total of 7 turns, and places 7 cards on his or her row of the game mat.
3. At the end of each round, players read their numbers aloud and compare them. Players record their numbers in standard notation and expanded form on the *Number Top-It* Record Sheet. The player with the larger number for the round scores 1 point. The player with the smaller number scores 2 points.
4. Play 5 rounds for a game. Shuffle the deck between each round. The player with the smaller total number of points at the end of 5 rounds wins the game.

Example

Andy and Barb played *Number Top-It*. Here is the result of one complete round of play.

	Millions	Hundred Thousands	Ten Thousands	Thousands	Hundreds	Tens	Ones
Andy	7,	6	4	5,	2	0	1
Barb	4,	9	7	3,	5	2	4

Andy's number is larger, so Andy scores 1 point for this round, and Barb scores 2 points.

Variation

Play with 3 or more players. The player with the largest number scores 1 point, the player with the next-largest number scores 2 points, and so on.

Over and Up Squares

Materials	☐ 1 *Over and Up Squares* Gameboard/Record Sheet (*Math Masters,* p. G25) ☐ 1 colored pencil for each player (different colors) ☐ 2 six-sided dice
Players	2
Skill	Plotting ordered pairs
Object of the Game	To score more points by plotting and connecting ordered pairs.

Directions

1. Players take turns. When it is your turn:
   - Roll both dice and use the numbers to make an ordered pair. For example, if you roll a 2 and a 3, you may choose to make either (3, 2) or (2, 3). If both possible ordered pairs are already marked on the grid, roll again.
   - Plot the point on the grid using your colored pencil. You earn 10 points.
   - If your point is next to another point on the same side of one of the grid squares, connect the points with a line segment. Sometimes more than one line segment can be drawn. If your line segments complete a square, color it in. You earn 10 points for each line segment drawn and 50 points for completing a square. Record your points on the record sheet.

Score	
ordered pair	10 points
line segment	10 points
square	50 points

2. Continue taking turns. The player with more points after 10 rounds wins.

Example

Round 1

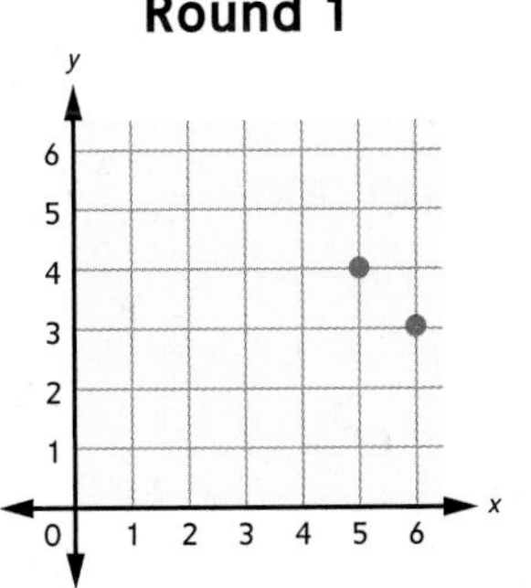

Li rolls a 3 and a 6. She marks the point (6, 3) on the grid in red. Joe rolls a 5 and a 4. He marks the point (5, 4) on the grid in blue. Each player scores 10 points.

Round 2

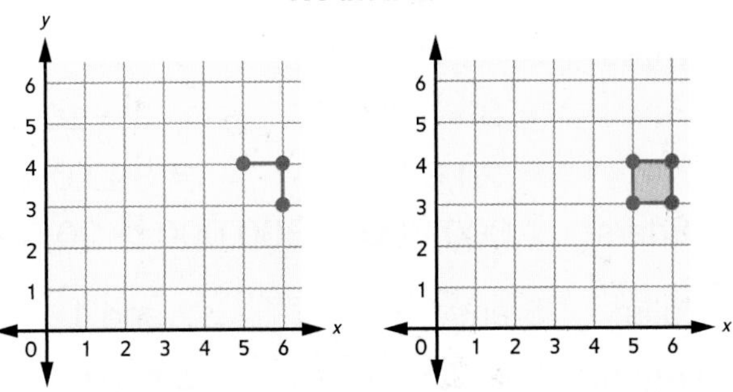

Li rolls a 6 and a 4. She marks (6, 4), scoring 10 points. She draws 2 red line segments, scoring 20 points. Her score for the round is 30 points. Then Joe rolls a 3 and a 5. He marks (5, 3), scoring 10 points. He draws 2 blue segments, scoring 20 points. His segments complete the square, so he colors it in and scores 50 points. His score for the round is 80 points.

Power Up

Materials ☐ 2 six-sided dice

☐ 1 *Power Up* Record Sheet for each player (*Math Masters,* p. G11)

Players 2

Skill Evaluating expressions containing powers of 10

Object of the Game To create the largest number.

Directions

1. Roll the dice. Choose one number as the exponent for a power of 10. Multiply the power of 10 by the other number. Write the expression on your record sheet.

Example

You roll a 5 and a 4. You can write $4 * 10^5$ or $5 * 10^4$.

2. Convert your power of 10 to standard notation. Then evaluate your expression. Write the resulting number in standard notation on your record sheet.
3. Play 3 rounds, taking turns with the other player. Then put your 3 numbers in order from largest to smallest.
4. Compare your list of numbers with the other player's list. The player who has the largest number wins. In case of a tie, play a fourth round.

Example

Ann	Rolls:	2 and 4	5 and 3	1 and 6
	Writes:	$2 * 10^4$	$3 * 10^5$	$1 * 10^6$
		$= 2 * 10{,}000$	$= 3 * 100{,}000$	$= 1 * 1{,}000{,}000$
		$= 20{,}000$	$= 300{,}000$	$= 1{,}000{,}000$
	Orders:	$1{,}000{,}000 > 300{,}000 > 20{,}000$		
Keith	Rolls:	5 and 5	2 and 1	4 and 3
	Writes:	$5 * 10^5$	$1 * 10^2$	$3 * 10^4$
		$= 5 * 100{,}000$	$= 1 * 100$	$= 3 * 10{,}000$
		$= 500{,}000$	$= 100$	$= 30{,}000$
	Orders:	$500{,}000 > 30{,}000 > 100$		

Ann's largest number is greater than Keith's largest number. Ann wins.

Prism Pile-Up

Materials ☐ 1 set of *Prism Pile-Up* Cards (*Math Journal 1*, Activity Sheets 3–4)

☐ 1 *Prism Pile-Up* Record Sheet for each player (*Math Masters*, p. G6)

☐ calculator (optional)

Players 2

Skill Finding volumes of rectangular prisms

Object of the Game To collect more cards.

Prism Pile-Up Cards 1–9

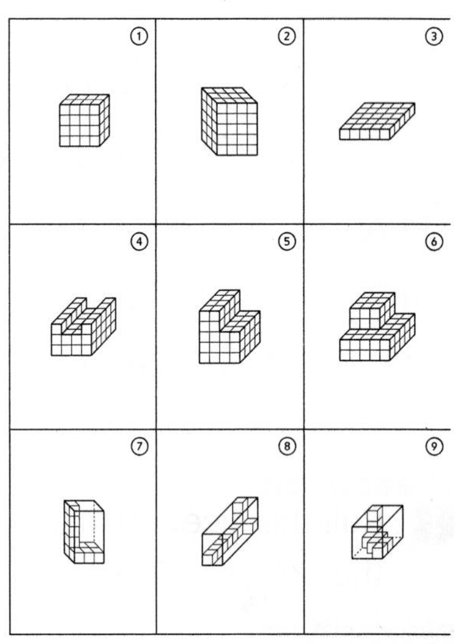

Prism Pile-Up Cards 10–18

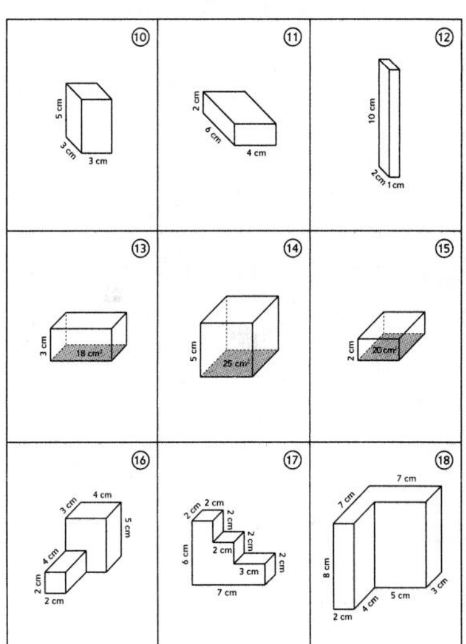

Directions

1. Shuffle the cards and place them facedown in a pile.
2. Each player draws a card from the pile and finds the volume of the figure. The cubes on cards 1–9 represent centimeter cubes, so their volume can be reported in cm^3. Players may use a calculator.
3. Players record the card number and volume of their figure on the *Prism Pile-Up* Record Sheet. They also record one or more number sentences for the volume.
4. Players check each other's work. The player whose figure has the greater volume takes both cards.
5. The game ends when there are no cards remaining. The player with more cards wins. If players have the same number of cards, they look through their cards to find the figure with the greatest volume. The player whose figure has the greatest volume wins.

Example

Libby draws Card 1. She finds the volume to be 32 cubic centimeters and writes 4 ∗ 2 ∗ 4 = 32.

Christopher draws Card 16. He calculates the volume to be 76 cubic centimeters and writes these number sentences: 2 ∗ 4 ∗ 2 = 16; 3 ∗ 4 ∗ 5 = 60; 16 + 60 = 76.

Christopher's figure has the greater volume, so he takes both cards.

Libby

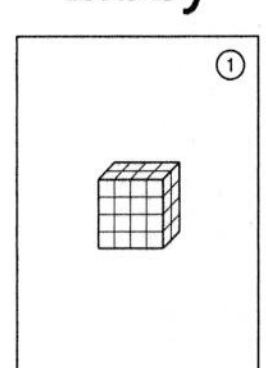

Christopher

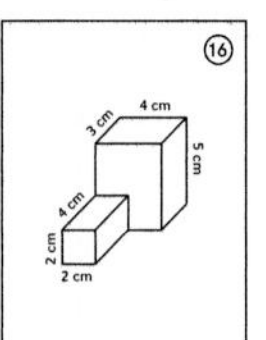

Property Pandemonium

Materials ☐ 1 *Property Pandemonium* Card Deck (*Math Journal 2*, Activity Sheet 20)
☐ 1 *Property Pandemonium* Record Sheet (*Math Masters*, p. G33)
☐ Quadrilateral Hierarchy Poster (or *Math Journal 2*, pp. 250–251)

Players 2

Skill Drawing and naming quadrilaterals with certain properties

Object of the Game To earn the fewest points.

Directions

1. Place the Property Cards and Quadrilateral Cards facedown in separate piles.
2. Players take turns. When it is your turn, draw one Property Card and one Quadrilateral Card. Record the information on the *Property Pandemonium* Record Sheet. If you draw a WILD card, you may choose a property or quadrilateral.
3. On the record sheet, draw an example of the quadrilateral given on your Quadrilateral Card that has the property or properties listed on your Property Card. You may look at the Quadrilateral Hierarchy to help you.
4. List all the other names you can give the quadrilateral you drew in the Additional Names column. Show your drawing and names to your partner. If your partner can write additional names for your quadrilateral, add them to your record sheet.
5. Your score for the round is the number of additional names recorded.
6. Play three rounds. The player with the fewest points wins.

Example

Nico and Rowan are playing *Property Pandemonium.* On his turn, Nico draws the "4 right angles" Property Card and the "trapezoid" Quadrilateral Card. He records his drawing on the record sheet. Nico writes that his trapezoid is also a quadrilateral, a rectangle, and a parallelogram. He scores 3 points for the round.

Round	Property	Quadrilateral	Drawing	Additional Names	Points
1	*4 right angles*	*trapezoid*		*quadrilateral* *rectangle* *parallelogram*	*3*

Rename That Mixed Number

Materials	☐ 1 *Rename That Mixed Number* Record Sheet for each player (*Math Masters,* p. G18) ☐ number cards 1–9 (4 of each) ☐ timer or stopwatch (optional)
Players	2
Skill	Renaming mixed numbers
Object of the Game	To find as many possible names for the starting mixed number without changing the denominator.

Directions

1. Shuffle the cards and place them number-side down in a pile.
2. Each player picks 3 cards and uses them to represent the 3 parts of a mixed number: whole number, numerator, and denominator. Players may place the cards in any position they wish.
3. Players write the mixed number in the Starting Mixed Number column for Round 1 on the record sheet. See the record sheet on the following page.
4. Players list as many other names for their starting mixed number as they can by making or breaking apart wholes. All names must have the same denominator as the starting mixed number. Players record each name in the Other Names column of the record sheet for Round 1. Players may choose to set a 2-minute time limit.
5. Players record the number of other names they found for their mixed number in the Points column.
6. Players exchange record sheets and check each other's work. If a player finds another name for their partner's mixed number, they may claim that point and record it in the Points column for their partner's number on their own record sheet.
7. If players disagree on any answer, draw a picture to check the answer. If necessary, cross out any incorrect names and adjust your points.
8. Repeat Steps 1–7 four more times. The player with the most points after 5 rounds wins. See the example on the following page.

Round	Starting Mixed Number	Other Names	Points from My Number	Points from My Partner's Number
1				
2				
3				
4				
5				
		Column Totals		
		Total Points		

Rename That Mixed Number Record Sheet

Example

Ally draws 5, 8, and 3. She makes the mixed number $5\frac{3}{8}$ and lists the names $\frac{43}{8}$, $4\frac{11}{8}$, $3\frac{19}{8}$, $2\frac{27}{8}$, and $1\frac{35}{8}$. She scores 5 points for her number.

Natalie draws 2, 7, and 5. She makes the mixed number $2\frac{7}{5}$. She lists the names $1\frac{12}{5}$ and $\frac{17}{5}$ and scores 2 points for her number. Ally notices that $3\frac{2}{5}$ is also a name for $2\frac{7}{5}$ and scores 1 point for Natalie's number.

Variation

For an additional challenge, play the game using number cards 1–20.

Spend and Save

Materials
- ☐ number cards 0–9 (4 of each)
- ☐ 1 coin
- ☐ 1 counter
- ☐ 1 *Spend and Save* Record Sheet for each player (*Math Masters,* p. G27)

Players 2

Skill Adding and subtracting money

Object of the Game To end the game with more money saved.

Directions

You are keeping track of the amount of money in your savings account. Each player starts Round 1 with a balance of $100.00.

1. Shuffle the deck and place it number-side down.
2. Players take turns. When it is your turn:
   - Write your starting amount in the first column of your record sheet.
   - Flip the coin. If it shows heads, you will spend money. If it shows tails, you will save money. Write "spend" or "save" on your record sheet. Remember, spending means you subtract money from your account. Saving means you add money to your account.
   - Draw 3 cards and use them to make a money amount. Use a counter as a decimal point. You can place the cards in any order. For example, if you draw 6, 8, and 2, you can make the amount $6.82, $8.62, $2.68, or any other amount that uses the same digits. Write that amount on your record sheet.
   - Add or subtract to find your new balance. Write it on your record sheet. This will also be your starting amount in the next round.
3. The game ends after 8 rounds. The player with more money is the winner.

Example

At the beginning of Round 4, Max has a starting amount of $72.19. His coin flip shows tails, which means he will save money, and he draws the number cards 2, 7, and 0. He shows the money amount $7.20. He adds $7.20 to his starting amount to get a new balance of $79.39. His record sheet for the round is shown below.

Round	Start Amount	Change		Balance (and next Start Amount)
		Spend or save?	How much?	
4	$72.19	save	$7.20	$79.39

Spoon Scramble

Materials ☐ 1 set of *Spoon Scramble* Cards (*Math Journal 2*, Activity Sheet 19)

☐ 3 spoons

Players 4

Skill Multiplying and dividing whole numbers, fractions, and decimals

Object of the Game To avoid getting all the letters in the word *SPOONS*.

$\frac{1}{4}$ of 24	$\frac{3}{4} * 8$	50% of 12	0.10 * 60
$\frac{1}{3}$ of 21	$3\frac{1}{2} * 2$	25% of 28	0.10 * 70
$\frac{1}{5}$ of 40	$2 * \frac{16}{4}$	1% of 800	0.10 * 80
$\frac{3}{4}$ of 12	$4\frac{1}{2} * 2$	25% of 36	0.10 * 90

Spoon Scramble Cards

Directions

1. Place the spoons in the center of the table.
2. The dealer shuffles the deck and deals 4 cards number-side down to each player.
3. Players look at their cards. If a player has 4 cards of equal value, proceed to Step 5 below. Otherwise, each player chooses a card to discard and passes it, number-side down, to the player on the left.
4. Each player picks up the new card and repeats Step 3. The passing of the cards should proceed quickly.
5. As soon as a player has 4 cards of equal value, the player places the cards number-side up on the table and grabs a spoon.
6. The other players then try to grab one of the 2 remaining spoons. The player left without a spoon is assigned a letter from the word *SPOONS,* starting with the first letter. If a player incorrectly claims to have 4 cards of equal value, that player receives a letter instead of the player left without a spoon.
7. Put the spoons back in the center and begin a new round. Play continues until 3 players each get all the letters in the word *SPOONS*. The player who does not have all the letters is the winner.

Variations

- For 3 players: Eliminate one set of 4 equivalent *Spoon Scramble* Cards. Use only 2 spoons.
- For a more challenging version of the game, players can make their own deck of *Spoon Scramble* Cards. Each player writes 4 computation problems that have equivalent answers on 4 index cards. Check to be sure the players have all chosen different values.

Top-It: Multiplication or Division

Materials ☐ number cards 0–9 (4 of each)

Players 2 to 4

Skill Multiplying or dividing 2- and 3-digit numbers

Object of the Game To collect the most cards.

Multiplication Top-It: Extended Facts

Directions

1. Shuffle the deck and place it number-side down on the table.
2. Each player turns over 2 cards. Attach a zero to the first card drawn and multiply by the second card. For example, if 7 is the first card drawn and 5 is the second card drawn, compute $70 * 5 = 350$.
3. The player with the largest product takes all cards. The game ends when there are not enough cards for all players to have another turn. The player with the most cards wins.

Multiplication Top-It: Larger Numbers

Directions

1. Shuffle the deck and place it number-side down on the table.
2. Each player turns over 4 cards. Choose 3 of them to form a 3-digit number, then multiply by the remaining number. Carefully consider how to form your 3-digit numbers. For example, $462 * 5 = 2{,}310$ while $256 * 4 = 1{,}024$.
3. The player with the largest product takes all cards. The game ends when there are not enough cards for all players to have another turn. The player with the most cards wins.

Variation Use the 4 cards to make two 2-digit numbers to multiply.

Division Top-It: Larger Numbers

Directions

1. Shuffle the deck and place it number-side down on the table.
2. Each player turns over 4 cards. Choose 3 of them to form a 3-digit number, then divide the 3-digit number by the remaining number. Ignore the remainder. Carefully consider how to form your 3-digit numbers. For example, 462 / 5 is greater than 256 / 4.
3. The player with the largest quotient takes all cards. The game ends when there are not enough cards for all players to have another turn. The player with the most cards wins.

What's My Attribute Rule?

Materials
- ☐ 1 set of attribute blocks
- ☐ 1 die
- ☐ 1 set of *What's My Attribute Rule?* Cards (*Math Masters,* p. G31)
- ☐ 2 sheets of paper

Players 2 to 4

Skill Sorting shapes according to attributes

Object of the Game To figure out rules that are used to sort shapes.

Directions

1. Label one sheet of paper: **These fit the rule**. Label another sheet of paper: **These do NOT fit the rule**. Shuffle and place the Attribute Rule Cards facedown.
2. Players roll the die. The player with the lowest number is the first "Rule Maker."
3. When you are the Rule Maker:
   - Turn over the top Attribute Rule Card, but don't show it to the other players.
   - Choose 3 or 4 attribute blocks that fit the rule. Place the blocks on the sheet labeled **These fit the rule**.
   - Choose 3 or 4 blocks that do NOT fit the rule. Place them on the sheet labeled **These do NOT fit the rule**.
4. The other players take turns choosing a block that they think might fit the rule and placing it on that sheet. If the Rule Maker says, "No," the player puts the block on the other sheet. If "Yes," the player gets to suggest what the rule might be. The Rule Maker tells the player whether the rule is correct.
5. The round continues until someone figures out the rule. That person becomes the Rule Maker for the next round.

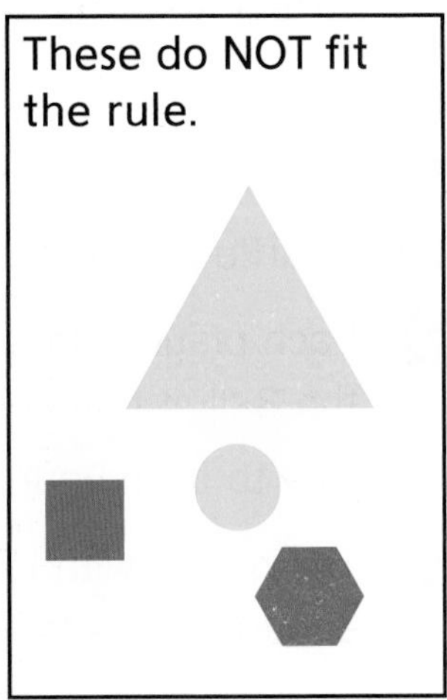

Attribute Rule: large shapes, but not triangles

Place-Value Chart

billions	hundred millions	ten millions	millions	hundred thousands	ten thousands	thousands	hundreds	tens	ones	.	tenths	hundredths	thousandths
1,000,000,000s	100,000,000s	10,000,000s	1,000,000s	100,000s	10,000s	1,000s	100s	10s	1s	.	0.1s	0.01s	0.001s
10^9	10^8	10^7	10^6	10^5	10^4	10^3	10^2	10^1	10^0	.	$\frac{1}{10^1}$	$\frac{1}{10^2}$	$\frac{1}{10^3}$
10 * 10 * 10 * 10 * 10 * 10 * 10 * 10 * 10	10 * 10 * 10 * 10 * 10 * 10 * 10 * 10	10 * 10 * 10 * 10 * 10 * 10 * 10	10 * 10 * 10 * 10 * 10 * 10	10 * 10 * 10 * 10 * 10	10 * 10 * 10 * 10	10 * 10 * 10	10 * 10	10	1	.	$\frac{1}{10}$	$\frac{1}{(10 * 10)}$	$\frac{1}{(10 * 10 * 10)}$

Multiplication and Divison Table

*, /	1	2	3	4	5	6	7	8	9	10
1	1	2	3	4	5	6	7	8	9	10
2	2	4	6	8	10	12	14	16	18	20
3	3	6	9	12	15	18	21	24	27	30
4	4	8	12	16	20	24	28	32	36	40
5	5	10	15	20	25	30	35	40	45	50
6	6	12	18	24	30	36	42	48	54	60
7	7	14	21	28	35	42	49	56	63	70
8	8	16	24	32	40	48	56	64	72	80
9	9	18	27	36	45	54	63	72	81	90
10	10	20	30	40	50	60	70	80	90	100

The numbers on the diagonal are square numbers.

Rules for Order of Operations

1. Do operations inside parentheses or other grouping symbols first.
2. Calculate all expressions with exponents (e.g., $10^2 = 10 * 10 = 100$).
3. Multiply and divide in order, from left to right.
4. Add and subtract in order, from left to right.

Prefixes

uni-	one	giga-	billion (10^9)
bi-	two	mega-	million (10^6)
tri-	three	kilo-	thousand (10^3)
quad-	four	hecto-	hundred (10^2)
penta-	five	deca-	ten (10^1)
hexa-	six	uni-	one (10^0)
hepta-	seven	deci-	tenth $\left(\frac{1}{10^1}\right)$
octa-	eight	centi-	hundredth $\left(\frac{1}{10^2}\right)$
nona-	nine	milli-	thousandth $\left(\frac{1}{10^3}\right)$
deca-	ten	micro-	millionth $\left(\frac{1}{10^6}\right)$
dodeca-	twelve	nano-	billionth $\left(\frac{1}{10^9}\right)$
icosa-	twenty		
tera-	trillion (10^{12})		

Metric System

Units of Length

1 kilometer (km) = 1,000 meters (m)
1 meter = 10 decimeters (dm)
= 100 centimeters (cm)
= 1,000 millimeters (mm)
1 decimeter = 10 centimeters
1 centimeter = 10 millimeters

Units of Area

1 square meter (m^2) = 100 square decimeters (dm^2)
= 10,000 square centimeters (cm^2)
1 square decimeter = 100 square centimeters
1 square kilometer (km^2) = 1,000,000 square meters

Units of Volume

1 cubic meter (m^3) = 1,000 cubic decimeters (dm^3)
= 1,000,000 cubic centimeters (cm^3)
1 cubic decimeter = 1,000 cubic centimeters

Units of Capacity and Liquid Volume

1 kiloliter (kL) = 1,000 liters (L)
1 liter = 1,000 milliliters (mL)
1 cubic centimeter = 1 milliliter

Units of Mass

1 metric ton (t) = 1,000 kilograms (kg)
1 kilogram = 1,000 grams (g)
1 gram = 1,000 milligrams (mg)

System Equivalents

1 inch is about 2.5 centimeters (2.54)
1 kilometer is about 0.6 mile (0.621)
1 mile is about 1.6 kilometers (1.609)
1 meter is about 39 inches (39.37)
1 liter is about 1.1 quarts (1.057)
1 ounce is about 28 grams (28.350)
1 kilogram is about 2.2 pounds (2.205)

U.S. Customary System

Units of Length

1 mile (mi) = 1,760 yards (yd)
= 5,280 feet (ft)
1 yard = 3 feet
= 36 inches (in.)
1 foot = 12 inches

Units of Area

1 square yard (yd^2) = 9 square feet (ft^2)
= 1,296 square inches ($in.^2$)
1 square foot = 144 square inches
1 acre (a.) = 43,560 square feet
1 square mile (mi^2) = 640 acres

Units of Volume

1 cubic yard (yd^3) = 27 cubic feet (ft^3)
1 cubic foot = 1,728 cubic inches ($in.^3$)

Units of Capacity and Liquid Volume

1 gallon (gal) = 4 quarts (qt)
1 quart = 2 pints (pt)
1 pint = 2 cups (c)
1 cup = 8 fluid ounces (fl oz)
1 fluid ounce = 2 tablespoons (tbs)
1 tablespoon = 3 teaspoons (tsp)

Units of Weight

1 ton (T) = 2,000 pounds (lb)
1 pound = 16 ounces (oz)

Units of Time

1 century (cent) = 100 years
1 decade = 10 years
1 year (yr) = 12 months
= 52 weeks (plus one or two days)
= 365 days (366 days in a leap year)
1 month (mo) = 28, 29, 30, or 31 days
1 week (wk) = 7 days
1 day (d) = 24 hours
1 hour (hr) = 60 minutes
1 minute (min) = 60 seconds (sec)

Equivalent Fractions					
	Multiply both the numerator and denominator by:				
Starting Fraction:	2	3	4	5	10
$\frac{1}{2}$	$\frac{2}{4}$	$\frac{3}{6}$	$\frac{4}{8}$	$\frac{5}{10}$	$\frac{10}{20}$
$\frac{3}{4}$	$\frac{6}{8}$	$\frac{9}{12}$	$\frac{12}{16}$	$\frac{15}{20}$	$\frac{30}{40}$
$\frac{5}{8}$	$\frac{10}{16}$	$\frac{15}{24}$	$\frac{20}{32}$	$\frac{25}{40}$	$\frac{50}{80}$

Fraction-Decimal Number Line

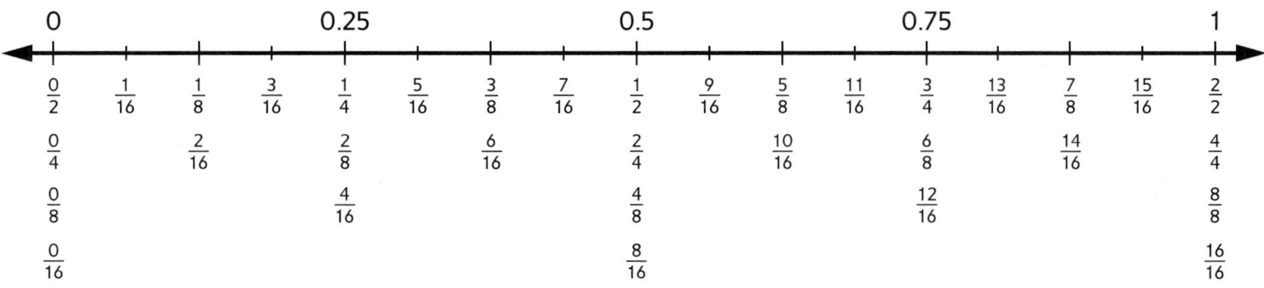

Formulas	Meaning of Variables
Rectangles	
• Perimeter: $p = 2l + 2w$, or $p = 2 * (l + w)$ • Area: $A = l * w$	p = perimeter; l = length; w = width A = area
Squares	
• Perimeter: $p = 4 * s$ • Area: $A = s^2$	p = perimeter; s = length of one side A = area
Regular Polygons	
• Perimeter: $p = n * s$	p = perimeter; n = number of sides; s = length of one side
Rectangular Prisms	
• Volume: $V = B * h$, or $V = l * w * h$	V = volume; B = area of the base; l = length; w = width; h = height

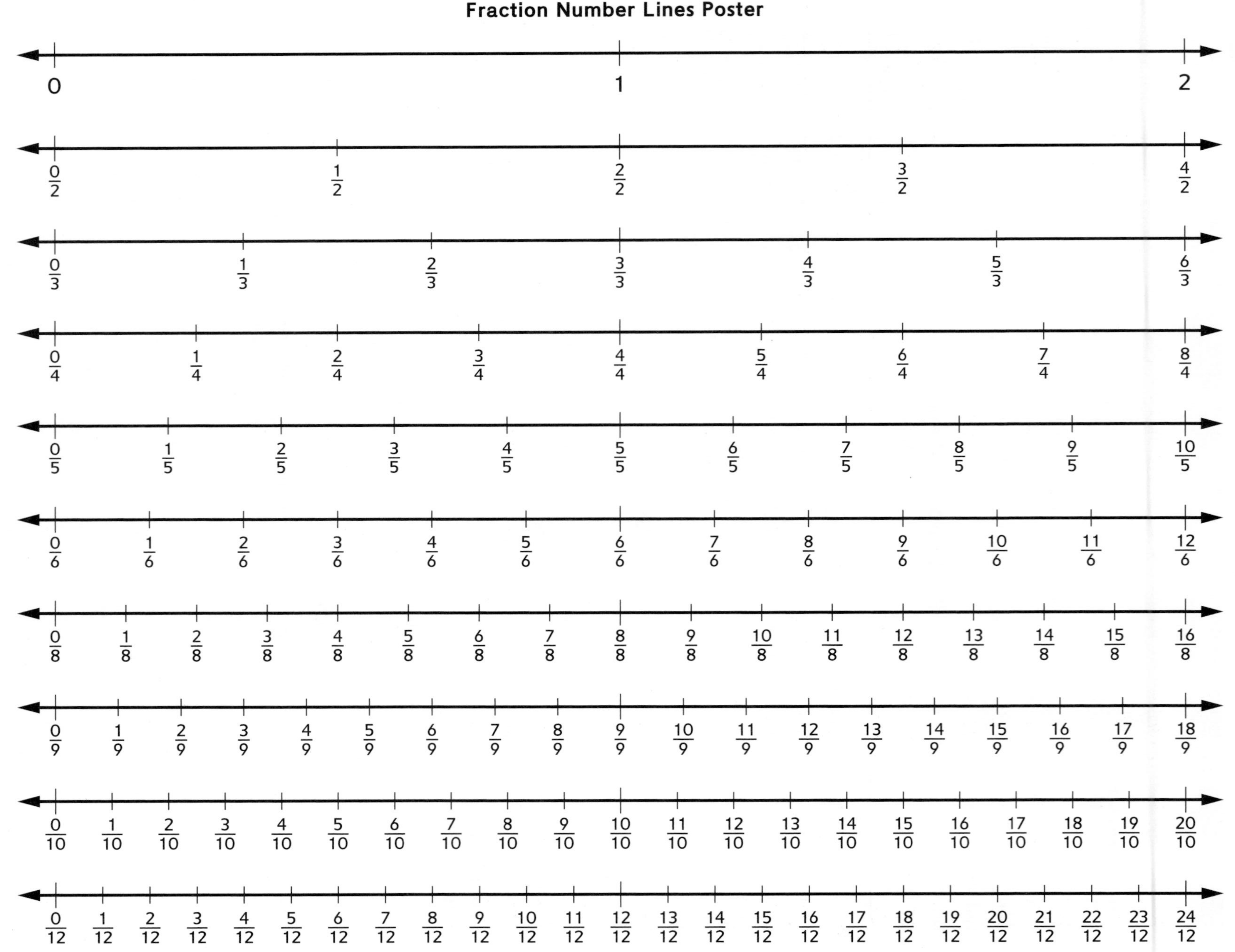
Fraction Number Lines Poster
0 1 2
0/2 1/2 2/2 3/2 4/2
0/3 1/3 2/3 3/3 4/3 5/3 6/3
0/4 1/4 2/4 3/4 4/4 5/4 6/4 7/4 8/4
0/5 1/5 2/5 3/5 4/5 5/5 6/5 7/5 8/5 9/5 10/5
0/6 1/6 2/6 3/6 4/6 5/6 6/6 7/6 8/6 9/6 10/6 11/6 12/6
0/8 1/8 2/8 3/8 4/8 5/8 6/8 7/8 8/8 9/8 10/8 11/8 12/8 13/8 14/8 15/8 16/8
0/9 1/9 2/9 3/9 4/9 5/9 6/9 7/9 8/9 9/9 10/9 11/9 12/9 13/9 14/9 15/9 16/9 17/9 18/9
0/10 1/10 2/10 3/10 4/10 5/10 6/10 7/10 8/10 9/10 10/10 11/10 12/10 13/10 14/10 15/10 16/10 17/10 18/10 19/10 20/10
0/12 1/12 2/12 3/12 4/12 5/12 6/12 7/12 8/12 9/12 10/12 11/12 12/12 13/12 14/12 15/12 16/12 17/12 18/12 19/12 20/12 21/12 22/12 23/12 24/12

About Calculators

You can use caluculators for working with whole numbers, fractions, and decimals. As with any mathematical tool or strategy, you need to think about when and how to use a calculator. It can help you compute quickly and accurately when you have many problems to do in a short time. Calculators can help you solve problems with very large and very small numbers that may be hard to do in your head or with pencil and paper. Whenever you use a calculator, estimation should be part of your work. Always ask yourself if the answer in the display makes sense.

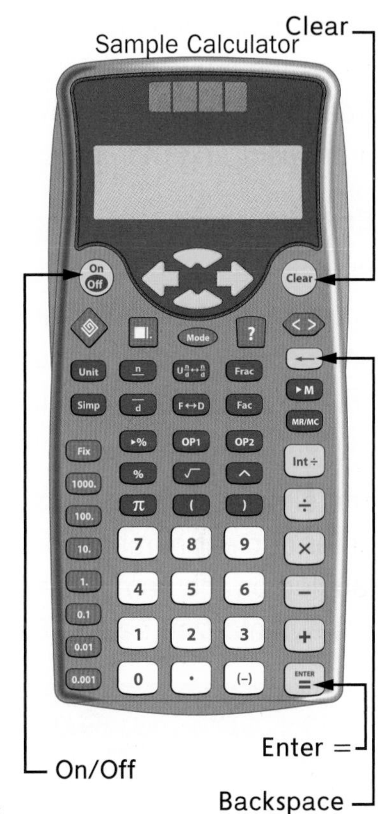

There are many different kinds of calculators. Four-function calculators do little more than add, subtract, multiply, and divide whole numbers and decimals. You can, however, use them to help you with more complex functions if you understand the processes of those functions. More advanced scientific calculators let you find powers and reciprocals and perform some operations with fractions. After elementary school, you may use graphing calculators that draw graphs, find data landmarks, and do even more complicated mathematics.

There are many calculators that work well with *Everyday Mathematics*. If the instructions in this book don't work for your calculator or the keys on your calculator are not explained here, you can refer to the directions that came with your calculator, look them up online, or ask your teacher for help.

Basic Operations on a Calculator

Many handheld calculators use light cells for power. If you press the ON key and see nothing on the display, hold the front of the calculator toward a light or a sunny window for a moment and then press ON again.

Note It is important to take proper care of your calculator. Dropping it, leaving it in the sun, or other kinds of carelessness may break it, ruin it, or make it less reliable.

Entering and Clearing

Pressing a key on a calculator is called *keying in,* or *entering.* In this book, calculator keys, except numbers and decimal points, are shown in rectangular boxes: (+), (=), and (×) and so on. A set of instructions for performing a calculation is called a *key sequence.*

The simplest key sequences turn the calculator on and enter or clear numbers or other characters. These keys are labeled on the sample calculator on the previous page and are summarized below.

Sample Calculator	
Key	Purpose
(On/Off)	Turn the display on.
(Clear) and (On/Off) at the same time	Clear the display and the short-term memory.
(Clear)	Clear only the display.
(←)	Clear the last digit.

Always clear both the display and the memory each time you turn on your calculator. It is also a good idea to clear both after you finish a problem and before you begin another problem.

Powers of 10 on a Calculator

Numbers like 10, 100, and 1,000 are called **powers of 10.** They are numbers that can be written as products of 10s. For example, 100 can be written as $10 * 10$ or 10^2. 1,000 can be written as $10 * 10 * 10$ or 10^3.

The number 1,000 is written in **standard notation** (or **standard form**). A number written with an exponent, such as 10^3, is in **exponential notation.** The raised digit is called an **exponent.** The number before the exponent is the **base.** The exponent tells how many times the base is used as a factor. For powers of 10, 10 is the base, so the exponent tells how many 10s are multiplied.

Exponential notation is a good way to represent very large or very small numbers. Some calculators have special keys for renaming numbers written in exponential notation in standard form.

Example

Use a calculator. Find the value of 10^5.

To rename 10^5 on the sample calculator, use this key sequence:

Sample Calculator	Key Sequence	Display
	10 [^] 5 [Enter =]	10^5 = 100000

$10^5 = 100,000$

You can check this by keying in 10 [×] 10 [×] 10 [×] 10 [×] 10 [=].

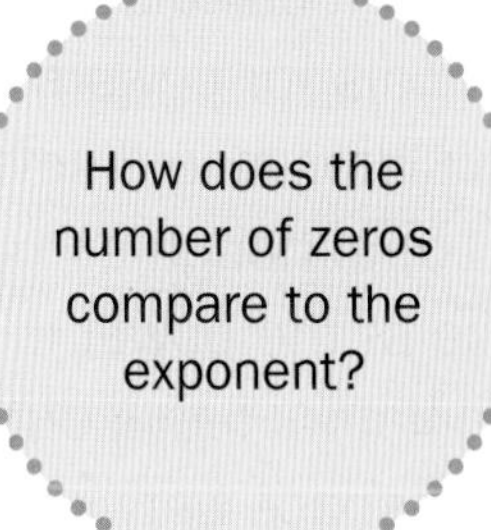

Look at your calculator to see which key it has for finding powers of ten.

- The key may look like [10^x] and is read as "10 to the *x*."
- The key may look like [^] as on the sample calculator and is called a *caret*.

Multiplying by Powers of 10

You can multiply whole numbers and decimals by powers of 10 on a calculator.

Example

Multiply $7 * 10^4$ and $4.35 * 10^5$. Show the products in standard notation.

Sample Calculator	Key Sequence	Display
	7 (×) 10 (^) 4 (Enter)	7 × 10^4 = 70000
	4.35 (×) 10 (^) 5 (Enter)	4.35 × 10^5 = 435000

$7 * 10^4 = 70{,}000$

$4.35 * 10^5 = 435{,}000$

Numbers multiplied by powers of 10 often have products with too many digits to fit on the display. Scientific calculators show the product as a number multiplied by a power of 10 in exponential notation.

Different calculators use different symbols for powers of 10. Your calculator may display raised exponents of 10, although some do not. Since the base of the power is always 10, some calculators leave out the 10 and simply put a space between the number and the exponent.

This calculator shows $9 * 10^9$.

Note Most calculators do not display large numbers in standard notation with a comma like you do with pencil and paper. Some use an apostrophe; others use no symbol at all.

How does your calculator display large numbers?

Dividing by Powers of 10

You can divide whole numbers and decimals by powers of 10 on a calculator.

Example

Divide 6 by 10^7, or $6 \div 10^7$.

Sample Calculator	Key Sequence	Display
	6 (÷) (( ) 10 (^) 7 ( )) (Enter)	6 ÷ (10^7) = 0.0000006

What do you notice about the location of the decimal point compared to the exponent?

Scientific calculators follow the order of operations and will calculate the expressions with exponents before multiplying or dividing, so parentheses are not needed around 10^7 in the example above.

Example

Use your calculator to show $12.8 \div 10^2$ in standard notation.

Sample Calculator	Key Sequence	Display
	12.8 (÷) 10 (^) 2 (Enter)	12.8 ÷ 10^2 = 0.128

$12.8 \div 10^2$ in standard notation is 0.128.

Did You Know?

Numbers divided by powers of 10 sometimes have products that are in the ten-millionths or smaller. These numbers have too many digits for the calculator to display. Scientific calculators show the product as a number multiplied by a negative power of 10 in exponential notation. You will learn more about negative powers of 10 in future grades.

Check Your Understanding

Use your calculator to convert the following to standard notation:

1. $5.8 * 10^4$ **2.** $7.6 * 10^7$ **3.** $4.389 \div 10^6$ **4.** $1.1 \div 10^5$

Check your answers in the Answer Key.

Division with Remainders on a Calculator

The answer to a division problem with whole numbers does not always result in whole number answers. When this happens, most calculators display the answer as a decimal. Some calculators also have a second division key that displays the whole number quotient with a whole number remainder.

Example

$39 \div 5 =$ **?**

Estimate. Since $40 \div 5 = 8$, the quotient will be a little less than 8.

Use the division key.

Sample Calculator	Key Sequence	Display
	39 (÷) 5 (=)	39 ÷ 5 = 7.8

$39 \div 5 =$ **7.8** This quotient is reasonable since 7.8 is a little less than 8.

$39 \div 5 =$ **?**

Use the division with remainder key.

Sample Calculator	Key Sequence	Display
	39 (Int÷) 5 (=)	39 ÷ 5 = 7r 4

$39 \div 5 \rightarrow$ **7 R4** This response makes sense because 7 with 4 left over is close to but not quite 8.

Note "Int" stands for "integer" on this calculator. Use (Int÷) because this kind of division is sometimes called "integer division."

Note Some calculators have a key that looks like (÷R) that means "divide with remainder." You can also divide positive fractions and decimals with (÷R).

Interpreting Division with Remainders

The way you interpret the quotient and remainder depends on the problem situation.

There are several ways to think about remainders:

- Ignore the remainder. Use the whole number quotient as the answer.
- Round the quotient up to the next whole number.
- Write the remainder as a fraction or decimal. Use this fraction or decimal as part of the answer.

Example

For all of the examples below, the calculator displays the answer either as a decimal 7.8 or as a quotient with a whole-number remainder "7 R4." You must decide which answer is appropriate for the situation.

Suppose 5 friends share 39 marbles equally. How many marbles will each friend get? Since you cannot split the remaining marbles among 5 people, ignore the remainder. Answer: Each friend will get 7 marbles, with 4 left over.	Suppose 39 postcards are placed in a scrapbook. How many pages are needed if 5 postcards can fit on a page? Round the quotient up to the next whole number. An eighth page is needed to include all of the postcards. The album will have some pages filled and other pages only partially filled. Answer: 8 pages are needed.	Suppose 5 classrooms share 39 pizzas. How much pizza does each class get? The answer, 7 R4, shows that each class receives 7 pizzas, and 4 pizzas are left over. Each of the remaining 4 pizzas can be divided into fifths. Each class receives 4 of these $\frac{1}{5}$-pizzas, or $\frac{4}{5}$-pizza. The remainder (4) is rewritten as a fraction ($\frac{4}{5}$). Use this as part of the answer. Answer: Each class will get $7\frac{4}{5}$ pizzas.

When the calculator shows the remainder as a fraction, the remainder is the **numerator** of the fraction, and the divisor is the **denominator** of the fraction.

Check Your Understanding

Solve. Use the division key or the division with remainder key on your calculator. Write a number model and the quotient from the calculator display. Then answer the question. Estimate to check that your answer is reasonable.

1. There are 357 students. Each classroom can have no more than 20 students. How many classrooms are needed?
2. It takes 3 pounds of blueberries to make 1 pie. How many pies can the bakery make with 100 pounds of blueberries?
3. A train travels an average of 96 miles in an hour. If a train traveled 984 miles, how many hours did it travel?

Check your answers in the Answer Key.

Fractions and Mixed Numbers on a Calculator

Some calculators let you enter, rewrite, and do operations with fractions. Once you know how to enter a fraction, you can add, subtract, multiply, or divide them just like whole numbers and decimals.

Entering Fractions and Mixed Numbers

Most calculators that let you enter fractions use similar key sequences. For fractions, always start by entering the numerator. Then press the key that tells the calculator to begin writing a fraction.

Example

Enter $\frac{5}{8}$ as a fraction in your calculator.

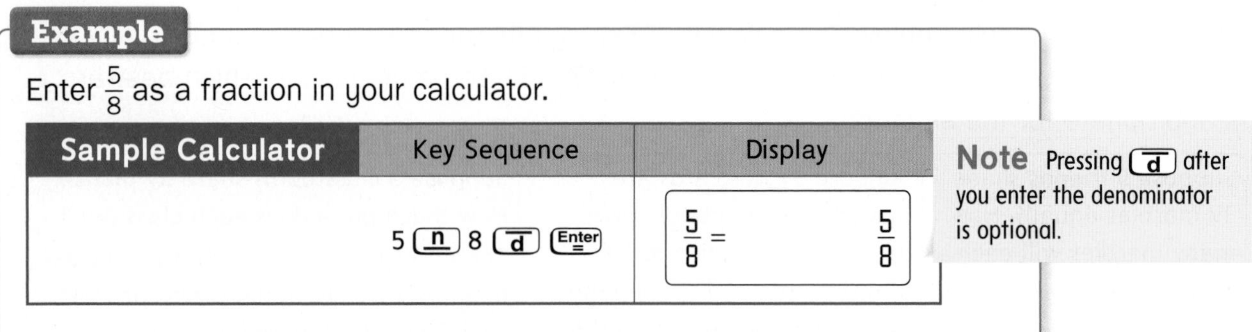

Sample Calculator	Key Sequence	Display
	5 (n) 8 (d) (Enter)	$\frac{5}{8} =$ $\frac{5}{8}$

Note Pressing (d) after you enter the denominator is optional.

To enter a mixed number, enter the whole number part first. Then press the fraction key or keys to enter the fraction part of the number.

Example

Enter $73\frac{2}{5}$ as a fraction in your calculator.

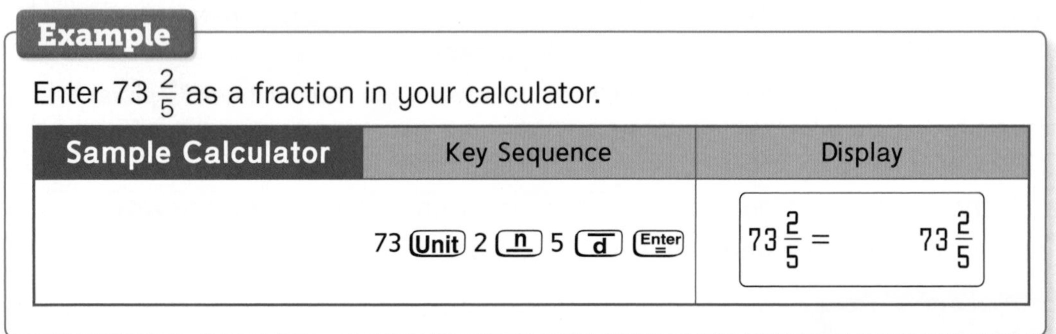

Sample Calculator	Key Sequence	Display
	73 (Unit) 2 (n) 5 (d) (Enter)	$73\frac{2}{5} =$ $73\frac{2}{5}$

Some calculators have a (a) key for entering the whole number part of a mixed number and a (b/c) key for entering the fraction part of a mixed number.

Try entering a mixed number in your calculator.

The keys to convert between mixed numbers and fractions greater than one are similar on most fraction calculators. Look for a key similar to [U n/d ↔ n/d] or [a b/c ↔ d/c].

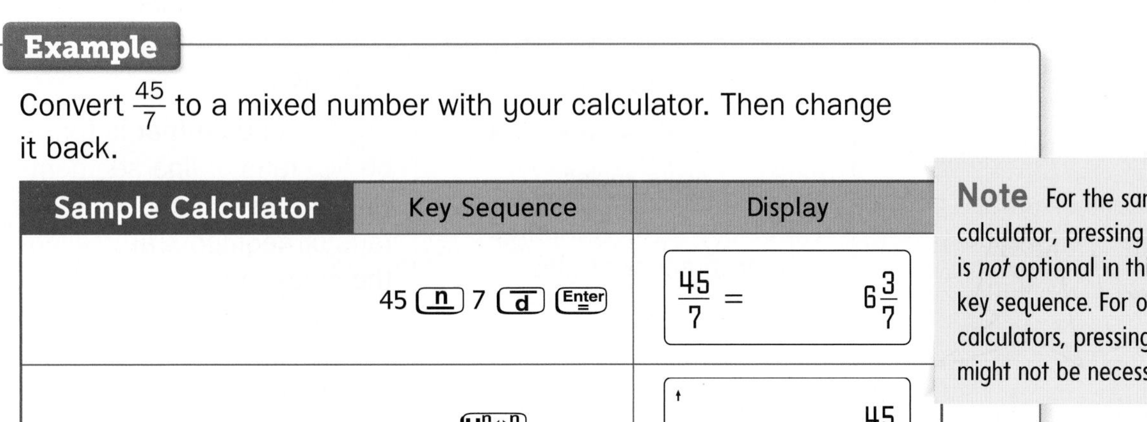

Example

Convert $\frac{45}{7}$ to a mixed number with your calculator. Then change it back.

Sample Calculator	Key Sequence	Display
	45 [n] 7 [d] [Enter =]	$\frac{45}{7} = 6\frac{3}{7}$
	[U n/d ↔ n/d]	$\frac{45}{7}$
	[U n/d ↔ n/d]	$6\frac{3}{7}$

Note For the sample calculator, pressing [Enter =] is *not* optional in this key sequence. For other calculators, pressing [=] might not be necessary.

Keys such as [U n/d ↔ n/d] and [a b/c ↔ d/c] toggle between mixed numbers and fractions greater than one.

2-dimensional
(2-D) Having *area* but not volume. A 2-dimensional *surface* can be flat like a piece of paper or curved like a dome.

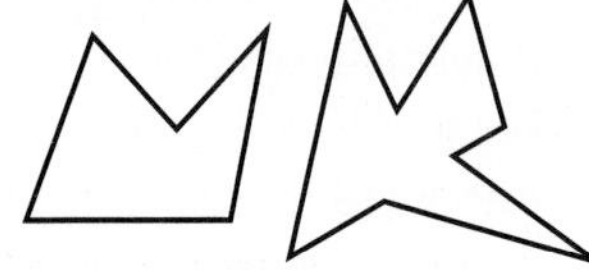

3-dimensional
(3-D) Having length, width, and thickness. Solid objects take up *volume* and are 3-dimensional. A figure whose points are not all in a single plane is 3-dimensional.

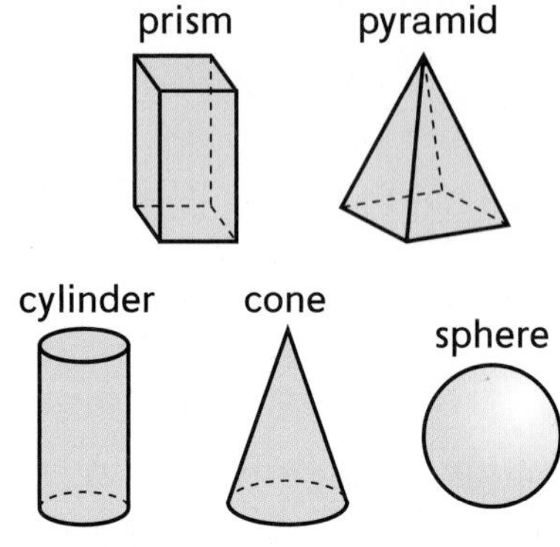

accurate As correct as possible for the situation.

acute angle An angle with a measure less than 90°.

Acute angles

acute triangle A triangle with three acute angles.

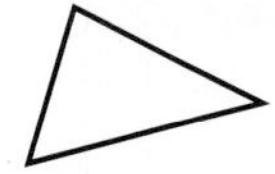
Acute triangle

addend Any one of a set of numbers that are added. For example, in $5 + 3 + 1 = 9$, the addends are 5, 3, and 1.

Additive Identity The number zero (0). The additive identity is the number that when added to any other number, gives that other number.

algebra The branch of mathematics that uses letters and other symbols to stand for quantities that are *unknown* or vary. Algebra is used to describe patterns, express numerical relationships, and model real-world situations.

algebraic expression An *expression* that contains a *variable*. For example, if Maria is 2 inches taller than Joe and if M represents Maria's height in inches, then the algebraic expression $M - 2$ represents Joe's height.

algorithm A set of step-by-step instructions for doing something, such as carrying out a computation or solving a problem.

angle A figure that is formed by two rays or line segments with a common endpoint. The rays or segments are called the sides of the angle. The common endpoint is called the *vertex* of the angle. Angles are measured in *degrees* (°).

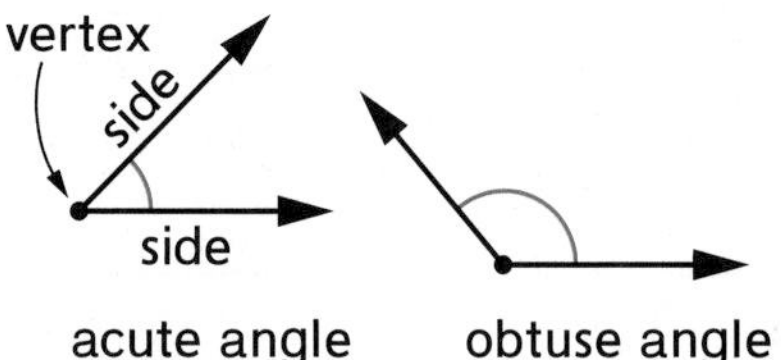

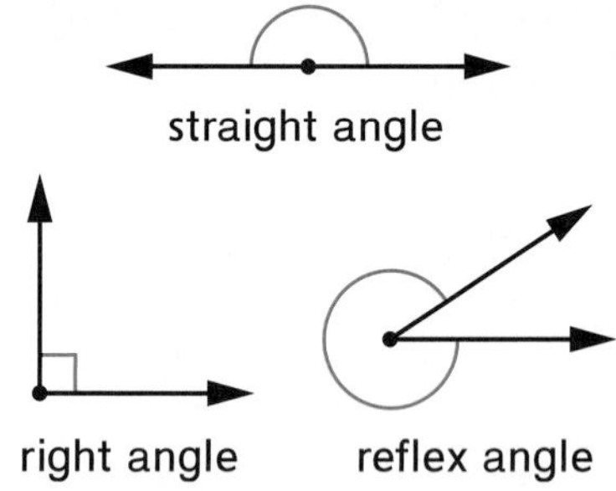

apex In a *pyramid* or cone, the point at the tip opposite the *base*. In a pyramid, all the nonbase faces meet at the apex.

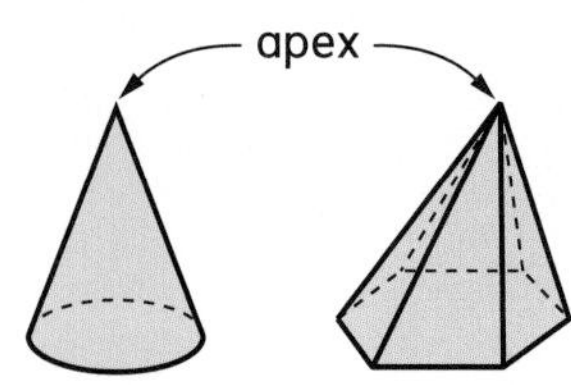

approximate Close to exact. In many situations it is not possible to get an exact answer, but it is important to be close to the exact answer.

area The amount of *surface* inside a *2-dimesional* shape. The measure of the area is how many units, such as square inches or square centimeters, cover the surface.

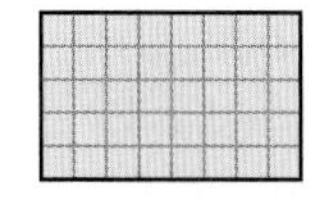

40 square units

about 21 square units

1 square centimeter

1 square inch

area model A *model* for multiplication problems in which the length and width of a *rectangle* represent the *factors,* and the *area* of the rectangle represents the *product.*

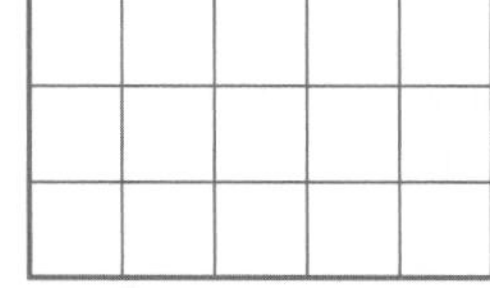

Area model for 3 * 5 = 15

array An arrangement of objects in a regular pattern, usually in rows and columns. In *Everyday Mathematics,* an array is a rectangular array unless specified otherwise.

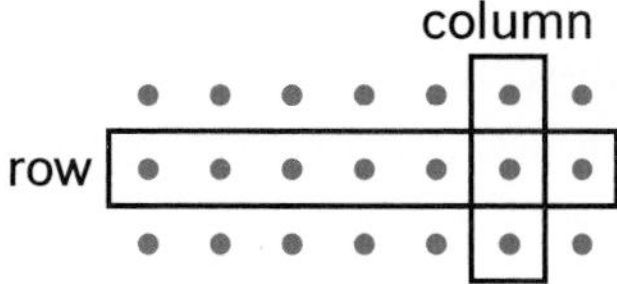

A rectangular array

Associative Property of Addition A *property* of addition (but not subtraction) that says that when you add three numbers, you can change the grouping without changing the *sum.* For example:
(4 + 3) + 7 = 4 + (3 + 7)

Associative Property of Multiplication A *property* of multiplication (but not division) that says that when you multiply three numbers, you can change the grouping without changing the *product.* For example:
(5 * 8) * 9 = 5 * (8 * 9)

axis of coordinate grid Either of the two *number lines* that intersect to form a *coordinate grid.*

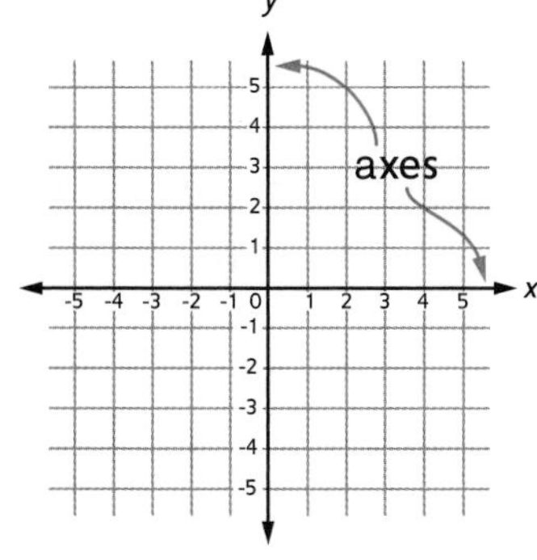

bar graph A graph that uses horizontal or vertical bars to represent *data.*

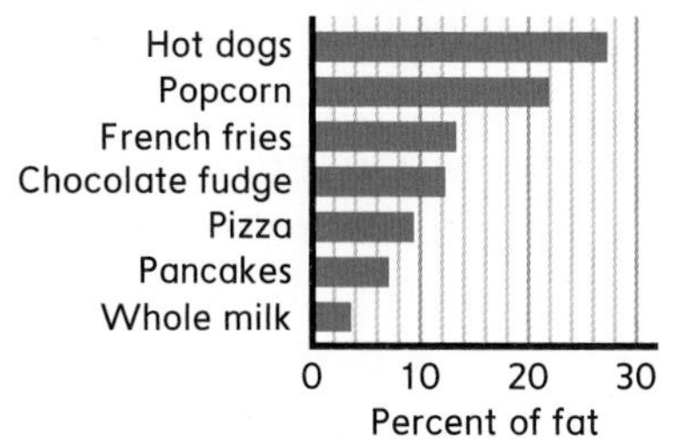

base The *side* of a *polygon or face* of a polyhedron from which the *height* is measured.

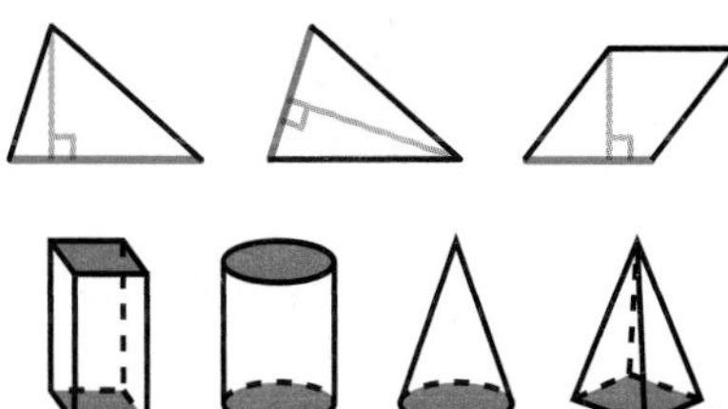

Bases are shown in red.

base (in exponential notation) The number that is raised to a power. For example, in 5^3, the base is 5. See *exponential notation* and *power of a number* n.

base ten Our system for writing numbers that uses 10 symbols called *digits.* The digits are 0, 1, 2, 3, 4, 5, 6, 7, 8, and 9. You can write any number using only these 10 digits. Each digit has a value that depends on its place in the number. In this system, moving a digit one place to the left makes that digit worth 10 times as much. And moving a digit one place to the right makes that digit worth one-tenth as much. See *place value.*

basic facts The addition facts (whole-number addends of 10 or less) and their related subtraction facts, and the multiplication facts (whole number factors of 10 or less) and their related division facts. Facts are organized into fact families.

benchmark A well-known number or measure that can be used to check whether other numbers, measures, or estimates make sense. For example, a benchmark for length is that the width of a man's thumb is about one inch. The numbers 0, $\frac{1}{2}$, 1, $1\frac{1}{2}$ may be useful benchmarks for fractions.

capacity (1) The amount a container can hold. Capacity is usually measured in units such as cups, fluid ounces, and liters. (2) The amount something can hold. For example, a computer hard drive may have a capacity of 64TB, or a scale may have a capacity of 400 lbs.

category A group whose members are defined by a shared attribute. For example, triangles are polygons that share the attribute of having three sides.

circle A 2-dimensional, closed, curved path whose points are all the same distance from a center point.

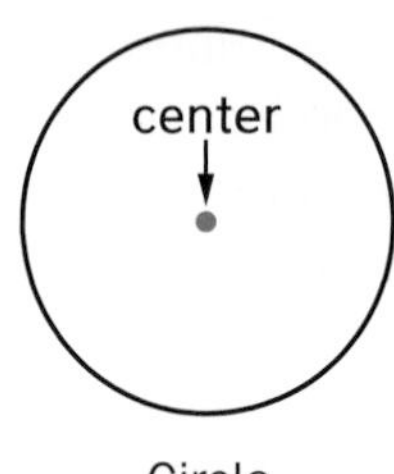

Circle

close-but-easier numbers Numbers that are close to the original numbers in the problem, but easier for solving problems. For example, to estimate 494 + 78, you might use the close-but-easier numbers 480 and 80.

column addition A method for adding numbers in which the *addends'* digits are first added in each *place-value* column separately, and then 10-for-1 trades are made until each column has only one digit. Lines are drawn to separate the place-value columns.

	100s	10s	1s
	2	4	8
+	1	8	7
	3	12	15
	3	13	5
	4	3	5

248 + 187 = 435

common denominator For two or more fractions, a number that is a multiple of both or all *denominators.* For example, the fractions $\frac{1}{2}$ and $\frac{2}{3}$ have the common denominators 6, 12, 18, and so on. If the fractions have the same denominator, that denominator is called a common denominator. See *quick common denominator.*

common multiple A number that is a multiple of two or more given numbers. For example, common multiples of 6 and 8 include 24, 48, and 72. See *least common multiple (LCM).*

common numerator Same as *like numerator.*

Commutative Property of Addition A *property* of addition (but not of subtraction) that says that changing the order of the numbers being added does not change the *sum.* This property is often called the *turn-around rule* in *Everyday Mathematics.* For example: 5 + 10 = 10 + 5

Commutative Property of Multiplication A *property* of multiplication (but not of division) that says that changing the order of the numbers being multiplied does not change the *product.* This property is often called the *turn-around rule* in *Everyday Mathematics.* For example: $3 * 8 = 8 * 3$

compose To make a number or shape by putting together smaller numbers or shapes. For example, you can compose a 10 by putting together ten 1s: $1 + 1 + 1 + 1 + 1 + 1 + 1 + 1 + 1 + 1 = 10$. You can compose a pentagon by putting together an equilateral triangle and a square.

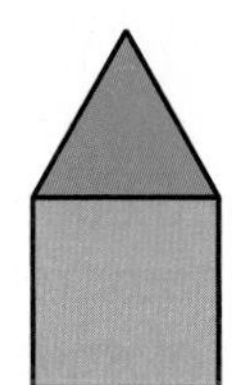

A composed pentagon

composite number A *counting number* that has more than two different *factors.* For example, 4 is a composite number because it has three factors: 1, 2, and 4.

composite unit A *unit* of measure made up of smaller units. For example, a foot is a composite unit of 12 inches, and a row of unit squares can be used to measure area.

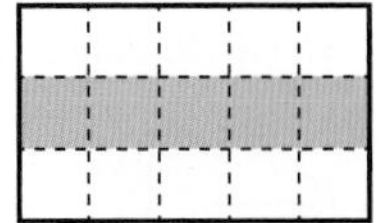

3 rows of 5 square units each have a total area of 15 square units.

concave polygon A *polygon* in which at least one vertex is "pushed in." At least one *angle* of a polygon measures greater than 180° inside the polygon. Same as nonconvex polygon.

congruent figures Figures that have the same shape and the same size. 2-dimensional figures will make an exact fit if one is placed on top of the other and lined up.

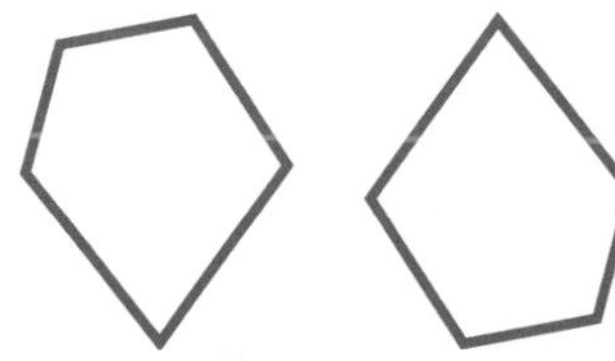

Congruent pentagons

conjecture A statement that is thought to be true based on information or mathematical thinking.

convex polygon A *polygon* in which all vertices are "pushed outward." Each *angle* of a convex polygon measures less than 180° inside the polygon.

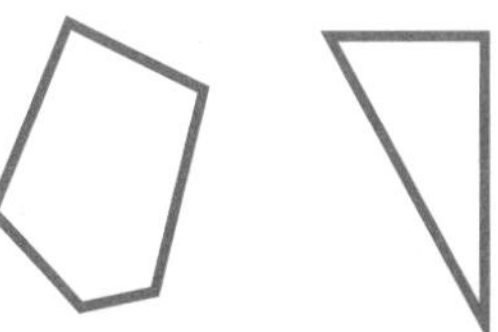

coordinate (1) One of the two numbers in an ordered pair. The number pair is used to locate a point on a coordinate grid. (2) A number used to locate a point on a number line.

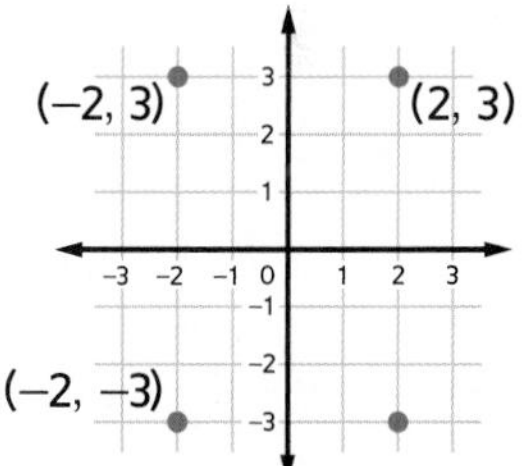

Coordinates

coordinate grid (rectangular coordinate grid) A grid formed by two number lines that intersect at their zero points and form right angles. Each number line is referred to as an *axis.* You can locate points on the grid with *ordered pairs* of numbers called coordinates.

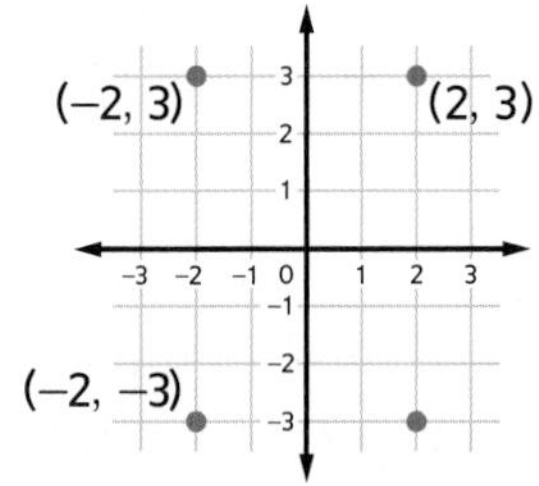

Coordinate grid

corresponding sides Sides in the same relative position in similar or congruent figures. Corresponding sides "match up."

corresponding terms Terms that are in the same position within two lists. For example, in the table below, the third term in the "in" column corresponds to the third term in the "out" column. They are corresponding terms.

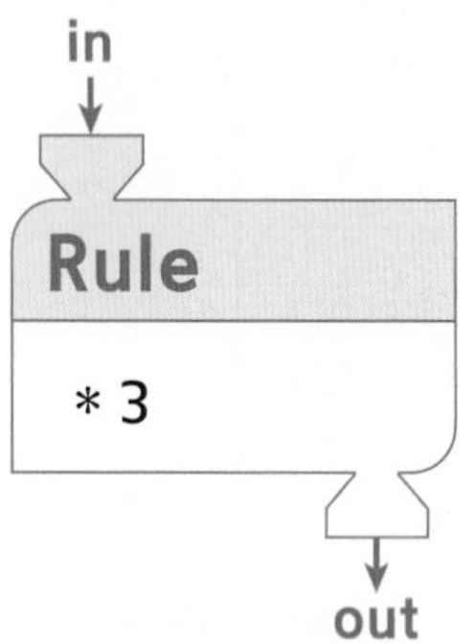

in	out
1	3
2	6
3	9
4	12

counting numbers The numbers used to count things. The set of counting numbers is {1, 2, 3, 4, ...}. Compare *whole numbers.*

counting-up subtraction A subtraction strategy in which you count up from the smaller to the larger number to find the *difference.* For example, to solve 16 – 9, count up from 9 to 16.

cube (1) A polyhedron with 6 square *faces.* A cube has 8 *vertices* and 12 *edges.* (2) The smallest base-10 block called a centimeter cube.

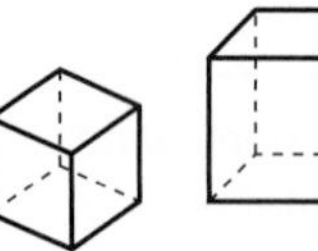
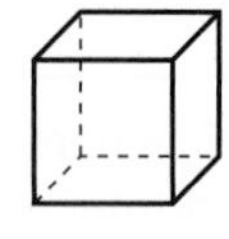
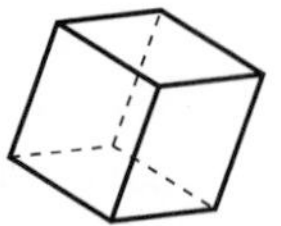

Cubes

cubic unit A unit used in measuring volume, such as a cubic centimeter or a cubic foot.

cubit An ancient unit of length, measured from the point of the elbow to the end of the middle finger. A cubit is about 18 inches.

cup A U.S. customary unit of *volume* or *capacity* equal to 8 fluid ounces or $\frac{1}{2}$ pint.

data Information that is gathered by counting, measuring, questioning, or observing.

decimal A number written in standard, *base-10* notation that contains a *decimal point,* such as 2.54. A whole number is a decimal, but is usually written without a decimal point.

decimal point A dot used to separate the ones and tenths places in *decimals.*

decompose To separate a number or shape into smaller numbers or shapes. For example, you can decompose 14 into 1 ten and 4 ones. You can decompose a square into two isosceles right triangles.

degree (°) (1) A unit of measure for *angles* based on dividing a *circle* into 360 equal parts. Latitude and longitude are measured in degrees, and these degrees are based on angle measures. (2) A unit of measure for *temperature.*

A small raised circle (°) can be used to show degrees, as in a 70° angle or 70°F for room temperature.

denominator The number below the line in a *fraction.* A fraction may be used to name part of a *whole.* If the whole is divided into equal parts, the denominator represents the number of equal parts into which the whole is divided. The denominator determines the size of each part. For example, in $\frac{3}{4}$, 4 is the denominator.

difference The result of subtracting one number from another.

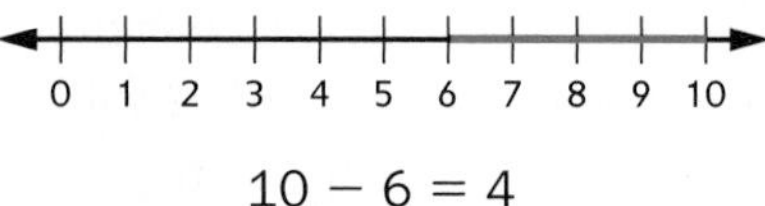

10 – 6 = 4

digit One of the number symbols 0, 1, 2, 3, 4, 5, 6, 7, 8, and 9 in the standard, *base-ten* system.

displacement method A method for measuring the *volume* of an object that involves submerging, or placing it under water, and measuring how much water it moves.

Distributive Property of Multiplication over Addition and Subtraction A *property* that relates multiplication and addition or subtraction. This property gets its name because it "distributes" a *factor* over terms inside parentheses. For example:

$2 * (5 + 3) =$
$(2 * 5) + (2 * 3) =$
$10 + 6 =$
16
and
$2 * (5 - 3) =$
$(2 * 5) - (2 * 3) =$
$10 - 6 =$
4

dividend The number in division that is being divided. For example, in $35 \div 5 = 7$, the dividend is 35.

divisible by If one *counting number* can be divided by a second counting number with a *remainder* of 0, then the first number is divisible by the second number. For example, 28 is divisible by 7 because 28 divided by 7 is 4, with a remainder of 0.

divisor In division, the number that divides another number. For example, in $35 \div 5 = 7$, the divisor is 5.

edge Any *side* of a *polyhedron's faces.*

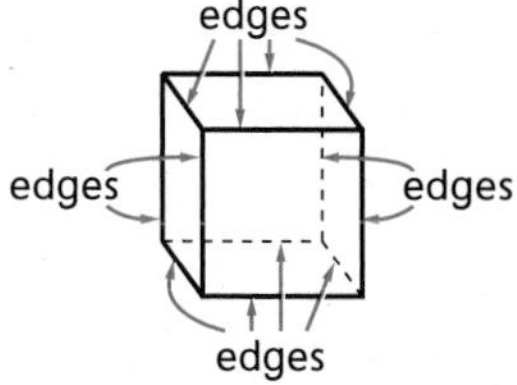

efficient strategy A method that can be applied easily and quickly.

equal See *equivalent.*

equal parts *Equivalent* parts of a *whole.* For example, dividing a pizza into 4 equal parts means each part is $\frac{1}{4}$ of the pizza and is equal in size to each of the other 3 parts.

equation A *number sentence* that contains an equal sign. For example, $15 = 10 + 5$ is an equation.

equilateral polygon A *polygon* in which all sides are the same length.

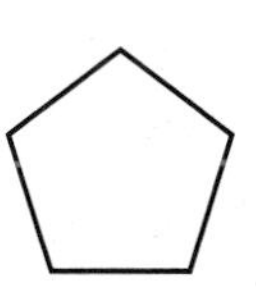

equilateral polygons

equilateral triangle A *triangle* with all three sides equal in length. In an equilateral triangle, all three angles have the same measure.

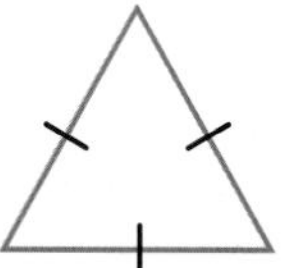

An equilateral triangle

equivalent *Equal* in value but possibly in a different form. For example, $\frac{1}{2}$, 0.5, and 50% are all equivalent.

equivalent fractions *Fractions* that name the same number. For example, $\frac{1}{2}$ and $\frac{4}{8}$ are equivalent fractions.

equivalent fractions rule A rule stating that if the *numerator* and *denominator* of a *fraction* are each multiplied or divided by the same nonzero number, the result is a fraction *equivalent* to the original fraction.

equivalent names Different ways to name the same number. For example, $2 + 6$, $12 - 4$, 2×4, $16 \div 2$, $5 + 1 + 2$, VIII, and eight are equivalent names for 8.

equivalent problem A division problem solved by writing an *equivalent expression.* For example, to solve 35.6/0.5, you may solve the equivalent problem 356/5.

estimate An answer close to an exact answer. To estimate means to give an answer that should be close to an exact answer.

evaluate a numerical expression To carry out the *operations* in a numerical *expression* to find a single value for the expression.

even number A *counting number* that can be divided by 2 with no remainder. The even numbers are 2, 4, 6, 8, and so on. 0, −2, −4, −6, and so on are also usually considered even.

expanded form A way of writing a number as the *sum* of the values of each *digit.* For example, in expanded form, 356 is written 300 + 50 + 6. Compare *standard form.*

exponent A number used in *exponential notation* to tell how many times the *base* is used as a *factor.* For example, in 5^3, the base is 5, the exponent is 3, and $5^3 = 5 * 5 * 5 = 125$. See *power of a number* n.

exponential notation A way to show repeated multiplication by the same *factor.* For example, 2^3 is exponential notation for $2 * 2 * 2$. The small raised 3 is the *exponent.* It tells how many times the number 2, called the *base,* is used as a factor.

expression A group of mathematical symbols that represents a number—or can represent a number if values are assigned to any *variables* in the expression. An expression may include numbers, variables, *operation symbols,* and *grouping symbols*—but *not relation symbols* (=, >, <, and so on). Any expression that contains one or more variables is called an algebraic expression.

extended facts Variations of basic facts involving multiples of 10, 100, and so on. For example, 30 + 70 = 100, 40 ∗ 5 = 200, and 560/7 = 80 are extended facts.

face A flat *surface* on a *3-dimensional* shape.

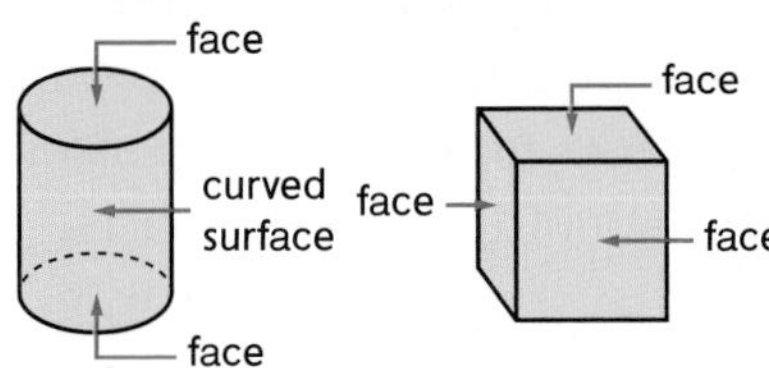

factor Whenever two or more numbers are multiplied to give a *product,* each of the numbers that is multiplied is called a factor. For example, in 4 ∗ 1.5 = 6, 6 is the product and 4 and 1.5 are called factors. Compare *factor of a counting number* n.

factor of a counting number *n* A *counting number* whose product with another counting number equals *n.* For example, 2 and 3 are *factors* of 6 because 2 ∗ 3 = 6. But 4 is not a *factor* of 6 because 4 ∗ 1.5 = 6 and 1.5 is not a counting number.

factor pair Two *factors* of a *counting number* whose *product* is the number. A number may have more than one factor pair. For example, the factor pairs for 18 are 1 and 18, 2 and 9, and 3 and 6.

factor tree A way to get the *prime factorization* of a *counting number.* Write the original number as a product of counting-number *factors.* Then write each of these factors as a product of factors, and so on, until the factors are all prime numbers. A factor tree looks like an upside-down tree, with the root (the original number) at the top and the leaves (the factors) beneath it.

false number sentence A *number sentence* that is not true. For example, 8 = 5 + 5 is a false number sentence.

fathom A unit used by people who work with boats and ships to measure depths underwater and lengths of cables. A fathom is now defined as 6 feet.

formula A general rule for finding the value of something. A formula is often written using letters, called *variables,* which stand for the quantities involved.
For example, the formula for the area of a rectangle may be written as $A = l * w$, where *A* represents the area of the rectangle, *l* represents its length, and *w* represents its width.

fraction A number in the form $\frac{a}{b}$ or a/b. Fractions can be used to name part of a whole or part of a collection. A fraction may also be used to represent division. For example, $\frac{2}{3}$ can be thought of as 2 divided by 3. See *numerator* and *denominator.*

Fraction Circle Pieces A set of colored circles each divided into equal-size slices, used to represent *fractions.*

Frames and Arrows A diagram used in *Everyday Mathematics* to show a number pattern or sequence.

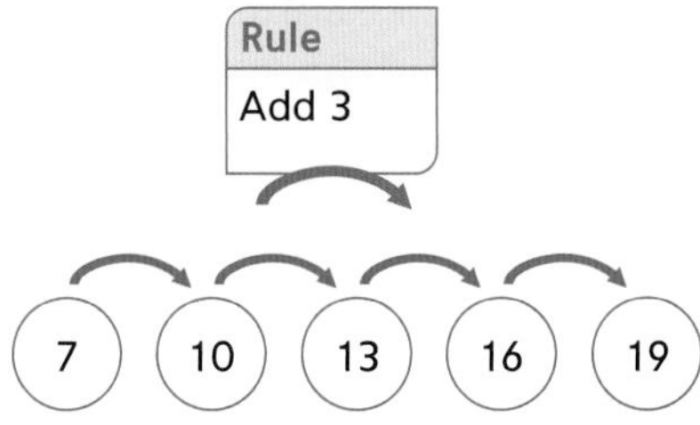

front-end estimation An estimation method that keeps only the left-most digit in the numbers and puts 0s in for all others. For example, the front-end *estimate* for 45,600 + 53,450 is 40,000 + 50,000 = 90,000.

function machine An imaginary machine that uses a rule to pair input numbers put in (inputs) with numbers put out (outputs). Each input is paired with exactly one output. Function machines are used in "What's My Rule?" problems.

geometric solid A *3-dimensional* shape, such as a *prism, pyramid,* cylinder, cone, or sphere. Despite its name, a geometric solid is hollow; it does not contain the points in its interior.

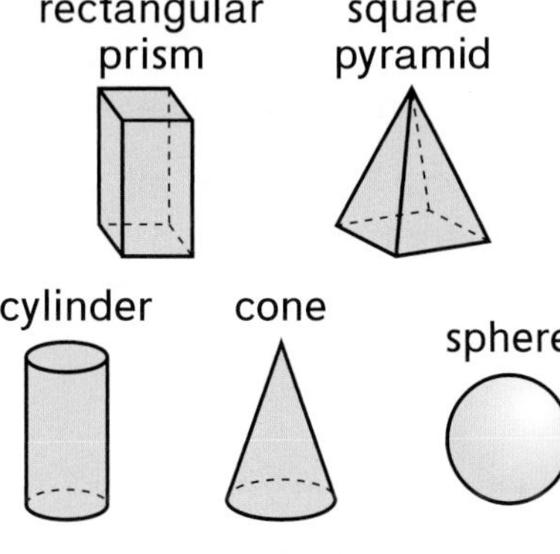

great span The distance from the tip of the thumb to the tip of the little finger (pinkie) when the hand is stretched as far as possible.

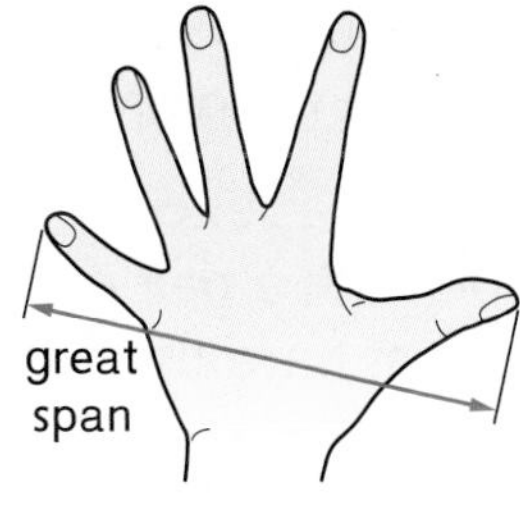

grouping symbols Symbols such as parentheses ( ), brackets [ ], and braces { } that tell the order in which operations in an *expression* are to be done. For example, in the expression (3 + 4) ∗ 5, the operation in the parentheses should be done first. The expression then becomes 7 ∗ 5 = 35.

height (1) The length of the shortest line segment connecting a *vertex* of a shape to the line containing the base opposite it. (2) The length of the shortest line segment connecting a vertex or apex of a solid to the plane containing the base opposite it. (3) The line segment itself.

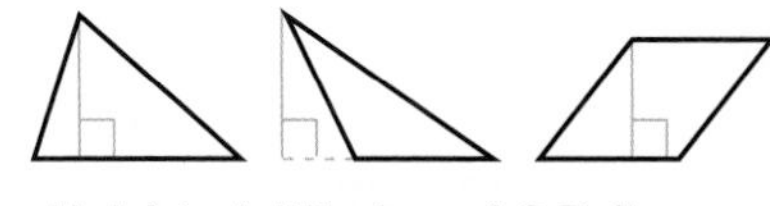

Heights/altitudes of 2-D figures are shown in blue.

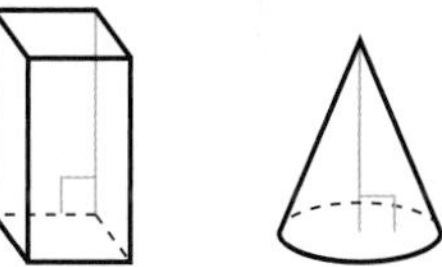

Heights/altitudes of 2-D figures are shown in blue.

hierarchy of shapes A way to organize shapes into categories and subcategories. All attributes for a category are attributes of each of its subcategories. Subcategories have more attributes. A hierarchy is often shown in a diagram with the most general category at the top and arrows or lines connecting categories to their subcategories.

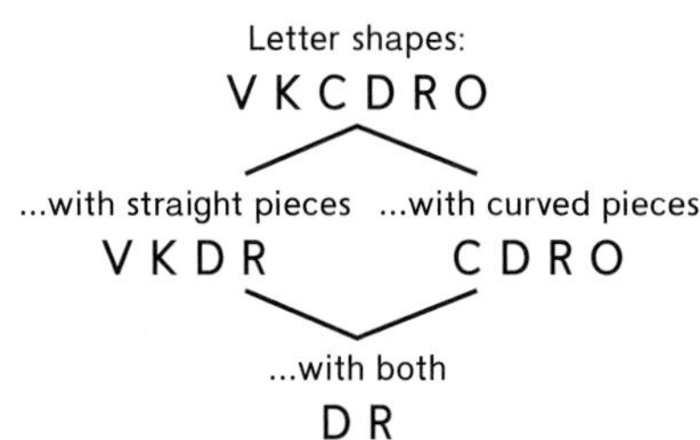

horizontal In a left-to-right orientation lined up with the horizon.

image The reflection of an object that you see when you look in a mirror. Also, a figure that is produced by a transformation (a reflection, translation, or rotation, for example) of another figure. Compare *preimage*.

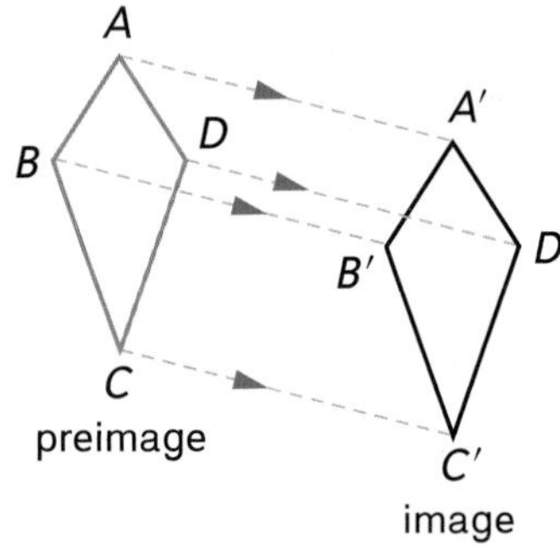

inequality A *number sentence* with $>$, $<$, $\geq$, $\leq$, or $\neq$. For example, the sentence $8 < 15$ is an inequality.

interior of a figure The inside of a closed figure. The interior is usually not considered to be part of the figure.

interpolate To *estimate* an unknown value between known values. Graphs are often useful tools for interpolation.

intersect To meet or cross.

Intersecting segments

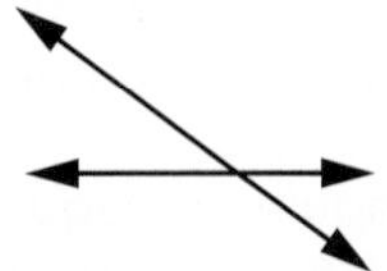

Intersecting lines

interval The set of all numbers between two numbers, *a* and *b*, which may include *a* or *b* or both.

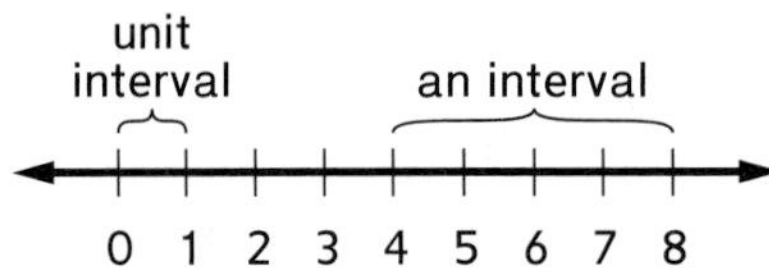

inverse operations Two *operations* that undo the effects of each other. Addition and subtraction are inverse operations, as are multiplication and division.

isosceles trapezoid A *trapezoid* with a pair of base angles that have the same measure.

isosceles trapezoids

isosceles triangle A *triangle* with at least two *sides* equal in length. In an isosceles triangle, at least two *angles* have the same measure. A triangle with all three sides the same length is an isosceles triangle, but is usually called an equilateral triangle.

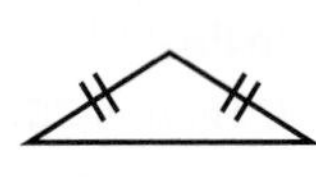

Isosceles triangles

iterate units To repeat a *unit* without gaps or overlaps in order to measure. For example, you can pack a shape with cubes to measure its volume.

kilogram A metric unit of mass equal to 1,000 grams. A bottle of water is usually 1 kilogram.

kite A *quadrilateral* with two pairs of adjacent equal length *sides*. The four sides can all have the same length, so a rhombus is a kite.

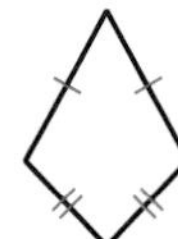
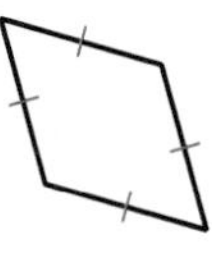
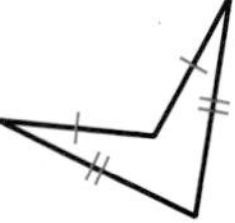

Kites

least common denominator (LCD) The *least common multiple* of the *denominators* of every *fraction* in a given collection. For example, the least common denominator of $\frac{1}{2}$, $\frac{4}{5}$, and $\frac{3}{8}$ is 40.

least common multiple (LCM) The smallest number that is a multiple of two or more numbers. For example, while some common multiples of 6 and 8 are 24, 48, and 72, the least common multiple of 6 and 8 is 24.

length The distance between two points along a path.

like Equal or the same.

like denominator Same as *common denominator.*

like numerator A number that is the numerator of two or more fractions. For example, the fractions $\frac{3}{11}$ and $\frac{3}{7}$ have a *common numerator* of 3.

line A straight path that goes on forever in both directions.

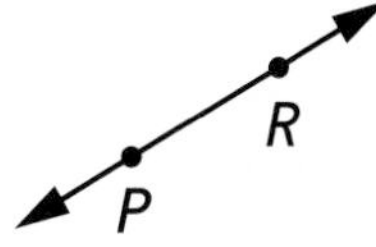

Line *RP* or *PR*

line graph A graph that uses line segments to connect *data* points. Line graphs are often used to show how something changed over a period of time.

line of reflection (mirror line) A line halfway between a figure (preimage) and its reflected image. In a reflection, a figure is "flipped over" the line of reflection.

line of symmetry A line drawn through a figure so that it is divided into two parts that are mirror images of each other. The two parts look alike but face in opposite directions. See *line symmetry.*

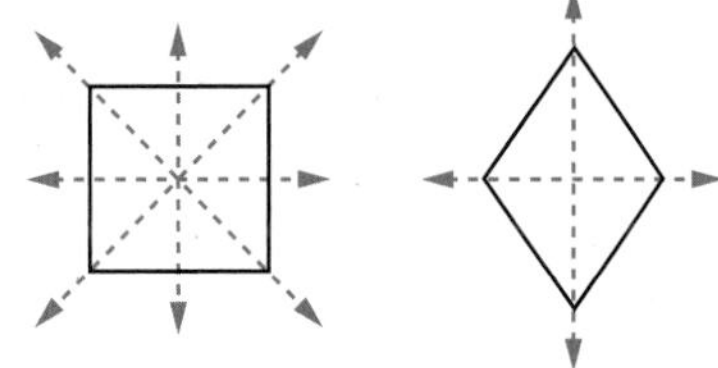

Lines of symmetry are shown in blue.

line plot A sketch of *data* in which check marks, Xs, or other marks above a labeled line show how many times each value appears in the set of data.

Test Scores

0	1	2	3	4	5
				x	
				x	
			x	x	
		x	x	x	
	x	x	x	x	
	x	x	x	x	x

Number Correct

line segment A straight path joining two points. The two points are called endpoints of the segment.

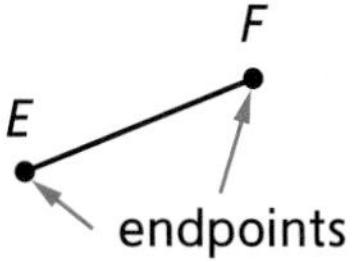

Line segment *EF* or *FE*

line symmetry A figure has line symmetry if a line can be drawn through it so that it is divided into two parts that are mirror images of each other. The two parts look alike but face in opposite directions. See *line of symmetry.*

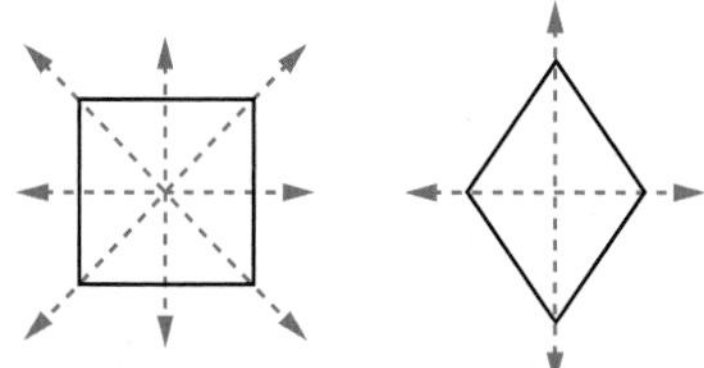

These figures have line symmetry.

liquid volume An amount of liquid measured in units such as liters and gallons. Units of liquid *volume* are frequently used to measure *capacity.*

mathematical argument An explanation of why a claim is true or false using words, pictures, symbols, or other representations. For example, if you claim that $\frac{1}{2} + \frac{3}{5} = \frac{4}{7}$ is not true, you say that $\frac{3}{5}$ is more than $\frac{1}{2}$, so the answer to $\frac{1}{2} + \frac{3}{5}$ is greater than 1. Since $\frac{4}{7}$ is less than 1, $\frac{1}{2} + \frac{3}{5} = \frac{4}{7}$ must not be true.

mathematical practices Ways of working with mathematics. Mathematical practices are habits or actions that help people use mathematics to solve problems.

mathematical structure A relationship among mathematical objects, operations, or relations; a mathematical pattern, category, or *property*. For example, the Distributive Property of Multiplication over Addition is a structure of arithmetic. The number grid illustrates some patterns and structures that exist in our number system.

measurement scale The spacing of the marks on a measuring device. The scales on this ruler are 1 millimeter on the left side and $\frac{1}{16}$ inch on the right side.

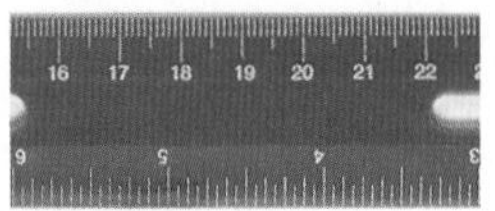

Scale of a number line

metric system A measurement system based on decimals and multiples of 10. The metric system is used by scientists and people in most countries around the world except the United States.

mixed number A number that is written using both a *whole number* and a *fraction.* For example, $2\frac{1}{4}$ is a mixed number equal to $2 + \frac{1}{4}$.

model A representation of a real-world object or situation. Number sentences, diagrams, and pictures can be models.

multiple of a number *n* A *product* of *n* and a *counting number.* For example, the multiples of 7 are 7, 14, 21, 28, and so on.

multiplication rule for equivalent fractions See equivalent fractions rule.

multiplicative identity The number 1. The multiplicative identity is the number that when multiplied by any other number is that other number.

name-collection box In *Everyday Mathematics*, a place to write *equivalent names* for a number.

50

100 ÷ 2   5 × 10

10 + 10 + 10 + 10 + 10

1 more than 49   25 + 25

fifty   *cincuenta*

negative numbers A number that is less than zero; a number to the left of zero on a horizontal *number line* or below zero on a vertical number line. The symbol – may be used to write a negative number. For example, "negative 5" is usually written as –5.

nonconvex polygon Same as *concave polygon.*

number line A *line* with numbers marked in order on it.

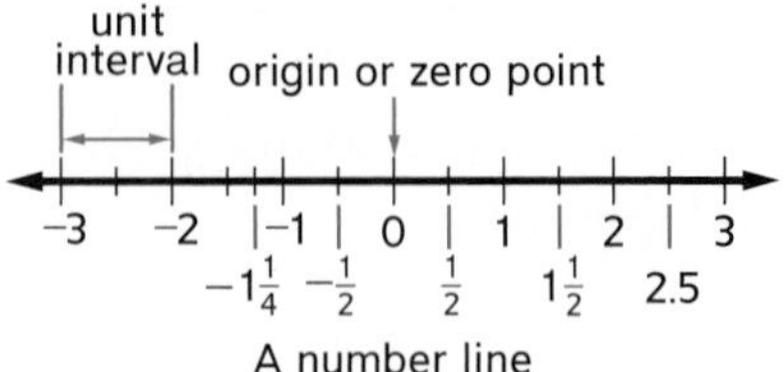

A number line

number model A *number sentence* or *expression* that models or fits a *number story* or situation. For example, the story *Sally had $5, and then she earned $8,* can be modeled as the number sentence 5 + 8 = 13, or as the expression 5 + 8.

number sentence At least two numbers or *expressions* separated by a *relation symbol* (=, >, < , ≥, ≤, ≠). Most number sentences contain at least one operation symbol (+, –, ×, ∗, ÷, or /). Number sentences may also have grouping symbols, such as parentheses and brackets.

number story A story with a problem that can be solved using arithmetic.

numeral A word, symbol, or figure that represents a number. For example, six, VI, and 6 are numerals that represent the same number.

numerator The number above the line in a *fraction.* A fraction may be used to name part of a *whole.* If the whole is divided into equal parts, the numerator represents the number of equal parts being considered. For example, in $\frac{3}{4}$, 3 is the numerator.

obtuse angle An angle with measure greater than 90° and less than 180°.

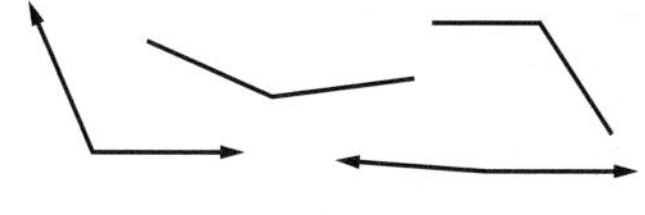

Obtuse angles

obtuse triangle A triangle with an obtuse angle.

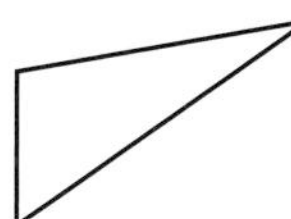

An obtuse triangle

open sentence A *number sentence* which has variables or missing numbers and is neither true nor false. For example, $5 + x = 13$ is an open sentence. The sentence is true if 8 is substituted for x. The sentence is false if 4 is substituted for x.

operation An action performed on numbers or *expressions* to produce other numbers or expressions. Addition, subtraction, multiplication, and division are the four basic arithmetic operations.

operation symbol A symbol used to stand for a mathematical operation. Common operation symbols are $+$, $-$, $\times$, $*$, $\div$, and $/$.

order of operations Rules that tell in what order to perform operations in arithmetic.

ordered pair Two numbers that are used to locate a point on a rectangular coordinate grid. The first number gives the position along the horizontal axis, and the second number gives the position along the vertical axis. The numbers in an ordered pair are called coordinates. Ordered pairs are usually written inside parentheses: (5, 3).

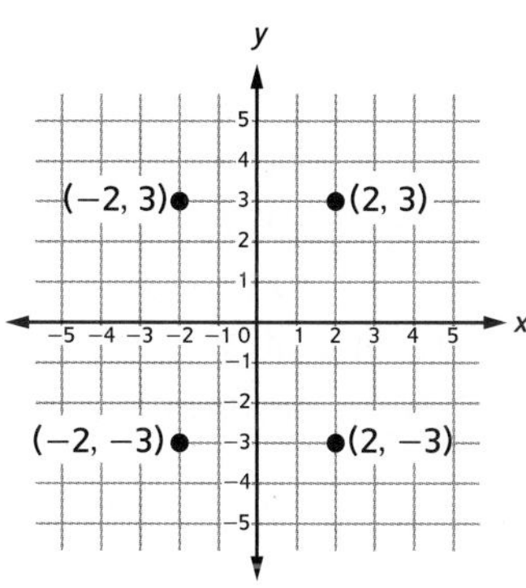

Ordered pairs

origin (1) The point (0, 0) where the two axes of a coordinate grid meet. (2) The 0 point on a *number line.*

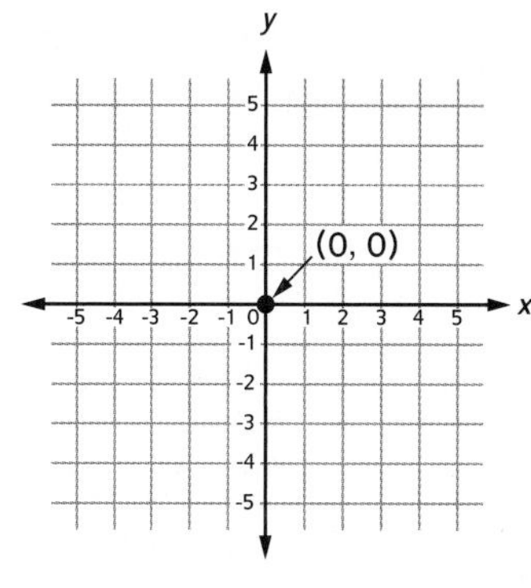

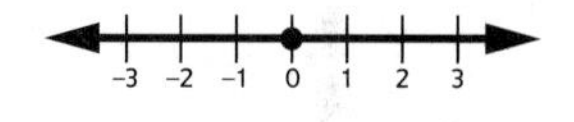

The points at (0, 0) and 0 are origins.

ounce (oz) A U.S. customary unit equal to $\frac{1}{16}$ of a pound.

parallel *Lines,* line segments, or rays in the same *plane* are parallel if they never cross or meet, no matter how far they are extended. Two planes are parallel if they never cross or meet. A line and a plane are parallel if they never cross or meet. The symbol || means is parallel to.

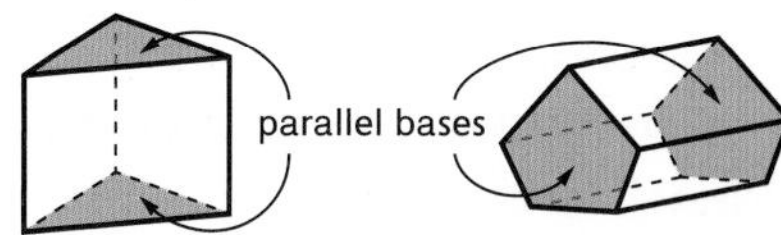

parallelogram A *quadrilateral* with two pairs of *parallel* sides. Opposite sides of a parallelogram are congruent. Opposite angles in a parallelogram have the same measure. All parallelograms are *trapezoids.*

Parallelogram

parentheses *Grouping symbols,* ( ), used to tell which parts of an expression should be calculated first.

partial-products multiplication (1) A way to multiply in which the value of each digit in one factor is multiplied by the value of each digit in the other factor. The final *product* is the sum of these partial products. (2) A similar method for multiplying mixed numbers.

partial-quotients division A way to divide in which the dividend is divided in a series of steps. The *quotients* for each step (called partial quotients) are added to give the final answer.

partial-sums addition A way to add in which *sums* are computed for each place (ones, tens, hundreds, and so on) separately. The partial-sums are then added to give the final answer.

perpendicular Being part of two lines that cross or meet at right angles. Planes that cross or meet at right angles are perpendicular. The symbol ⊥ means "is perpendicular to."

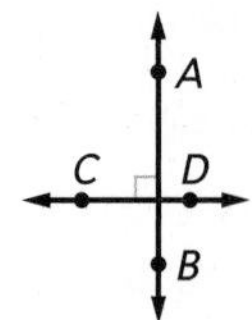

Perpendicular lines

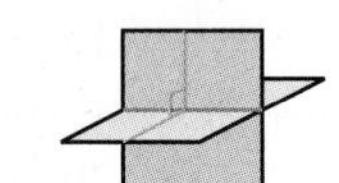

Perpendicular planes

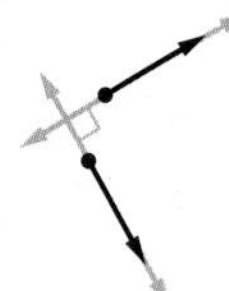

Perpendicular rays

place value A system that gives a *digit* a value according to its position in a number. In our *base-ten* system for writing numbers, moving a digit one place to the left makes that digit worth 10 times as much. And moving a digit one place to the right makes that digit worth one-tenth as much. For example, in the number 456, the 4 in the hundreds place is worth 400; but in the number 45.6, the 4 in the tens place is worth 40.

thousands	hundreds	tens	ones

A place-value chart

plane A flat *surface* that extends forever.

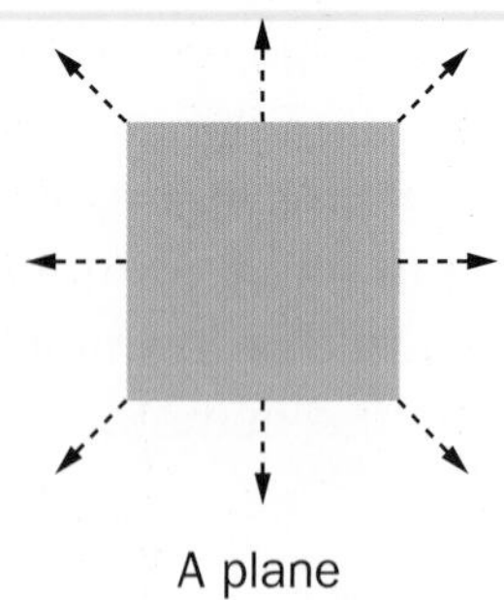

A plane

plane figure A set of points that is entirely contained in a single plane

plot To draw on a number line, coordinate grid, or graph. The points plotted may come from *data.*

point An exact location in space. The center of a circle is a point. Lines have an unlimited number of points on them.

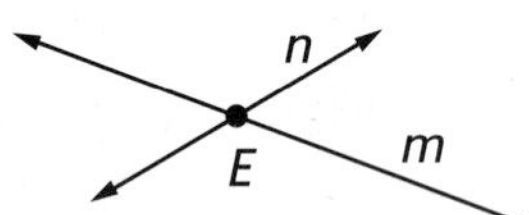

Lines *m* and *n* intersect at point *E*.

polygon A 2-dimensional figure that is made up of *line segments* joined end to end to make one closed path. The line segments of a polygon may not cross.

Polygons

polyhedron A *3-dimensional* figure whose surfaces (*faces*) are all flat and formed by *polygons.* Each face consists of a polygon and the interior of that polygon. The faces may meet but not cross. A polyhedron does not have any curved surface.

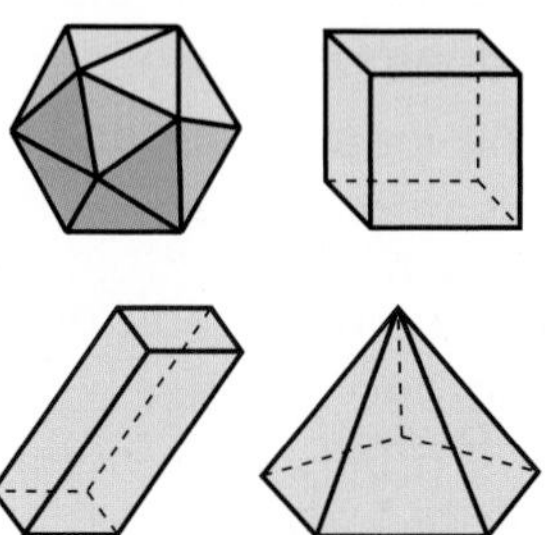

Polyhedra

positive numbers Numbers that are greater than zero. Positive numbers are usually written to the right of zero on a horizontal *number line* or above zero on a vertical number line. A positive number may be written using the + symbol, but is usually written without it. For example, +10 = 10.

pound (lb) A U.S. customary unit equal to 16 ounces. A small can of soup weighs about 1 pound.

power of 10 (1) A whole number that can be written as a product of 10s. For example, 100 is equal to 10 ∗ 10, or 10^2. 100 is called "the second power of 10," "10 to the second power, " or "10 squared." (2) A number that can be written as a product of $\frac{1}{10}$s is also a power of 10.

precise Exact. The smaller the *unit* used in measuring, the more precise the measurement is. For example, a measurement to the nearest inch is more precise than a measurement to the nearest foot. A ruler with $\frac{1}{8}$-inch markings is more precise than a ruler with $\frac{1}{4}$ -inch markings.

preimage A geometric figure that is changed (by stretching or moving, for example) to produce another figure. Compare *image.*

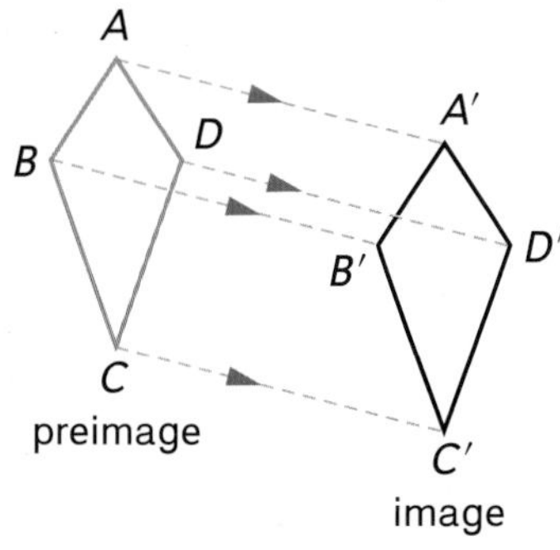

prime factorization A *counting number* written as a product of prime *factors.* Every counting number greater than 1 has a unique prime factorization. For example, the prime factorization of 24 is 2 ∗ 2 ∗ 2 ∗ 3. The factorization of a prime number is that number. For example, the prime factorization of 13 is 13.

prime number A *counting number* that has exactly two different *factors:* itself and 1. For example, 5 is a prime number because its only factors are 5 and 1. The number 1 is not a prime number because 1 has only a single factor, the number 1 itself.

prism A polyhedron with two *parallel* faces, called *bases* that are the same size and shape. The other *faces* connect the bases and are shaped like *parallelograms.* The edges that connect the bases are parallel. Prisms get their name from the shape of their bases.

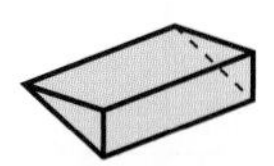

Triangular prism

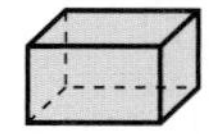

Rectangular prism

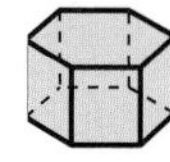

Hexagonal prism

product The result of multiplying two or more numbers, called *factors.* For example, in 4 ∗ 3 = 12, the product is 12.

property (1) A general statement about a mathematical relationship, such as the turn-around rule or the Distributive Property of Multiplication over Addition. (2) Same as attribute.

protractor A tool used to measure and draw *angles.* A half-circle protractor can be used to measure and draw angles up to 180°; a full-circle protractor, to measure angles up to 360°.

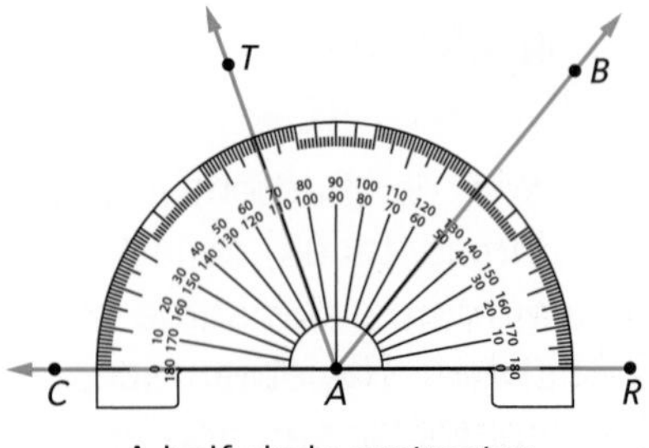

A half-circle protractor

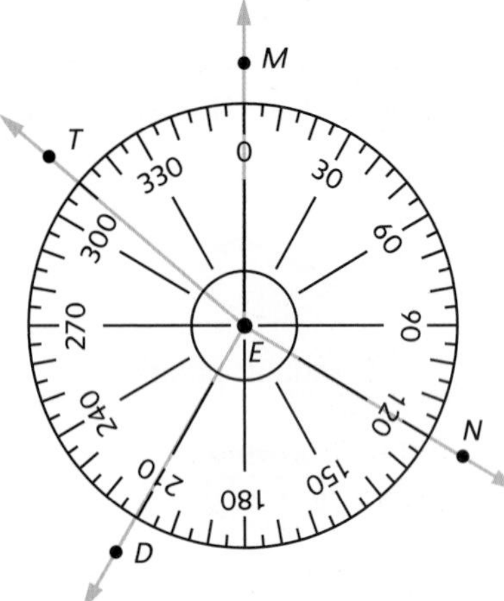

A full-circle protractor

pyramid A *polyhedron* in which one *face,* the *base,* may have any polygon shape. All of the other faces are triangular and come together at a point called the *apex.* A pyramid takes its name from the shape of its base.

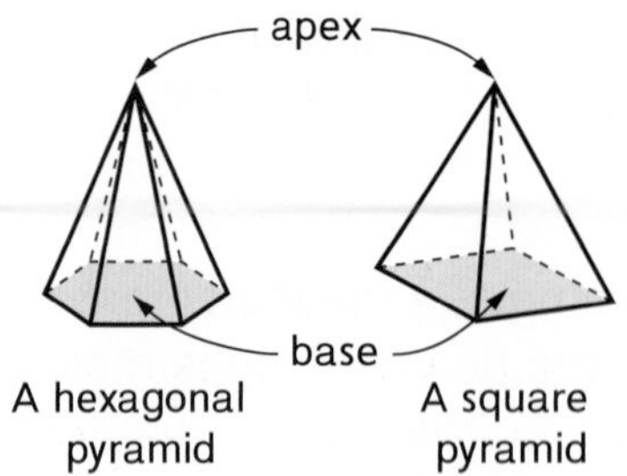

Q

quadrangle A polygon that has four angles. Same as *quadrilateral.*

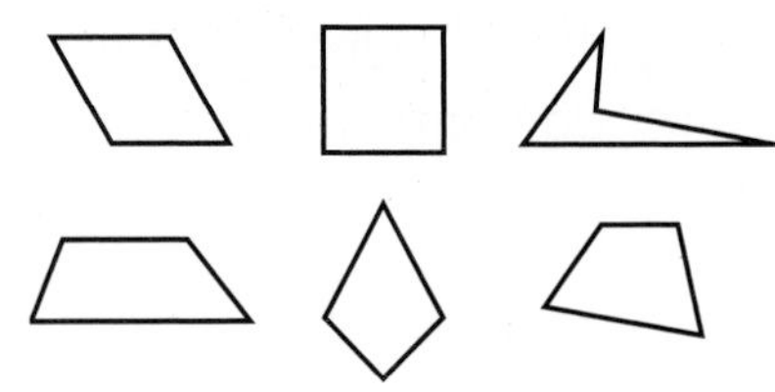

quadrilateral A *polygon* that has four sides. Same as *quadrangle.*

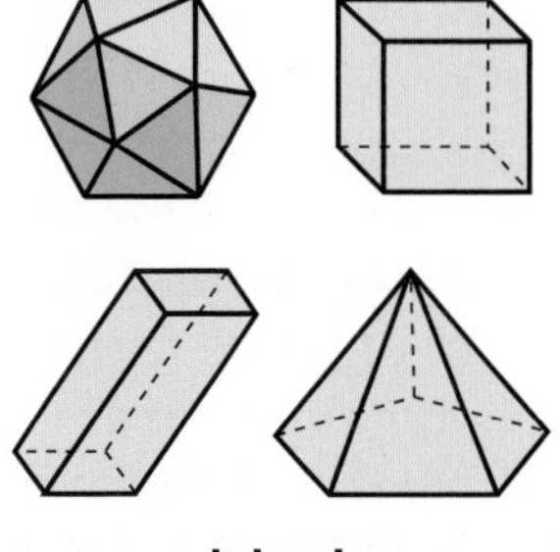
polyhedra

quantity A number with a unit, usually a measurement or count.

quart A U.S. customary unit of *volume* or *capacity* equal to 32 fluid ounces, 2 pints, or 4 cups.

quick common denominator (QCD) The *product* of the *denominators* of two or more *fractions.* For example, the quick common denominator of $\frac{1}{4}$ and $\frac{3}{6}$ is $4 * 6$, or 24. As the name suggests, this is a quick way to get a *common denominator* for a collection of fractions, but it does not necessarily give the *least common denominator.*

quotient The result of dividing one number by another number. For example, in $35 \div 5 = 7$, the quotient is 7.

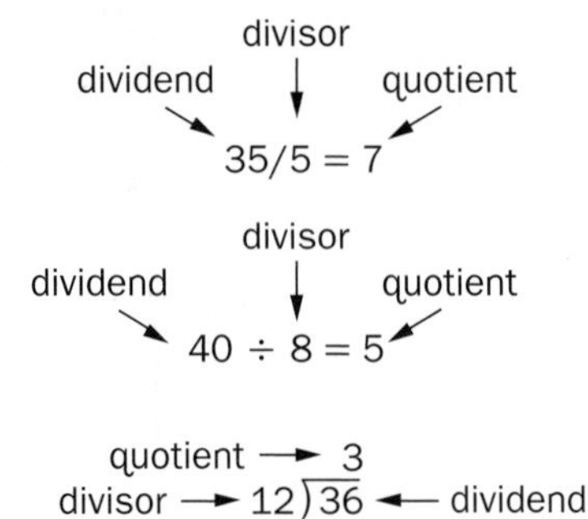

R

ray A straight path that starts at one endpoint and goes on forever in one direction.

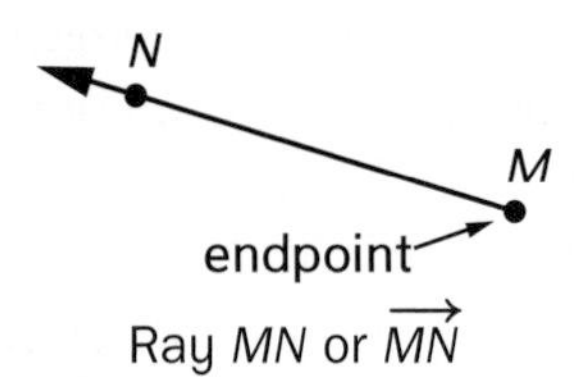

Ray *MN* or $\overrightarrow{MN}$

rectangle A *parallelogram* whose corners are all *right angles.*

Rectangles

rectangle method A method for finding area in which rectangles are drawn around a figure or parts of a figure. The rectangles form regions with boundaries that are rectangles or triangular halves of rectangles. The area of the original figure can be found by adding or subtracting the areas of these regions.

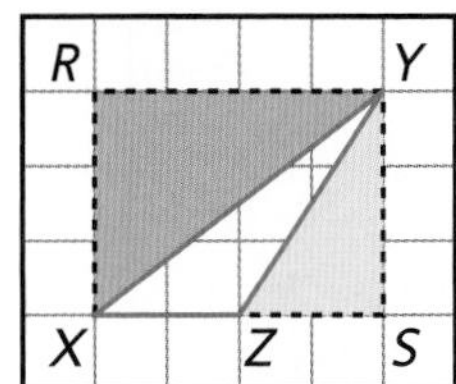

Area of $\triangle XYZ$ = area of rectangle $RYSX$ – area of $\triangle XRY$ – area of $\triangle YSZ$

rectangular prism A *prism* with rectangular *bases.* The four *faces* that are not bases are either *rectangles* or other *parallelograms.* A rectangular prism may model a shoebox.

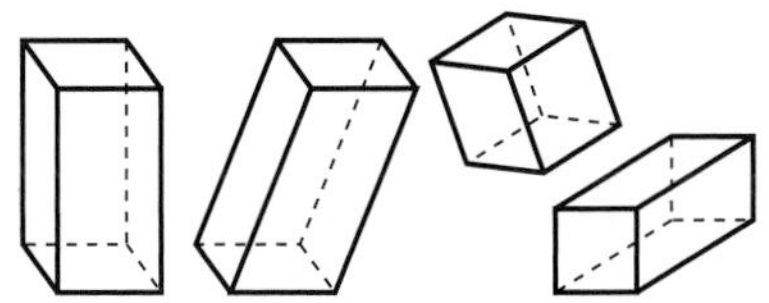

Rectangular prisms

rectangular pyramid A *pyramid* with a rectangular *base.*

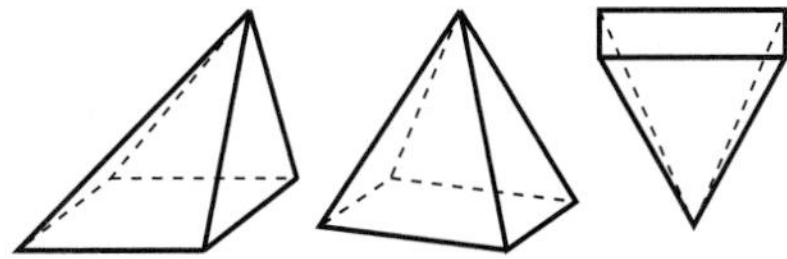

Rectangular pyramids

reflection The "flipping" of a figure over a line (the line of reflection) so that its image is the mirror image of the original figure (preimage). A reflection of a solid figure is a mirror-image "flip" over a plane.

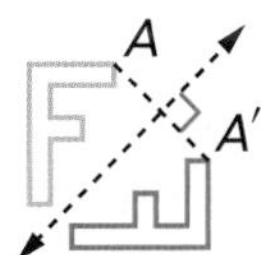

A reflection

regular polygon A *polygon* whose sides are all the same length and whose interior *angles* are all the same measure. For example, a *square* is a regular polygon.

regular polyhedron A *polyhedron* whose *faces* are formed by congruent *regular polygons.*

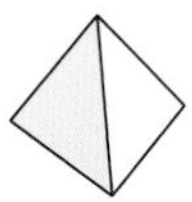

A tetrahedron (4 equilateral triangles)

A cube (6 squares)

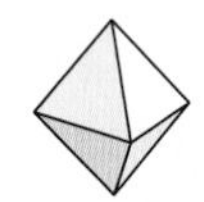

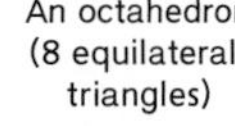

An octahedron (8 equilateral triangles)

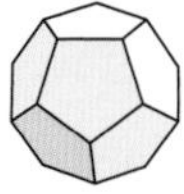

A dodecahedron (12 regular pentagons)

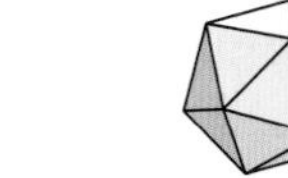

An icosahedron (20 equilateral triangles)

relation symbol A symbol used to express a relationship between two quantities. Some relationship symbols are <, >, and =.

remainder An amount left over when one number is divided by another number. For example, if 38 books are divided into 5 equal piles, there will be 7 books per pile, with 3 books left over. We may write 38 ÷ 5 → 7 R3, where R3 stands for the remainder.

represent To show, symbolize, or stand for something. For example, numbers can be represented using base-10 blocks, spoken words, or written numerals.

rhombus A *quadrilateral* whose sides are all the same length. All rhombuses are *parallelograms* and *kites.* Every square is a rhombus, but not all rhombuses are squares.

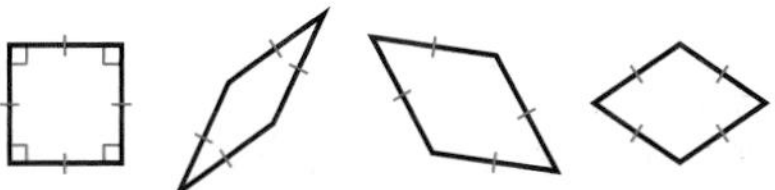

Rhombuses

right angle An *angle* with a measure of 90°.

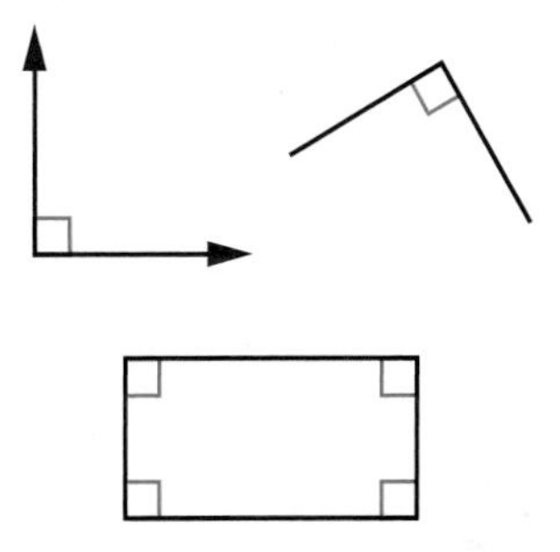

Right angles

right triangle A triangle that has a right angle (90°).

Right triangle

rotation A movement of a figure around a fixed point; a "turn."

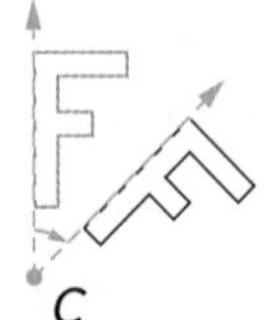

A rotation

round To change a number slightly to make it easier to work with or to make it better reflect the level of precision of the data. Often numbers are rounded to the nearest multiple of 10, 100, 1,000, and so on. For example, 12,964 rounded to the nearest thousand is 13,000.

rubric A tool used to rate work based on its quality.

scale (1) A comparison between the number of units in a picture or model and the actual number of units. A picture graph may show 1 smiley face to stand for 10 people. (2) See *scale of a number line or measurement scale.* (3) A tool for measuring *weight* or mass.

scale of a number line The spacing of the marks on a *number line.* The scale of the number line below is halves.

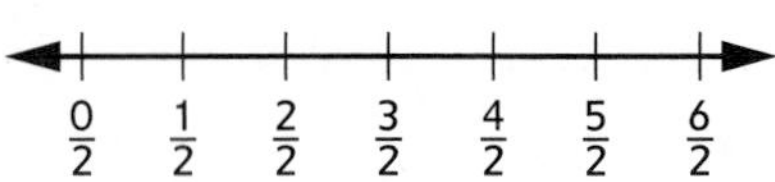

scalene triangle A *triangle* with *sides* of three different lengths.

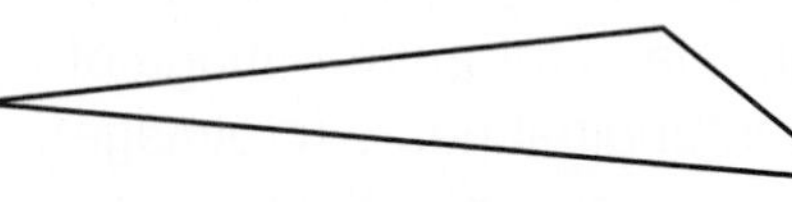

Scalene triangle

scientific notation A system for writing numbers in which a number is written as the product of a *power of 10* and a number that is at least 1 and less than 10. Scientific notation allows you to write big and small numbers with only a few symbols. For example, $4 * 10^{12}$ is scientific notation for 4,000,000,000,000.

sequence A list of numbers, often created by a rule that can be used to extend the list. Frames-and-Arrows diagrams can represent sequences.

side (1) One of the *line segments* of a *polygon.* (2) One of the rays or segments that make up an *angle.* (3) One of the *faces* of a 3-dimensional figure.

situation diagram In *Everyday Mathematics,* a diagram used to organize information in a problem situation.

Total	
7	
Part	Part
2	5

Suzie has 2 pink balloons and 5 yellow balloons. She has 7 balloons in all.

solid See geometric solid.

solution of an open sentence A value that makes an open sentence true when it is substituted for the *variable.* For example, 7 is a solution of $5 + n = 12$.

square A *rectangle* whose sides are all the same length. A rectangle that is also a *rhombus.*

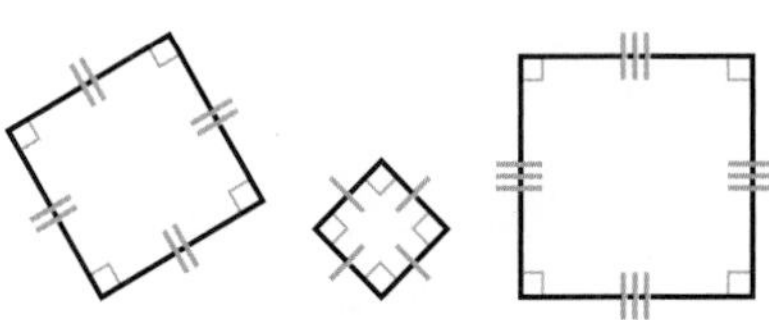

Squares

square number A number that is the *product* of a counting number with itself. For example, 25 is a square number because $25 = 5 * 5$. The square numbers are 1, 4, 9, 16, 25, and so on. A square number can be represented by a square array.

square of a number *n* The product of a number with itself. For example, 81 is the square of 9 because $81 = 9 * 9$. And 0.64 is the square of 0.8 because $0.64 = 0.8 * 0.8$.

square unit A *unit* used in measuring area, such as a square centimeter or a square foot.

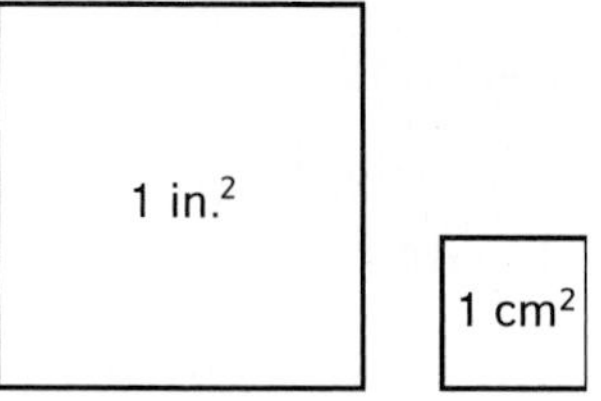

Square units

standard form The most familiar way of representing numbers. In standard form, numbers are written using the *base-ten place-value* system. For example, standard form for three hundred fifty-six is 356. Compare *expanded form.*

standard notation Same as *standard form.*

standard unit Measurement *units* that are the same size no matter who uses them and when or where they are used.

subcategory A more specific category within a given category. Subcategories are usually defined by an attribute shared by some, but not all, of the members of the larger category. For example, triangles are a subcategory of the larger category of polygons because some, but not all, polygons have three sides.

sum The result of adding two or more numbers. For example, in $5 + 3 = 8$, the sum is 8.

surface (1) The boundary of a *3-dimensional* object. Common surfaces include the top of a body of water, the outermost part of a ball, and the topmost layer of ground that covers Earth. (2) Any *2-dimensional* layer, such as a *plane* or the *faces* of a polyhedron.

symmetry A figure has line symmetry if a line can be drawn through it so that it is divided into two parts that are mirror images of each other. The two parts look alike but face in opposite directions. More generally, a figure has symmetry if parts of it look alike but are in different positions.

A figure with line symmetry

A figure with rotation symmetry

temperature A measure of how hot or cold something is.

term In an expression, a number or a product of a number and one or more *variables.* In the equation $5y + 3k = 8$, the terms are $5y$, $3k$, and 8. (2) An element in a *sequence.* In the sequence of multiples of 10, the terms are 10, 20, 30, 40, and so on.

tile (verb) To cover a *surface* completely with shapes without overlaps or gaps. Tiling with same-size squares is a way to measure area.

tool Anything that can be used for performing a task. Calculators, rulers, fraction circle pieces, and number grids are examples of mathematical tools.

trade-first subtraction A subtraction method in which all trades are done before any subtractions are carried out.

translation A movement of a figure along a straight line; a slide. In a translation, each point of the figure slides the same distance in the same direction.

trapezoid A *quadrilateral* that has at least one pair of *parallel* sides.

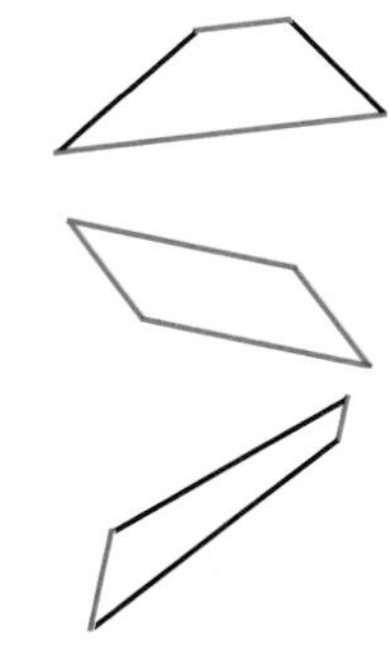
Trapezoids with parallel bases marked in the same color

triangle A *polygon* that has 3 sides and 3 angles.

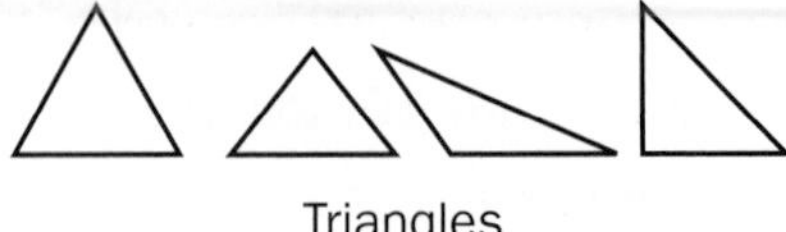

Triangles

triangular prism A *prism* whose *bases* are triangles.

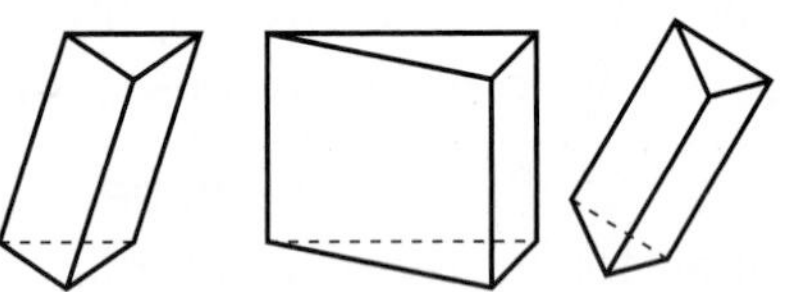

Triangular prisms

true number sentence A *number sentence* in which the relation symbol accurately connects the two sides. For example, $15 = 5 + 10$ and $25 > 20 + 3$ are both true number sentences.

turn-around rule A rule for solving addition and multiplication problems based on the Commutative Property. For example, if you know that $6 * 8 = 48$, then, by the turn-around rule, you also know that $8 * 6 = 48$.

U.S. traditional addition algorithm An addition method that involves adding digits by place-value columns starting at the right.

$$\begin{array}{r} {}^{1}\,{}^{1} \\ 3\,4\,8 \\ +\;2\,6\,3 \\ \hline 6\,1\,1 \end{array}$$

U.S. traditional multiplication algorithm A multiplication method that produces partial sums based on multiplying the values of each digit starting from the right.

$$\begin{array}{r} 1\;\;1\;\; \\ 1\;\; \\ 1\;2\;2 \\ *\;\;\;7\;5 \\ \hline 6\;1\;0 \\ +\;8\;5\;4\;0 \\ \hline 9\;1\;5\;0 \end{array}$$

U.S. traditional subtraction algorithm A subtraction method that involves subtracting digits by place-value columns starting from the right, making 10-for-1 trades as needed.

$$\begin{array}{r} 3\;\;\;\;\;\;\;\;\; \\ 10\;\;\not{14}\;\;\; \\ \not{4}\;\;\not{4}\;\;\not{7} \\ -\;\;1\;\;6\;\;5 \\ \hline 2\;\;8\;\;2 \end{array}$$

unit A label used to put a number in context. In measuring length, for example, the inch and the centimeter are units. In a problem about 5 apples, *apple* is the unit.

unit conversion A change from one measurement *unit* to another using a fixed relationship such as 1 yard = 3 feet or 1 inch = 2.54 centimeters.

unit cube A *cube* with edge lengths of 1.

unit fraction A *fraction* whose *numerator* is 1. For example, $\frac{1}{2}$, $\frac{1}{3}$, $\frac{1}{8}$, and $\frac{1}{20}$ are unit fractions.

unit square A *square* with side lengths of 1.

unknown A quantity whose value is not known. An unknown is sometimes represented by a ______, a ?, or a letter.

unlike Unequal or not the same.

variable (1) A letter or other symbol that can be replaced by any number. In the *number sentence* $5 + n = 9$, any number may be substituted for *n*, but only 4 makes the sentence true. See *unknown.* (2) A number or data set that can have many values is variable.

vertex The point where the *sides* of an *angle,* the sides of a polygon, or the *edges* of a *polyhedron* meet. Plural is vertexes or vertices.

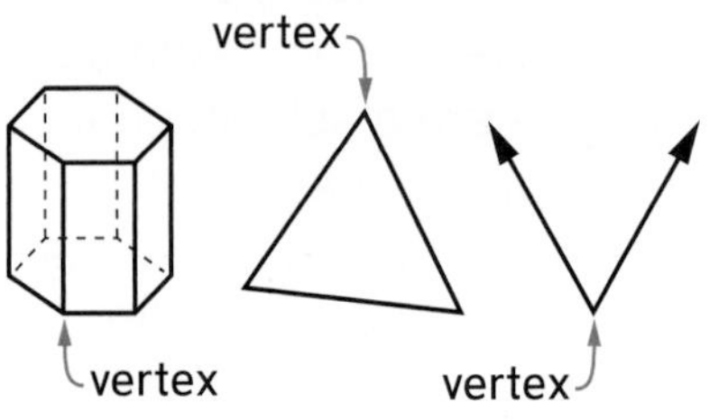

vertical Upright; perpendicular to the horizon.

volume A measure of how much space a solid object takes up. Volume is often measured in liquid units, such as liters, or cubic units, such as cubic centimeters or cubic inches. The volume or capacity of a container is a measure of how much the container will hold.

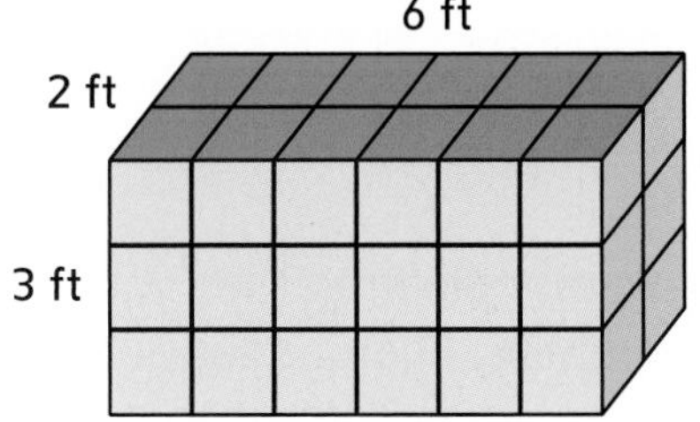

36 cubic feet

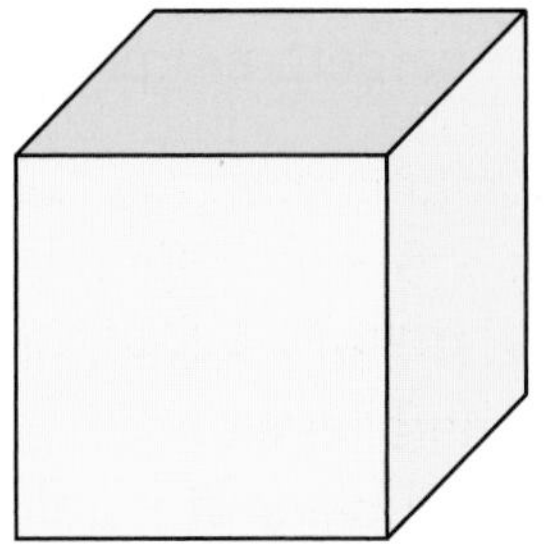
1 cubic inch

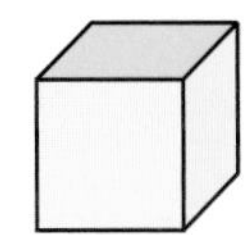
1 cubic centimeter

If the cubic centimeter were hollow, it would hold exactly 1 milliliter ($\frac{1}{1000}$ liter).

"What's My Rule?" A type of problem with "in" numbers, "out" numbers, and a rule that changes the in numbers to the out numbers. Sometimes you have to find the rule. Other times, you use the rule to figure out the in or out numbers. Sometimes, you can graph in and out numbers as ordered pairs.

whole An entire object, collection of objects, or quantity being considered.

whole numbers The *counting numbers,* together with 0. The set of whole numbers is {0, 1, 2, 3, ...}

x-axis In a *coordinate grid,* the horizontal number line.

x-coordinate The first number in a pair of numbers used to locate a point on a *coordinate grid.* The *x*-coordinate gives the position of the point along the horizontal axis. For example, 2 in the point (2, 3) is the *x*-coordinate.

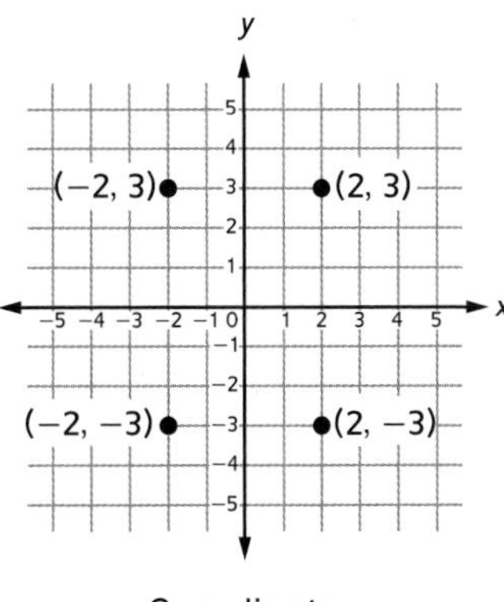

Coordinates

y-axis In a *coordinate grid,* the vertical number line.

y-coordinate The second number in a pair of numbers used to locate a point on a *coordinate grid.* The *y*-coordinate gives the position of the point along the vertical axis. For example, 3 in the point (2, 3) is the *y*-coordinate.

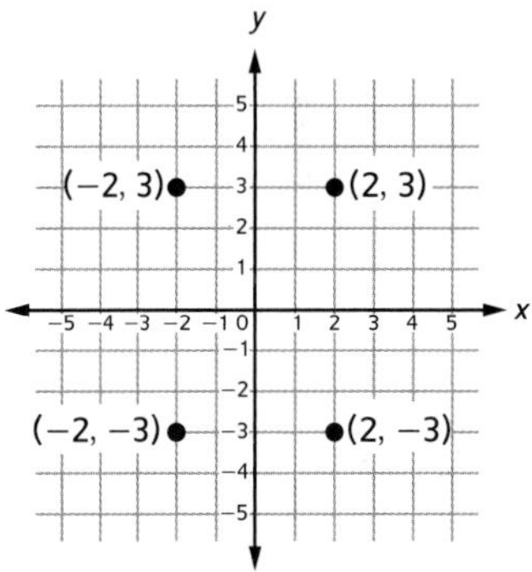

Coordinates

zero (1) The number representing no amount. (2) To adjust a scale or balance before use so that it reads 0 when no object is being weighed. You need to zero a scale for *accurate* measurements.

Page 11

Sample answer: Josh's conjecture is false, because $\frac{1}{2} * \frac{2}{3} = \frac{1}{3}$ (one-half of two-thirds would be one-third). The answer of $\frac{1}{3}$ is smaller than each of the two factors ($\frac{1}{2}$ and $\frac{2}{3}$).

Page 14

1. 187 square feet models the area of the wall that needs to be painted.
2. Each 25 models the area that one pint of paint covers, and the 100 models the area that 4 pints of paint cover.
3. The numbers on the right side of the problem model the number of pints of paint needed to cover each smaller area subtracted from the total area of the wall.
4. The numbers representing pints of paint on the right side of the problem can be added to find the total number of pints of paint needed to cover most of the wall. One more pint of paint would be needed to cover the remaining 12 square feet of the wall. $1 + 1 + 1 + 4 + 1 = 8$ pints of paint.
5. Sample answers: This model is similar to Emily's picture and Ava's table because it shows that 1 pint of paint covers 25 square feet. It is different because it subtracts 25s from the total instead of counting up by 25s to reach the total. It is also different because it groups 4 pints of paint together to cover 100 square feet rather than counting each 25 square-foot area separately.

Page 21

1. **a.** city blocks or miles
   b. square yards
2. Sample answers: I pictured the situation in real life. Then I used personal references to determine what units made the most sense for each measurement. I've heard on TV commercials that carpets are priced per square yard.

Page 26

1. Sample answers: D and E each only have one pair of parallel sides.
   F does not have any pairs of parallel sides. G has 5 sides instead of 4.
2. Mr. Bates would need 18 square tables.

Page 29

1. Sample answer: The rules are alike because both rules are based on counting the number of zeros in the factors and realizing that the total number of zeros in the factors is the number of zeros in the product.
2. Sample answer: The rules are different because Sophia counts all of the zeros in the factors, while Josh just writes the first factor and then counts on and writes the number of zeros in the second factor.
3. Yes. Sample answer: Yes, both rules work because in both cases the total number of zeros in the factors is the same as the number of zeros in the product. Both rules lead to a correct product and are based on the number of tens in the factors that are multiplied together to get the product.

Page 40

1. True
2. False
3. False
4. True
5. True
6. False

Page 42

1. $25 - (15 + 10) = 0$
2. $100 = (9 + 1) * 10$
3. $(4 + 1) / (10 - 5) = 1$

Page 43

1. 12
2. 25
3. 0
4. True
5. False

Page 45

1. $\$50 + \$4.50 * 26 = t$
2. $3\frac{1}{2} - 2\frac{1}{3} = f$ or $2\frac{1}{3} + f = 3\frac{1}{2}$

Page 47

1. >
2. >
3. =

Page 54

Rule: "$\frac{1}{4}$ of," or "$\div$ 4"

Page 56

1. $1\frac{1}{2}$ hours
2. $120

Page 67

1. twenty-five thousand, three hundred eight; 5,000
2. seventy-four million, five hundred forty-six thousand, two; 500,000
3. six hundred forty-three thousand, fifty-seven; 50
4. two million, four hundred fifty thousand, six hundred nine; 50,000
5. 1,234,788
6. 3,399

Page 69

1. 8,000,000
2. 760,000
3. 12,000,000
4. 490
5. $5 * 10^2$
6. $44 * 10^3$, or $4.4 * 10^4$
7. $9 * 10^8$

Page 70

1. 83,405
2. 2,673,952
3. Sample answer: $(8 * 1{,}000) + (7 * 100) + (4 * 10) + (4 * 1)$
4. Sample answer: $(1 * 10^6) + (4 * 10^5) + (5 * 10^4) + (6 * 10^3) + (9 * 10^2)$

Page 72

1. Sample answers: 12, 24, 36, ...
2. Sample answers: 30, 60, 90, ...
3. Sample answers: 45, 90, ...
4. Sample answers: 42, 84, ...
5. Sample answer: No, multiples of 10 and powers of 10 are not all the same. A power of 10 is a product of 10s only, such as 10, 100, 1,000, and so on. The multiples of 10 are a product of 10 and another number, such as 10 = 10 * 1; 20 = 10 * 2, 300 = 10 * 30; 2,400 = 10 * 240.

Page 73

1. 1, 3, 5, 15
2. 1, 2, 4, 8
3. 1, 2, 4, 7, 14, 28
4. 1, 2, 3, 4, 6, 9, 12, 18, 36
5. 1, 11
6. 1, 2, 4, 5, 10, 20, 25, 50, 100

Page 74

Divisible by 2: 4,470; 616; 14,580

Divisible by 3: 705; 4,470; 621; 14,580

Divisible by 5: 705; 4,470; 14,580

Divisible by 6: 4,470; 14,580

Divisible by 9: 621; 14,580

Divisible by 10: 4,470; 14,580

Page 75

The prime factors can be given in any order.

1. 3 * 5
2. 2 * 2 * 5
3. 2 * 2 * 2 * 5
4. 2 * 2 * 3 * 3
5. 17
6. 2 * 2 * 5 * 5

Page 78

1. Emily is not correct. The front-end estimate is 800 + 200 + 700 = 1,700. She should check her work.
2. Luis is not correct. The front-end estimate is 900 / 30 = 30. He should check his work.

Page 82

1. 25,300
2. 30,000
3. 25,300
4. 25,000
5. Sample estimate: 150 + 700 + 200 + 400 = 1,450
6. Sample estimate: 70 * 80 = 5,600

Page 84

1. Sample estimate: 5,000 / 20 = 250
2. Sample estimates: 250 * 40 = 10,000; 300 * 50 = 15,000
3. Sample estimates: 3,200 / 16 = 200; 3,000 / 15 = 200

Page 88

1. 887; Sample estimate: 350 + 530 = 880
2. 133; Sample estimate: 45 + 90 = 135
3. 321; Sample estimate: 280 + 40 = 320
4. 1,023; Sample estimate: 675 + 350 = 1,025
5. 863; Sample estimate: 300 + 550 = 850
6. 830; Sample estimate: 750 + 100 = 850

Page 90

1. 456; Sample estimate: 500 − 60 = 440
2. 517; Sample estimate: 750 − 225 = 525
3. 283; Sample estimate: 450 − 150 = 300
4. 2,708; Sample estimate: 3,100 − 400 = 2,700

Page 92

1. 35; Sample estimate: 125 − 90 = 35
2. 101; Sample estimate: 1,000 − 900 = 100
3. 130; Sample estimate: 2,000 − 1,900 = 100

Page 94

1. 38; Sample estimate: 70 − 30 = 40
2. 382; Sample estimate: 850 − 450 = 400
3. 366; Sample estimate: 650 − 300 = 350
4. 257; Sample estimate: 700 − 450 = 250
5. 4,279; Sample estimate: 7,300 − 3,000 = 4,300

Page 96

1. 42,000
2. 345,700
3. 20,000
4. 320,000

Page 98

1. 4,500
2. 240,000
3. 4,800,000
4. 4,000,000

Page 101

1. 795; Sample estimate: 250 ∗ 3 = 750
2. 2,814; Sample estimate: 40 ∗ 70 = 2,800
3. 2,320; Sample estimate: 40 ∗ 60 = 2,400
4. 4,482; Sample estimate: 80 ∗ 50 = 4,000
5. 18,600; Sample estimate: 400 ∗ 50 = 20,000

Page 103

1. 13,984; Sample estimate: 40 ∗ 400 = 16,000
2. 43,108; Sample estimate: 800 ∗ 50 = 40,000

Page 105

1. 90
2. 120
3. 20
4. 100

Page 106

1. 30
2. 80
3. 20
4. 50

Page 112

1. 17 R3; An estimated quotient will be less than $4\overline{)80}$ = 20.
2. 29 R10; An estimated quotient will be less than 750 / 25 = 30.
3. 85 R2; An estimated quotient will be greater than 320 / 4 = 80.
4. 21 R2; An estimated quotient will be less than $30\overline{)690}$ = 23.

Page 114

1. 3 R1; Each station will get $3\frac{1}{3}$ watermelons.
2. 6 R2; 7 picnic tables are needed.
3. 7 R1; Abner can buy 7 tickets.

Page 117

1. 0.08; eight hundredths
2. 0.9; nine tenths
3. $\frac{70}{100}$; seventy hundredths
4. $4\frac{506}{1,000}$ or $\frac{4,506}{1,000}$; four and five hundred six thousandths
5. $24\frac{68}{100}$ or $\frac{2,468}{100}$; twenty-four and sixty-eight hundredths
6. $\frac{14}{1,000}$; fourteen thousandths

Page 120

1. **a.** twenty thousands
   b. two hundredths
   c. two thousandths
2. 0.003
3. 0.9 or 0.90

Page 123

1. 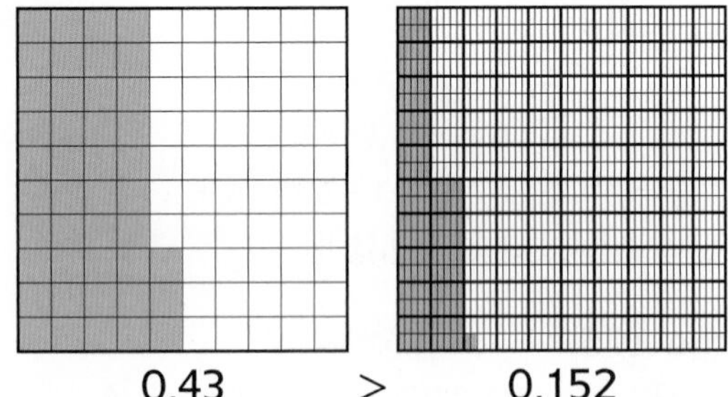

0.43 > 0.152

2. 2.8 > 2.102, since 2.800 > 2.102
3. 1.402 < 1.41

1s ones	.	0.1s tenths	0.01s hundredths	0.001s thousandths
1	.	4	0	2
1	.	4	1	

The ones digit and the tenths digits are the same. The 0 is worth 0 hundredths and 1 is worth 1 hundredth, so 1.41 is greater than 1.402.

4. 0.853

Page 127

1. nearest tenth: 5.4 + 2.1 = 7.5; nearest one: 5 + 2 = 7
2. nearest one: 14 + 21 = 35
3. nearest dollar: \$4 + \$1 = \$5; nearest dime: \$3.60 + \$1.10 = \$4.70

Page 128

1. Sample estimate: 15 − 13 = 2
2. Sample estimate: 20 * 2 = 40
3. Sample estimate: 30 / 15 = 2

Page 132

1. 1.02; Sample estimate: 0.6 + 0.4 = 1.0
2. 14.58; Sample estimate: 13 + 1 = 14
3. 0.023; Sample estimate: 2.40 − 2.38 = 0.02

Page 133

1. 456; Sample estimate: 100 * 4 = 400
2. 2,800; Sample estimate: 0.28 is close to 0.25 or $\frac{1}{4}$. $\frac{1}{4}$ of 10,000 is 2500.
3. \$4,500; Sample estimate: 1,000 * \$4 = \$4,000 and 1,000 * \$5 = \$5,000, so the product will be between \$4,000 and \$5,000.
4. 10.4; Sample estimate: 1 * 10 = 10

Page 135

1. 9.69; Sample estimate: $2 * 6 = 12$
2. 19.572; Sample estimate: $2 * 8 = 16$
3. 2.4644; Sample estimate: $1 * 4 = 4$
4. 0.0063; Sample estimate: $0.3 * 10^1 = 3$ and $0.02 * 10^2 = 2$; $3 * 2 = 6$. $6/(10^1 * 10^2) = 0.006$

Page 136

1. 5.67
2. 0.0047
3. \$0.29
4. 0.006

Page 137

1. 24.8; Sample estimate: $120/6 = 20$
2. 2.11; Sample estimate: $24/12 = 2$
3. 1.3; Sample estimate: 45/35 is greater than 1, but less than 2.

Page 141

1. $8.01/0.3 \approx 27$; An equivalent expression for 8.01/0.3 is 80.1/3. Sample estimate: $81/3 = 27$
2. 300; An equivalent expression for 3.9/0.013 is $3{,}900/13 = 300$
3. 300; Sample estimate: 0.45 is close to $\frac{1}{2}$. Dividing 135 by $\frac{1}{2}$ is the same as multiplying by 2. $135 * 2 = 270$

Page 159

$\frac{11}{6}$ and $1\frac{5}{6}$

Page 160

1.
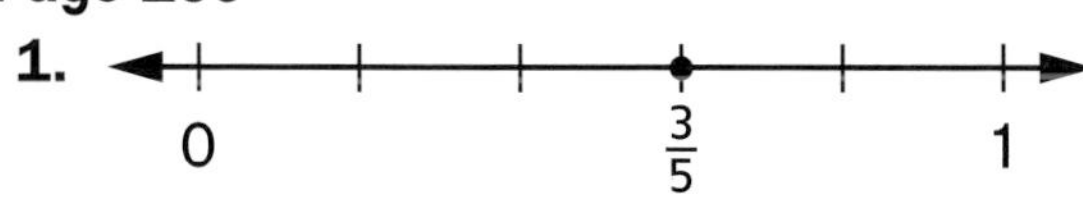

2. 0, 1, 2, $2\frac{2}{3}$, 3

3. 0, 1, $\frac{5}{3}$, 2

Page 167

1. Sample answers: Point A: $\frac{3}{16}$; Point B: $\frac{7}{8}$, $\frac{14}{16}$; Point C: $4\frac{1}{4}$, $4\frac{4}{16}$, $\frac{17}{4}$
2. **a.** $\frac{3}{4}$ in. **b.** $\frac{6}{8}$ in. **c.** $\frac{12}{16}$ in.

Page 169

Sample answers:

1. **a.** $\frac{9}{12}, \frac{12}{16}, \frac{6}{8}$ **b.** $\frac{6}{16}, \frac{9}{24}, \frac{15}{40}$
   c. $\frac{4}{10}, \frac{6}{15}, \frac{8}{20}$ **d.** $\frac{8}{14}, \frac{12}{21}, \frac{16}{28}$
   e. $\frac{16}{6}, \frac{24}{9}, \frac{32}{12}$ **f.** $\frac{22}{24}, \frac{33}{36}, \frac{44}{48}$
2. **a.** $\frac{2}{3}$ or $\frac{4}{6}$
   b. $\frac{2}{5}$
   c. $\frac{2}{3}, \frac{4}{6}, \frac{6}{9}, \frac{8}{12}$, or $\frac{12}{18}$
   d. $\frac{3}{4}, \frac{9}{12}$, or $\frac{15}{20}$
   e. $\frac{3}{2}, \frac{6}{4}, \frac{9}{6}$
   f. $\frac{25}{2}$ or $\frac{50}{4}$

Page 170

1. **a.** True **b.** True
   c. True **d.** False
2. Sample answer: $\frac{4}{6}, \frac{6}{9}, \frac{8}{12}, \frac{10}{15}, \frac{12}{18}$

Page 173

1. Sample answer: $12\frac{3}{4}$
2. Sample answer: $8\frac{2}{3}$
3. Sample answer: $6\frac{4}{5}$
4. Sample answer: $3\frac{12}{16}$
5. Sample answer: $\frac{19}{4}$
6. Sample answer: $\frac{11}{3}$
7. Sample answer: $\frac{29}{6}$
8. Sample answer: $\frac{7}{3}$
9. $5\frac{5}{3}, 4\frac{8}{3}, 3\frac{11}{3}, 2\frac{14}{3}, 1\frac{17}{3}, \frac{20}{3}$

Page 180

1. $\frac{2}{8}$ melon, or $\frac{1}{4}$ melon; Sample answer:

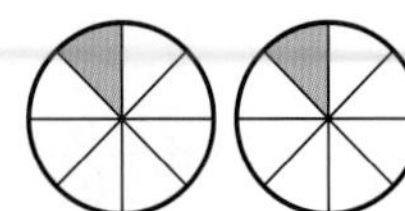

2. $\frac{7}{2}$ inches or $3\frac{1}{2}$ inches; Sample answer:

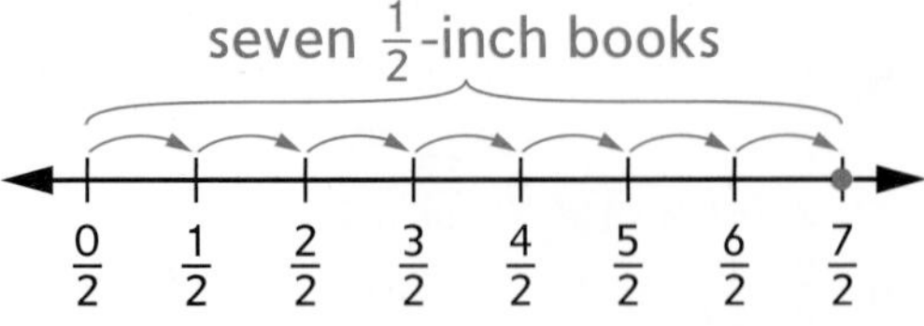

Page 184

Sample estimates:

1. about 3
2. about 2
3. about 10
4. about 50
5. Sample answer: I rounded $2\frac{1}{4}$ down to 2 and $\frac{5}{6}$ up to 1. Then added 2 and 1 to get 3.

Page 185

1. Sample answer: No, $\frac{3}{4}$ is too small. $2\frac{1}{8}$ is more than 2, and $\frac{4}{5}$ is less than 1. Subtracting less than 1 from more than 2 should give a result that is more than 1.
2. Sample answer: No, $3\frac{3}{5}$ is too small. $\frac{2}{3}$ is greater than $\frac{1}{2}$, and $3\frac{1}{2} + \frac{1}{2} = 4$, so the sum must be greater than 4.

Page 186

1. Sample estimate: $\frac{1}{4}$ is less than $\frac{1}{2}$ and $\frac{2}{4}$ is equal to $\frac{1}{2}$. The answer should be more than $\frac{1}{2}$, but less than 1. Exact sum: $\frac{3}{4}$
2. Sample estimate: $\frac{4}{6}$ is twice the size of $\frac{2}{6}$, so the answer should be half of $\frac{4}{6}$, or $\frac{2}{6}$. Exact difference: $\frac{2}{6}$
3. Sample estimate: $\frac{2}{3}$ is just over $\frac{1}{2}$ and $\frac{1}{3}$ is just under $\frac{1}{2}$, so together they should be close to two-halves, or 1. Exact sum: $\frac{3}{3}$, or 1
4. Sample estimate: $\frac{9}{10}$ is almost 1 whole and $\frac{3}{10}$ is less than $\frac{1}{2}$, so the answer should be just a little more than $\frac{1}{2}$. Exact difference: $\frac{6}{10}$

Page 191

1. Sample estimate: $2\frac{1}{2} + 1 = 3\frac{1}{2}$; Exact sum: $3\frac{3}{4}$
2. Sample estimate: $4 + 7 = 11$; Exact sum: $10\frac{13}{12}$, or $11\frac{1}{12}$
3. Sample estimate: $5 + 3 = 8$. $1 + \frac{2}{3} = 1\frac{2}{3}$. $8 + 1\frac{2}{3} = 9\frac{2}{3}$; Exact sum: $8\frac{9}{6}$, $9\frac{3}{6}$, or $9\frac{1}{2}$
4. Sample estimate: 5 and 3 is 8. $\frac{5}{6}$ is close to 1, 1 and $\frac{2}{3}$ is $1\frac{2}{3}$. The sum of $5\frac{5}{6}$ and $3\frac{2}{3}$ is about $9\frac{2}{3}$.

Page 194

1. $\frac{1}{6} + \frac{1}{6} + \frac{1}{6} + \frac{1}{6} + \frac{1}{6} = \frac{5}{6}$; $5 * \frac{1}{6} = \frac{5}{6}$
2. Sample answer:

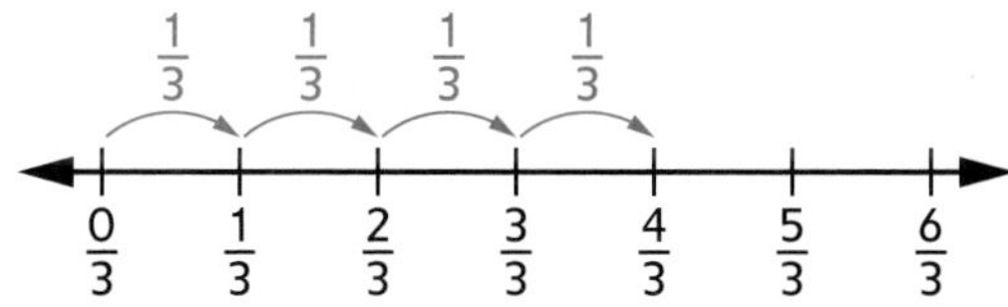

Page 196

1. Sample estimate: $\frac{1}{4}$ is part of 32, so the answer is less than 32. $\frac{1}{4}$ is less than $\frac{1}{2}$, so the answer is also less than half of 32, or 16. Exact answer: 8
2. Sample estimate: $\frac{3}{4}$ is only part of 32, but it is more than half of 32, so the answer is greater than 16 and less than 32. Exact answer: 24
3. Sample estimate: $\frac{1}{3}$ is less than half of 25, so the answer is less than $12\frac{1}{2}$. Exact answer: $8\frac{1}{3}$
4. Sample estimate: $\frac{2}{3}$ is twice that of $\frac{1}{3}$. Since $\frac{1}{3}$ of 25 is $8\frac{1}{3}$ (see Problem 3) or about 8, $\frac{2}{3}$ of 25 is twice 8, or 16; Exact answer: $16\frac{2}{3}$
5. Sample estimate: $\frac{4}{5}$ is greater than $\frac{1}{2}$, so Rita gets more than \$10. $\frac{1}{5}$ is less than $\frac{1}{2}$, so Hunter gets less than \$10. Exact answer: Rita gets \$16 and Hunter gets \$4.

Page 198

1. $<$ 2. $>$ 3. $>$ 4. $=$

Page 200

1. Sample estimate: $\frac{3}{4}$ is close to but less than 1; so the product of 8 times a number just less than 1 will be close to but less than 8. Exact answer: $\frac{24}{4}$, or 6
2. Sample estimate: $\frac{2}{5}$ less than $\frac{1}{2}$; so the product will be less than $\frac{1}{2}$ of 3, or $1\frac{1}{2}$; Exact answer: $\frac{6}{5}$, or $1\frac{1}{5}$
3. Sample estimate: since $\frac{7}{8}$ is less than 1, the fraction of 16 should be less than 16; Exact answer: 14, or $\frac{112}{8}$
4. 16, or $\frac{96}{6}$

Page 201

1. $\frac{4}{10}$, or $\frac{2}{5}$; Sample answer:

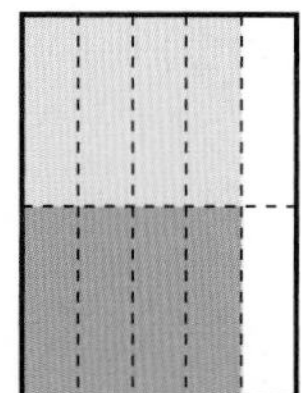

2. $\frac{1}{6}$; Sample answer: I showed $\frac{2}{3}$ of a whole circle with 2 orange pieces. I found I could cover the 2 orange pieces with 4 light blue pieces. So each light blue piece is $\frac{1}{4}$ of 2 orange pieces. One light blue piece is $\frac{1}{6}$ of the whole circle.

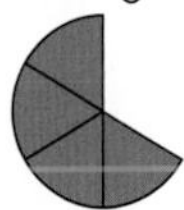

3. $\frac{3}{24}$, or $\frac{1}{8}$; Sample answer:

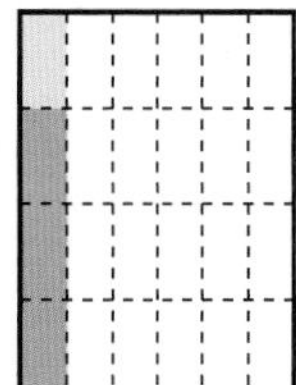

Page 203

1. Sample estimate: $\frac{1}{2}$ of $\frac{2}{3}$ is less than $\frac{2}{3}$; Exact answer: $\frac{2}{6}$, or $\frac{1}{3}$
2. Sample estimate $\frac{2}{5} * 1 = \frac{2}{5}$; Exact answer: $\frac{6}{20}$, or $\frac{3}{10}$
3. Sample estimate: a fraction of $\frac{3}{8}$ is less than $\frac{3}{8}$; Exact answer: $\frac{15}{48}$, or $\frac{5}{16}$
4. Sample estimate: a fraction of $\frac{1}{3}$ is less than $\frac{1}{3}$; Exact answer: $\frac{1}{12}$
5. Sample answer: Since I was finding a fraction of $\frac{3}{4}$, I knew the answer would be less than $\frac{3}{4}$, or $\frac{15}{20}$.

Page 206

1. Sample estimate: $1\frac{2}{3}$ is close to but less than 2. $6 * 2 = 12$, so the product should be close to but less than 12; Exact answer: $\frac{30}{3}$, or 10
2. Sample estimate: $1\frac{1}{2}$ is between 1 and 2. $\frac{1}{4}$ of 1 is $\frac{1}{4}$, and $\frac{1}{4}$ of 2 is $\frac{2}{4}$, so the product should be between $\frac{1}{4}$ and $\frac{2}{4}$; Exact answer: $\frac{3}{8}$
3. Sample estimate: $3\frac{2}{5}$ is close to 3, and $2\frac{1}{2}$ is close to 3. $3 * 3 = 9$, so the product should be close to 9. Exact answer: $\frac{17}{2}$, or $8\frac{1}{2}$
4. Sample estimate: $4\frac{5}{6}$ is close to 5. Half of 5 is $2\frac{1}{2}$, so the product should be close to $2\frac{1}{2}$. Exact answer: $\frac{29}{12}$, or $2\frac{5}{12}$
5. Sample answer: I thought that $4\frac{5}{6}$ is almost 5. Then I took $\frac{1}{2}$ of 5 and got $2\frac{1}{2}$. The answer should be close to $2\frac{1}{2}$.

Page 210

1. 20
2. $\frac{1}{15}$
3. $\frac{1}{32}$
4. 40

Page 214

1. foot, pound, inch, yard
2. $\frac{1}{1{,}000}$
3. 300 centimeters

Page 216

1. 5,000 meters 2. 2 cups

Page 217

1. 108 inches
2. 4 pints

Page 220

1. 15 cm
2. 27 feet, or 324 inches

Page 223

1. Sample answer: Square inches, because the area of the desk is probably a lot less than a 1 square yard.
2. Sample answer: Square meters, because the measurement in square centimeters would be a really big number and the number of square meters could more easily be understood or visualized.

Page 225

1. $7\frac{1}{2}$ cm^2 2. $22\frac{1}{2}$ in.^{2} 3. 25 cm^2

Page 227

1. Sample answer:

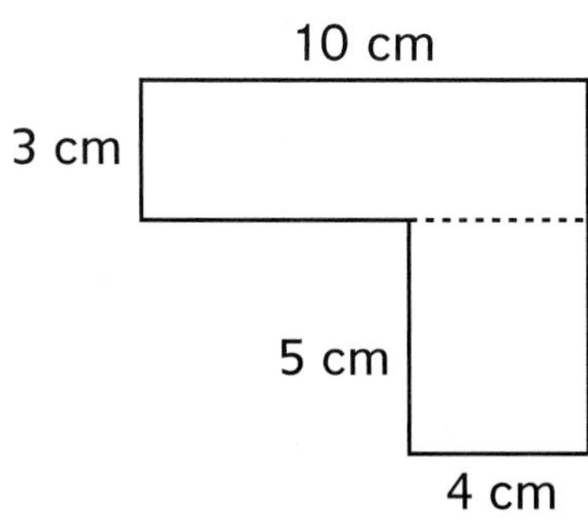

$3 \text{ cm} * 10 \text{ cm} = 30 \text{ cm}^2$ and $5 \text{ cm} * 4 \text{ cm} = 20 \text{ cm}^2$. Area of shape $= 30 \text{ cm}^2 + 20 \text{ cm}^2 = 50 \text{ cm}^2$.

2. Sample answer:

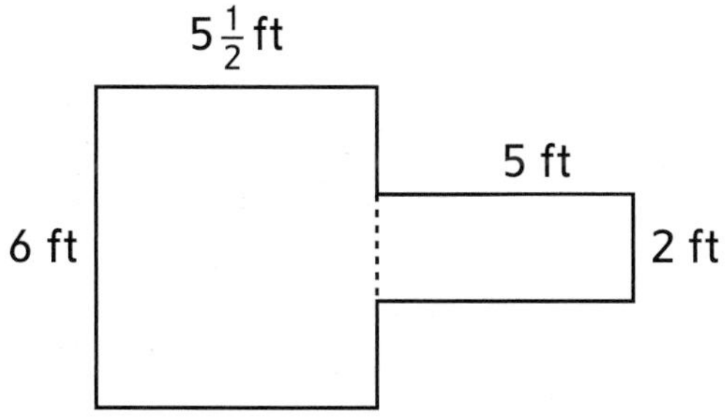

$6 \text{ ft} * 5\frac{1}{2} \text{ ft} = 33 \text{ ft}^2$ and $2 \text{ ft} * 5 \text{ ft} = 10 \text{ ft}^2$. Area of shape $= 33 \text{ ft}^2 + 10 \text{ ft}^2 = 43 \text{ ft}^2$.

Page 229

1. 8 units^2 2. 15 units^2 3. 20 units^2

Page 231

1. 9 cm^3 2. 10 in.^{3} 3. 60 ft^3

Page 234

1. You could decompose the figure into top and bottom rectangular prisms. The volume of the top prism $= 1 \text{ cm} * 2 \text{ cm} * 2 \text{ cm} = 4 \text{ cm}^3$. The volume of the bottom prism $= 4 \text{ cm} * 2 \text{ cm} * 2 \text{ cm} = 16 \text{ cm}^3$. The volume of the total figure is $4 \text{ cm}^3 + 16 \text{ cm}^3 = 20 \text{ cm}^3$.
2. 78 in.^{3}
3. 54 ft^3

Page 235

1. Sample answer: Cubic centimeters, because a pencil box is less than a meter long, less than a meter wide, and less than a meter tall. All of those dimensions can be measured in centimeters.
2. Sample answer: Cubic yards, because the measure in cubic inches would be a really big number that would be hard to picture. The number of cubic yards would be easier to visualize.

Page 238

1. 470 mL, or 470 cm^3
2. 60 mL, or 60 cm^3

Page 241

1. 35 years 2. 23 days

Page 245

1. a. $3\frac{1}{2}$, $3\frac{3}{4}$, $3\frac{3}{4}$, $3\frac{3}{4}$, $3\frac{3}{4}$, 4, 4, 4, $4\frac{1}{2}$, $4\frac{1}{2}$; smallest: $3\frac{1}{2}$ inches; largest: $4\frac{1}{2}$ inches

   b. Sunflower Heights After 2 Weeks

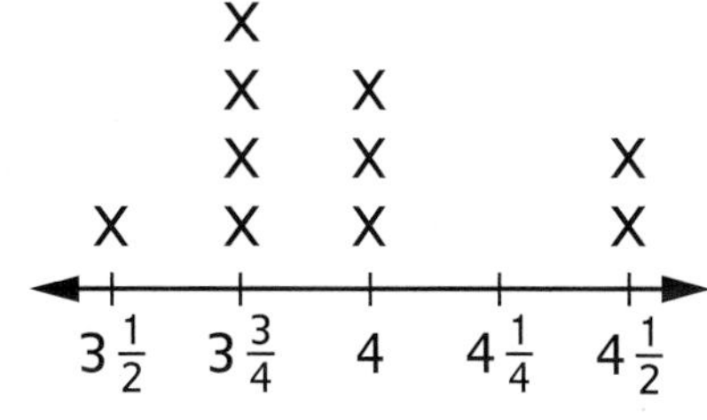

Length (inches)

   c. Each interval is $\frac{1}{4}$ inch.

   d. Yes. $4\frac{1}{4}$ inches. Chad did not have any seedling measuring $4\frac{1}{4}$ inches.

Page 251

1.

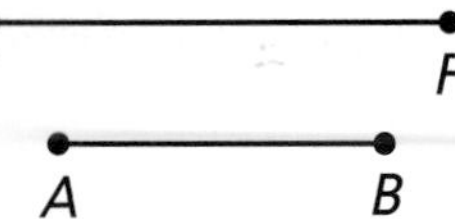

2.

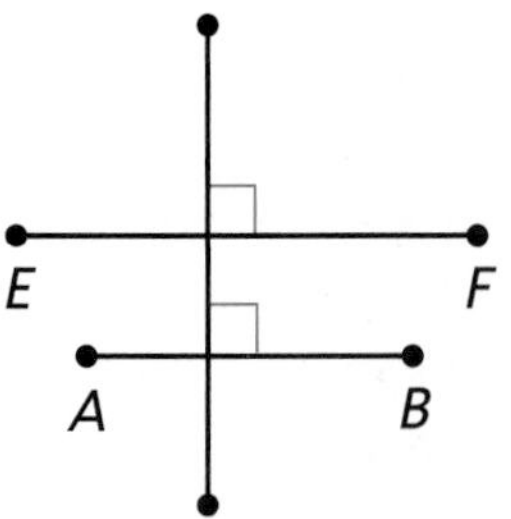

Page 262

1. a. hexagon
   b. Sample answers: quadrilateral, parallelogram, square, rectangle, trapezoid
   c. octagon
2. Sample answers:

   a.

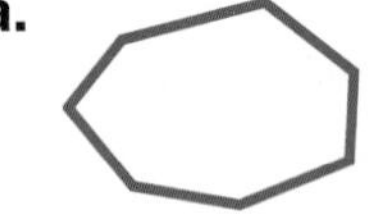

   b.

Page 263

Sample answers:

1.

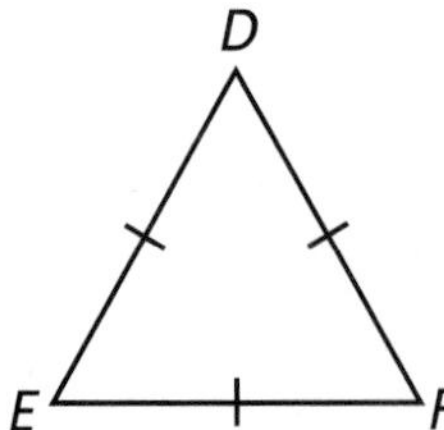

△*DFE*
△*EDF*
△*EFD*
△*FDE*
△*FED*

2.

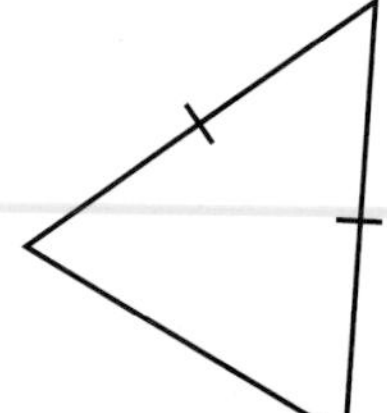

3.

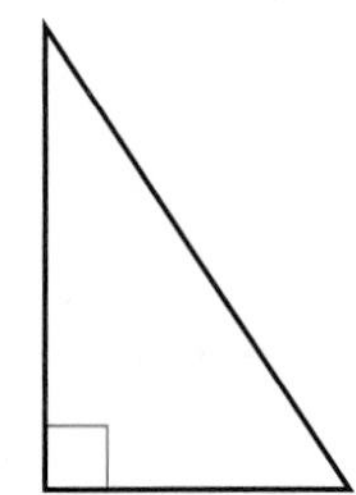

Page 265

1.

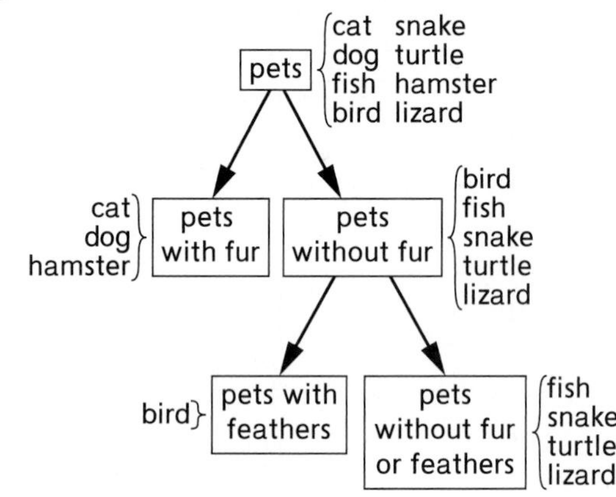

Page 271

Sample answers:

1. **a.** Cylinders and cones each have at least one circular face. They each have a curved surface.
   b. A cylinder has three surfaces; a cone has two. A cylinder has 2 circular bases, 2 edges, and no vertices. A cone has 1 circular base, 1 edge, and 1 vertex.
2. **a.** Pyramids and cones each have at least one vertex. They each have a flat base.
   b. A cone has a curved surface; the surfaces of a pyramid are all flat surfaces (faces). A cone has 1 circular face; the faces of a pyramid are all shaped like polygons. A cone has only one vertex. A pyramid has at least four vertices.

Page 272

1. **a.** 5 **b.** 1
2. **a.** 5 **b.** 2

Page 273

1. **a.** 8 **b.** 18 **c.** 12
2. decagonal prism

Page 274

1. regular tetrahedron, regular octahedron, regular icosahedron
2. **a.** 12 edges **b.** 6 faces
3. Sample answers:
   a. Their faces are same-size equilateral triangles
   b. The tetrahedrons have 4 faces, 6 edges, and 4 vertices. The octahedrons have 8 faces, 12 edges, and 6 vertices.

Page 278

All of the methods (a, b, and c) would make a congruent copy of the square.

Page 335

1. 58,000
2. 76,000,000
3. 0.000004389
4. 0.000011

Page 337

1. $357 \div 20 = 17.85$, or 17 R17; 18 classrooms
2. $100 \div 3 = 33.333333333$, or 33 R1; 33 pies
3. $984 \div 96 = 10.25$, or 10 R24; $10\frac{1}{4}$ hours, or 10 hours 15 minutes

A

B

D

F

G

M

N

S

T